D1281634

FUEL CELLS

THE ELECTROCHEMICAL SOCIETY SERIES

ECS-The Electrochemical Society
65 South Main Street
Pennington, NJ 08534-2839
http://www.electrochem.org

A complete list of the titles in this series appears at the end of this volume.

FUEL CELLS

Problems and Solutions

VLADIMIR S. BAGOTSKY
A.N. Frumkin Institute of Electrochemistry
and Physical Chemistry
Russian Academy of Sciences
Moscow, Russia

WILEY

A JOHN WILEY & SONS, INC., PUBLICATION

Published by John Wiley & Sons, Inc., Hoboken, New Jersey.
Published simultaneously in Canada.

Library of Congress Cataloging-in-Publication Data:

Bagotsky, V. S. (Vladimir Sergeevich)
Fuel cells: problems and solutions/Vladimir Bagotsky
p. cm.
Includes index.
ISBN 978-0-470-23289-7 (cloth)
1. Fuel cells. I. Title.

TK2931.B35 2008
621.31′2429–dc22

2008033276

Printed in the United States of America

10 9 8 7 6 5 4 3 2 1

CONTENTS

PREFACE

When fuel cells were first suggested and discussed, in the nineteenth century, it was firmly hoped that distinctly higher efficiencies could be attained with them when converting the chemical energy of natural fuels to electric power. Now that the world supply of fossil fuels is seen to be finite, this hope turns into a need: into a question of maintaining advanced standards of living. Apart from conversion efficiency, fuel cells have other aspects that make them attractive: Their conversion process is clean, they may cogenerate useful heat, and they can be used in a variety of fields of application. One worker in the field put it this way: "Fuel cells have the potential to supply the electricity powering a wristwatch or a large city, replacing a tiny battery or an entire power generating station."

With some important achievements made in the past, fuel cells today are a subject of vigorous R&D, engineering, and testing conducted on a broad international scale in universities, research centers, and private companies in various sectors of the economy. Combining engineers, technicians, and scientists, several 10,000 workers contribute their efforts and skills to advancing the field.

Progress in the field is rapid. Each month hundreds of publications report new results and discoveries. Important synergies exist with work done to advance the concepts of a hydrogen economy.

The book is intended for people who have heard about fuel cells but ignore the detailed potential and applications of fuel cells to focus on the information they need: engineers in civil, industrial, and military jobs; R&D people of diverse profile; investors; decision makers in government, industry, trade, and all levels of administration; journalists; school and university teachers and students; and hobby scientists. The work is also intended for people in industry and research who in their professional work are concerned with various special aspects of the development and applications of fuel cells and want to gain an overview of fuel cell problems and their economic and scientific significance.

The aim of this book is to provide readers across trades and lifestyles with a compact, readable introduction and explanation of what fuel cells do, how they do it, where they are important, what the problems are, and how they will continue in the field: what they could do against air pollution and for portable devices. All this is done with a critical attitude based on a detailed and advanced presentation. Problems and achievements are discussed at the level attained by the end of 2007.

Contradictions and a lack of consensus have existed in the field, along with ups and downs. In a field where the subject may range in size from milliwatt to megawatt output, and where many technical systems compete, this will not come as a surprise. To guide the reader through the maze, a sampling of literature references is provided. Unfortunately, a lot of work just as important as the work cited had to be omittted. Selection was also made difficult because of the strongly interdisciplinary character of fuel cell work.

The presentation is made against the historical background, and looks at future prospects, including those of a synergy with a potential future hydrogen economy. Where views diverge, they are presented as such. Some of the ideas offered may well be open to further discussion.

My sincere thanks are due Dr. Felix Büchi of the Paul Scherrer Institute in Villigen, Switzerland, who contributed the important chapter on the modeling of fuel cells. My gratitude goes to my colleagues the late Dr. Nina Osetrova and to Dr. Alexander Skundin, of Moscow, for their help in selecting relevant literature, and to Timophei Pastushkin for preparing graphical representations. My thanks also go to Dr. Klaus Müller, formerly at the Battelle Institute of Geneva, who transformed chapters written in Russian into English, contributed Section 18.2, and made a number of very valuable suggestions.

I sincerely hope that what has inspired me during more than 50 years of research and teaching at the Moscow Quant Power Sources Institute and the A.N. Frumkin Institute of Electrochemistry and Physical Chemistry, Russian Academy of Sciences, will continue to inspire current and future specialists and people in general who work to improve our lives and solve our problems.

VLADIMIR SERGEEVICH BAGOTSKY

Moscow, Russia and Mountain View, California
May 2008
E-mail:vbag@mail.ru

SYMBOLS

Symbol	Meaning	Dimensions (values)	Section*
ROMAN SYMBOLS			
c_j	concentration	mol/dm^3	
D_j	diffusion coefficient	cm^2/s	
E	electrode potential	V	1.4.3
E^0	equilibrium electrode potential	V	1.4.3
F	Faraday constant	94850 C/mol	7.2
G	Gibbs energy	kJ/mol	1.1.2
H	enthalpy	kJ/mol	1.1.2
i	current density	mA/cm^2	1.4.3
i^0	exchange current density	mA/cm^2	1.4.3
I	current	A, mA	1.4.3
M	mass	kg	
	molar concentration	mol/dm^3	
n	number of electrons in the reaction's elementary act	none	1.4.2
p	power density	W/kg	1.5.5
	power	W, kW	1.5.2
q	heat (in eV)	eV	1.4.2
Q	heat, thermal energy	J, kJ	1.1. 1

*Section where this symbol is used for the first time and/or where it is defined.

R	resistance	Ω	1.4.3
	molar gas constant	8.314 J/mol·K	7.2
S	entropy	kJ/K	1.1.2
	surface area	cm^2	
T	absolute temperature	K	1.1.1
U	cell voltage	V	1.4.4
w	energy density	kWh/kg	1.5.5
W	work, useful energy	W, kW	1.1.2

GREEK SYMBOLS

γ	roughness factor	none	
δ	thickness	cm	
$\mathcal{E}^0$	electromotive force	V	1.4.4
λ_e	amount of coulombs	none	1.5.3
η	efficiency	none, %	
σ	conductivity	S/cm^2	

SUBSCRIPTS

ads	adsorbed
app	apparent
e	electrical
exh	exhaust
ext	external
h.e.	hydrogen electrode
i	under current
j	any ion, substance
loss	energy loss
o.e.	oxygen electrode
ox	oxidizer
S	per unit area
red	reducer
V	per unit volume
0	without current
+	cation
−	anion

ACRONYMS AND ABBREVIATIONS*

ac	alternating current
AFC	alkaline fuel cell
APU	auxiliary power unit
ATR	autothermal reforming
CD	current density
CHP	combined heat and power
CNT	carbon nanotube
CTE	coefficient of thermal expansion
DBHFC	duirect borohydride fuel cell
dc	direct current
DCFC	direct carbon fuel cell
DEFC	direct ethanol fuel cell
DFAFC	direct formic acid fuel cell
DHFC	direct hydrazine fuel cell
DLFC	direct liquid fuel cell
DMFC	direct methanol fuel cell
DSA	dimensionally stable anode
EMF	electromotive force
EPS	electrochemical power source
ET-PEMFC	elevated-temperature PEMFC
FCI	Fuel Cells International
FCV	fuel cell vehicle

* These acronyms and abbreviations are used in most chapters. Acronyms for oxide materials used as electrolytes and electrodes in solid-oxide fuel cells are given in Chapter 8.

GDL gas-diffusion layer
GLDL gas–liquid diffusion layer
ICV internal combustion vehicle
IRFC internal reforming fuel cell
IT-SOFC interim-temperature SOFC
LHV lower heat value
LT-SOFC low-temperature SOFC
MCFC molten arbonate fuel cell
MEA membrane–electrode assembly
OCP open-circuit potential
OCV open-circuit voltage
ORR oxygen reduction reaction
Ox, ox oxidized form
PAFC phosphoric acid fuel cell
PBI polybenzimidazole
PCB printed circuit board
PD potential difference
PEEK polyether ether ketone
PEMFC proton-exchange membrane fuel cell (polymer electrolyte membrane fuel cell)
PFSA perfluorinated sulfonic acid
POX partial oxidation (reforming by)
PVD physical vapor deposition
Red, red reduced form
SHE standard hydrogen electrode
SOFC solid-oxide fuel cell
SR steam reforming
URFC unitized regenerative fuel cell
UCC Union Carbide Corporation
UTC United Technologies Corporation
WGSR water-gas shift reaction

PART I

INTRODUCTION

INTRODUCTION

> Fuel cells have the potential to supply electricity to power a wrist watch or a large city, replacing a tiny battery or a power generating station.
>
> — George Wand, Fuel cell history, Part 1, *Fuel Cells Today*, April 2006

What Is a Fuel Cell? Definition of the Term

A *fuel cell* may be one of a variety of electrochemical power sources (EPSs), but is more precisely a device designed to convert the energy of a chemical reaction directly to electrical energy. Fuel cells differ from other EPSs: the primary galvanic cells called *batteries* and the secondary galvanic cells called *accumulators* or *storage batteries*, (1) in that they use a supply of gaseous or liquid reactants for the reactions rather than the solid reactants (metals and metal oxides) built into the units; (2) in that a continuous supply of the reactants and continuous elimination of the reaction products are provided, so that a fuel cell may be operated for a rather extended time without periodic replacement or recharging.

Possible reactants or fuels for the current-producing reaction are natural types of fuel (e.g., natural gas, petroleum products) or products derived by fuel processing, such as hydrogen produced by the reforming of hydrocarbon fuels

Fuel Cells: Problems and Solutions, By Vladimir S. Bagotsky

or water gas (syngas) produced by treating coal with steam. This gave rise to their name: fuel cells.

Significance of Fuel Cells for the Economy

In this book we show that fuel cells, already used widely throughout the economy, offer:

- Drastically higher efficiency in the utilization of natural fuels for large-scale power generation in megawatt power plants, and a commensurate decrease in the exhaust of combustion products and contaminants into the atmosphere from conventional thermal power plants
- Improved operation of power grids by load leveling with large-scale plants for temporary power storage
- A widely developed grid of decentralized, silent, local power plants with a capacity of tens to hundreds of kilowatts for use as a power supply or as a combined power and heat supply in remote locations, buildings, or installations not hooked up to the grid, such as stations for meteorological and hydrological observation; and for use as an emergency power supply in individual installations such as hospitals and control points
- Traction power plants with a capacity of tens of kilowatts for large-scale introduction of electric cars, leading to an important improvement in the ecological situation in large cities and densely populated regions
- Installations for power supply to spacecraft and submarines or other underwater structures, in addition to supplying crews with drinking water
- Small power units with a capacity of tens of watts or milliwatts, providing energy for extended continuous operation of portable or transportable devices used in daily life, such as personal computers, videocameras, and mobile communication equipment, or in industrial applications such as signaling and control equipment

For all these reasons, the development of fuel cells has received great attention since the end of the nineteenth century. In the middle of the twentieth century, interest in fuel cells became more general and global when dwindling world resources of oil and more serious ecological problems in cities were recognized. Space exploration provided a singular stimulus from the 1950s onward. An additional push was felt toward the end of the twentieth century in connection with the advent of numerous portable and other small devices used for civil and military purposes, that required an autonomous power supply over extended periods of use.

Today, numerous fuel cell–based power plants have been built and operated successfully, on a scale of both tens of megawatts and tens or hundreds of kilowatts. A great many small fuel cell units are in use that output between a few milliwatts and a few watts. Fuel cells are already making an important contribution to solving economic and ecological problems facing humankind. There can be no doubt that this contribution will continue to increase.

Large-scale research and development (R&D) efforts concerning the development and application of fuel cells are conducted today in many countries, in national laboratories, in science centers and universities, and in industrial establishments. Several hundred publications in the area of fuel cells appear every month in scientific and technical journals.

CHAPTER 1

THE WORKING PRINCIPLES OF A FUEL CELL

1.1 THERMODYNAMIC ASPECTS

1.1.1 Limitations of the Carnot Cycle

Up to the middle of the twentieth century, all human energy needs have been satisfied by natural fuels: coal, oil, natural gas, wood, and a few others. The thermal energy Q_{react} set free upon combustion (a chemical reaction of oxidation by oxygen) of natural fuels is called the *reaction enthalpy* or *lower heat value* (LHV): "lower" because the heat of condensation of water vapor as one of the reaction products is usually disregarded. A large part of this thermal energy serves to produce mechanical energy in heat engines (e.g., steam turbines, various types of internal combustion engines).

According to one of the most important laws of nature, the *second law of thermodynamics*, the conversion of thermal to mechanical energy W_m is always attended by the loss of a considerable part of the thermal energy. For a heat engine working along a Carnot cycle within the temperature interval defined by an upper limit T_2 and a lower limit T_1, the highest possible efficiency, $\eta_{theor} \equiv W_m/Q_{react}$, is given by

$$\eta_{theor} = \frac{T_2 - T_1}{T_2} \tag{1.1}$$

Fuel Cells: Problems and Solutions, By Vladimir S. Bagotsky

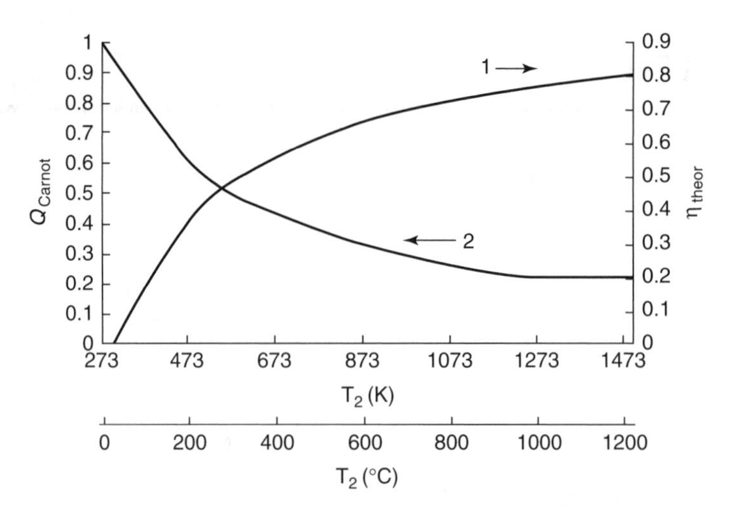

FIGURE 1.1 Limitations of the Carnot cycle. Theoretical efficiency η_{theor} (1) and the Carnot heat Q_{Carnot} (2) as functions of the upper operating temperature T_2 of the heat engine at a lower temperature T_1 of 298 K (25°C).

T_2 and T_1 being the temperatures (in kelvin) of the working fluid entering into and leaving the heat engine, respectively. The Carnot heat Q_{Carnot} (or *irretrievable heat*), for thermodynamic reasons known as the *Carnot-cycle limitations* is given by $Q_{Carnot} = (T_1/T_2)Q_{react}$. There is no way to reduce this loss. For a steam engine operating with superheated steam of 350°C ($T_2 = 623$ K) and release of the exhausted steam into a medium having an ambient temperature of 25°C ($T_1 = 298$ K), the maximum efficiency according to Eq. (1.1) is about 50%, so half of the thermal energy is irretrievably lost. As a matter of fact, the efficiency that can be realized in practice is even lower because of various other types of thermal losses Q_{loss} (e.g., heat transfer out of the engine, friction of moving parts); the total losses ($Q_{exh} = Q_{Carnot} + Q_{loss}$) are even higher. The efficiency η_{theor} can be raised by working with a higher value of T_2 (Figure 1.1), but losses due to nonideal heat transfer will also increase.

In part, the mechanical energy produced in heat engines is used, in turn, to produce electrical energy in the generators of stationary and mobile power plants. This additional step of converting mechanical into electrical energy involves additional energy losses, but these could be as low as 1 to 2% in a large modern generator. Thus, for a modern thermal power generating plant, a total efficiency η_{total} of about 40% is regarded as a good performance figure.

1.1.2 Electrochemical Energy Conversion

Until about 1850, the only source of electrical energy was the galvanic cell, the prototype of modern storage and throwaway batteries. In such cells, an electric current is produced through a chemical reaction involving an oxidizing agent and a reducing agent, which are sometimes quite expensive. In mercury primary cells, the current is generated through an overall reaction between mercuric oxide (HgO) and metallic zinc (Zn). In the cell, this *redox* (reducing and oxidizing) *reaction* occurs via an electrochemical mechanism that is fundamentally different from ordinary chemical mechanisms. In fact, in a reaction following chemical mechanisms, the reducing agent (here, Zn) reacts directly with the oxidizing agent (here, HgO):

$$Zn + HgO \rightarrow ZnO + Hg \tag{1.2}$$

the reaction involving a change in the valence states of the metals:

$$Zn + Hg^{2+} \rightarrow Zn^{2+} + Hg \tag{1.2a}$$

or electron transfer from Zn to Hg (the oxygen simply changing partners). If one were to mix zinc and mercuric oxides as powders in a reaction vessel and cause them to react, the electron transfers between the reacting particles would occur chaotically throughout the space taken up by the reactants, and no electron flow in any particular direction would be observed from the outside. For this reason, all of the chemical energy set free by the reaction would be evolved in the form of heat.

When an electrochemical mechanism is realized, then in the present example, electrons are torn away from the zinc at *one* electrode by making zinc dissolve in an aqueous medium:

$$Zn - 2e^- + 2OH^- \rightarrow ZnO + H_2O \tag{1.3}$$

or, essentially,

$$Zn - 2e^- \rightarrow Zn^{2+} \tag{1.3a}$$

and are added to mercuric oxide (HgO or Hg^{2+}) at *the other* electrode, by making the mercury deposit onto the electrode:

$$HgO + 2e^- + H_2O \rightarrow Hg + 2OH^- \tag{1.4}$$

or, essentially,

$$Hg^{2+} + 2e^- \rightarrow Hg \tag{1.4a}$$

the overall reaction occurring spatially separately at two different electrodes contacting the (aqueous) medium or electrolyte. Reaction (1.3) is zinc oxidation occurring as the anodic reaction at the anode. Reaction (1.4) is mercury reduction occurring as the cathodic reaction at the cathode. These two electrode reactions taken together yield the same products as those in chemical reaction (1.2).

Reactions (1.3) and (1.4) will actually proceed only when the two electrodes are connected outside the cell containing them. Electrons then flow from the zinc anode (the negative pole of the cell) to the mercuric oxide cathode (the positive pole). The cell is said to undergo *discharge* while producing current. Within the cell, the hydroxyl ions (OH^-) produced by reaction (1.4) at the cathode are transferred (migrate) to the anode, where they participate in reaction (1.3). The ions and electrons together yield a closed electrical circuit.

Of the total thermal energy of these two processes, Q_{react} [the *reaction enthalpy* ($-\Delta H$)], a certain part [called the *Gibbs reaction energy* ($-\Delta G$)] is set free as electrical energy W_e (the energy of the current flowing in the external part of the cell circuit). The remaining part of the reaction energy is evolved as heat, called the latent heat of reaction Q_{lat} [or reaction entropy ($-T\,\Delta S$)] (the latent heat in electrochemical reactions is analogous to the Carnot heat in heat engines):

$$Q_{react} = W_e + Q_{lat} \tag{1.5}$$

In summary, in the electrochemical mechanism, a large part of the chemical energy is converted directly into electrical energy without passing through thermal and mechanical energy forms. For this reason, and since the value of Q_{lat} usually (if not always) is small compared to the value of Q_{react}, the highest possible theoretical efficiency of this conversion mode,

$$\eta_{theor} = \frac{Q_{react} - Q_{lat}}{Q_{react}} \tag{1.6}$$

is free of Carnot cycle limitations and may approach unity i.e., 100%).* Even in this case, of course, different losses Q_{loss} have the effect that the practical efficiency is lower than the theoretical maximum, yet the efficiency will always be higher than that attained with a heat engine. The heat effectively exhausted in the electrochemical mechanism is the sum of the two components mentioned: $Q_{exh} = Q_{lat} + Q_{loss}$.

Toward the end of the nineteenth century, after the invention of the electric generator in 1864, thermal power plants were built in large numbers, and grid power gradually displaced the galvanic cells and storage batteries that had been used for work in laboratories and even for simple domestic devices. However, in 1894, a German physical chemist, Wilhelm Ostwald, formulated the idea that the electrochemical mechanism be used instead for the combustion (chemical oxidation) of natural types of fuel, such as those used in thermal power plants, since in this case the reaction will bypass the intermediate stage of heat generation. This would be cold combustion, the conversion of chemical

* For certain reactions, Q_{lat} is actually negative, implying that latent heat is absorbed by the system from the surrounding medium rather than being given off into the surrounding medium. In this case, the theoretical efficiency may even have values higher than 100%.

energy of a fuel to electrical energy not being subject to Carnot cycle limitations. A device to perform this direct energy conversion was named a *fuel cell*.

The electrochemical mechanism of cold combustion in fuel cells has analogies in living beings. In fact, the conversion of the chemical energy of food by humans and other living beings into mechanical energy (e.g., blood circulation, muscle activity) also bypasses the intermediate stage of thermal energy. The physiological mechanism of this energy conversion includes stages of an electrochemical nature. The average daily output of mechanical energy by a human body is equivalent to an electrical energy of a few tens of watthours.

The work and teachings of Ostwald were the beginning of a huge research effort in the field of fuel cells.

1.2 SCHEMATIC LAYOUT OF FUEL CELL UNITS

1.2.1 An Individual Fuel Cell

Fuel cells, like batteries, are a variety of galvanic cells, devices in which two or more *electrodes* (electronic conductors) are in contact with an *electrolyte* (the ionic conductor). Another variety of galvanic cells are *electrolyzers*, where electric current is used to generate chemicals in a process that is the opposite of that occurring in fuel cells, involving the conversion of electrical to chemical energy.

In the simplest case, a fuel cell consists of two metallic (e.g., platinum) electrodes dipping into an electrolyte solution (Figure 1.2). In an operating fuel cell, the negative electrode, the anode, produces electrons by "burning" a fuel. The positive electrode, the cathode, absorbs electrons in reducing an oxidizing agent. The fuel and the oxidizing agent are each supplied to its electrode. It is important at this point to create conditions that exclude direct mixing of the reactants or that supply to the "wrong" electrode. In these two undesirable cases, direct chemical interaction of the reactants would begin and would yield thermal energy, lowering or stopping the production of electrical energy completely.

So as to exclude accidental contact between anode and cathode (which would produce an internal short of the cell), an electronically insulating porous separator (holding an electrolyte solution that supports current transport by ions) is often placed into the gap between these electrodes. A solid ionically conducting electrolyte may serve at once as a separator. In any case, the cell circuit continues to be closed.

For work by the fuel cell to continue, provisions must be made to realize a continuous supply of reactant to each electrode and continuous withdrawal of reaction products from the electrodes, as well as removal and/or utilization of the heat being evolved.

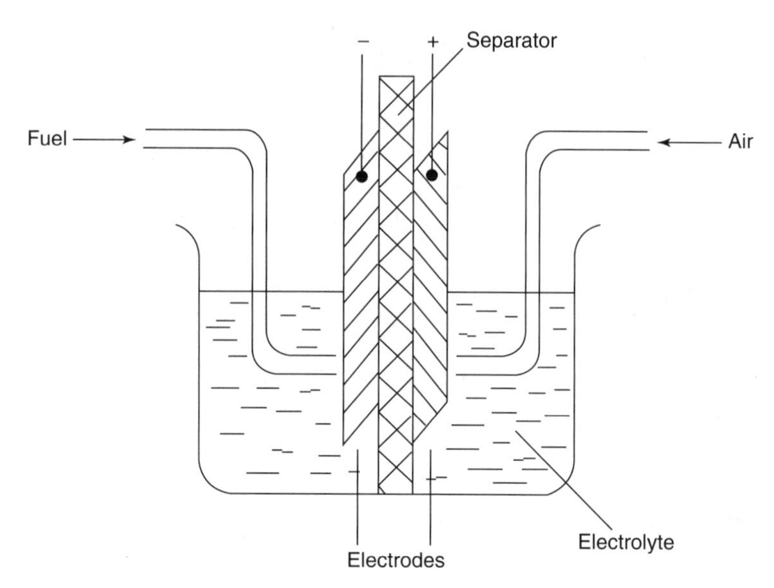

FIGURE 1.2 Schematic of an individual fuel cell.

1.2.2 Fuel Cell Stacks

As a rule, any individual fuel cell has a low working voltage of less than 1 V. Most users need a much higher voltage: for example, 6, 12, or 24 V or more. In a real fuel cell plant, therefore, the appropriate number of individual cells is connected in series, forming *stacks* (*batteries*).* A common design is the *filter-press design* of stacks built up of bipolar electrodes, one side of such electrodes working as the anode of one cell and the other side working as the cathode of the neighboring cell (Figure 1.3). The active (catalytic) layers of each of these electrodes face the separator, whose pores are filled with an electrolyte solution. A bipolar fuel cell electrode is generally built up from two separate electrodes, their backs resting on opposite sides of a separating plate known as the *bipolar plate*. These plates are electronically conducting and function as cell walls and intercell connectors (i.e., the current between neighboring cells merely crosses this plate, which forms a thin wall that has negligible resistance). This implies considerable savings in the size and mass of the stack. The bipolar plates alternate with electrolyte compartments, and both must be carefully sealed along the periphery to prevent electrolyte overflow and provide reliable separation of the electrolyte in neighboring compartments. The stacks formed from the bipolar plates (with their electrodes) and the electrolyte compartments (with their separators) are compressed and tightened with the aid of end plates and tie bolts. Sealing is achieved with the aid of gaskets compressed when

* A dc–dc transformer could be used to produce higher output voltage, but would introduce efficiency loss.

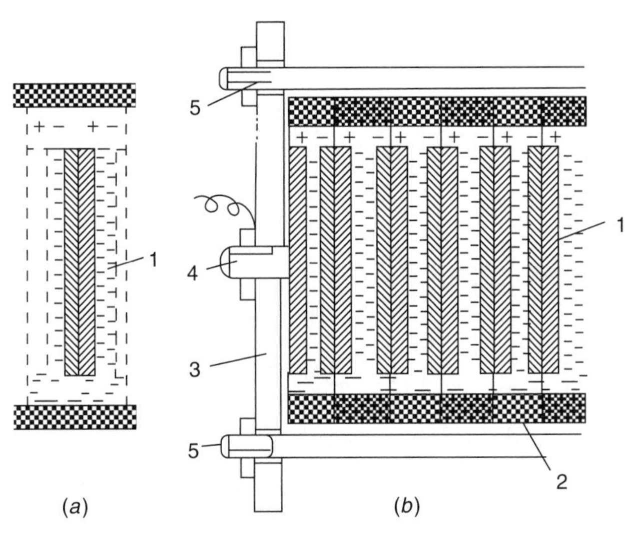

FIGURE 1.3 Fuel cell components: (a) bipolar electrode; (b) filter-press battery: 1, bipolar electrode; 2, gaskets; 3, end plate; 4, positive current collector; 5, tie bolts.

tightening the assembly. After sealing, the compartments are filled with electrolyte via manifolds and special narrow channels in the gaskets or electrode edges. Gaseous reactants are supplied to the electrodes via manifolds and grooves in the bipolar plates.

1.2.3 Power Plants Based on Fuel Cells

The heart of any fuel cell power plant (electrochemical generator or direct energy converter) is one or a number of stacks built up from individual fuel cells. Such plants include a number of auxiliary devices needed to secure stable, uninterrupted working of the stacks. The number or type of these devices depends on the fuel cell type in the stacks and the intended use of the plant. Below we list the basic components and devices. An overall layout of a fuel cell power plant is presented in Figure 1.4.

1. *Reactant storage containers.* These containers include gas cylinders, recipients, vessels with petroleum products, cryogenic vessels for refrigerated gases, and gas-absorbing materials among others.
2. *Fuel conversion devices.* These devices have as their purpose (a) the reforming of hydrocarbons, yielding technical hydrogen; (b) the gasification of coal, yielding water gas (syngas); or (c) the chemical extraction of the reactants from other substances, including devices for reactant purification, devices to eliminate harmful contaminants, and devices to separate particular reactants from mixtures. These are considered in greater detail in Chapter 11.

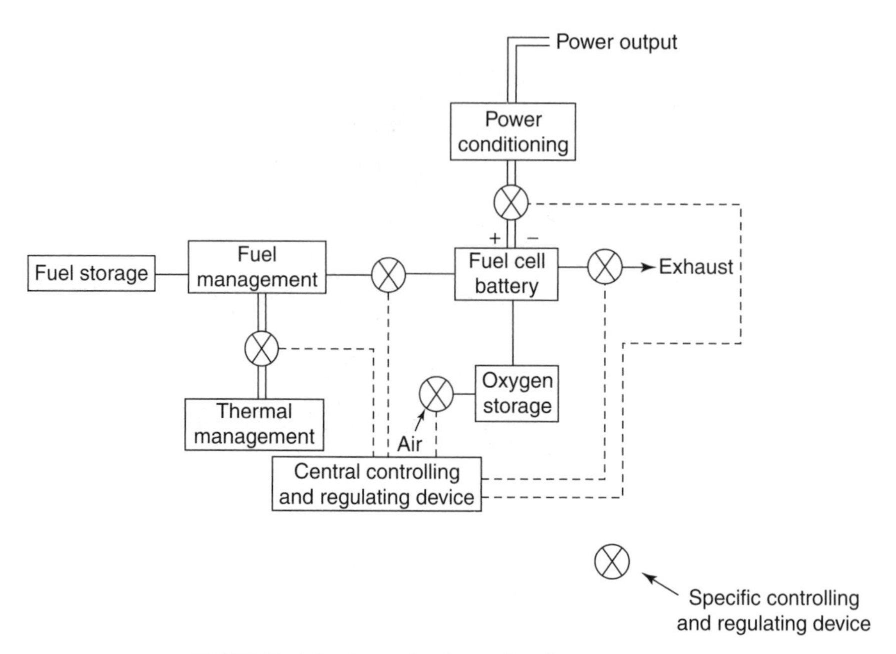

FIGURE 1.4 Overall schematic of a power plant.

3. *Devices for thermal management.* In most cases, the working temperature is distinctly above ambient temperature. In these cases the working temperature is maintained by exhaust (Q_{exh}) of the heat evolved during fuel cell operation. A cooling system must be provided when excess heat is evolved in fuel cell stacks. Difficulties arise when starting up the plant while its temperature is below the working temperature (such as after interruptions). In these cases, external heating of the fuel cell stack must be made possible. In certain cases, sufficient heat may be generated in the stack by shorting with a low-resistance load, where heating is begun at a low current and leads to a larger current producing more heat, and so on, until the working temperature is attained.
4. *Regulating and monitoring devices.* These devices have as their purpose (a) securing an uninterrupted reactant supply at the required rate and amount, (b) securing product removal (where applicable, with a view to their further utilization), (c) securing the removal of excess heat and maintaining the correct thermal mode, and (d) maintaining other operating fuel cell parameters needed in continuous operation.
5. *Power conditioning devices.* These devices include voltage converters, dc–ac converters, and electricity meters, among others.
6. *Internal electrical energy needs.* Many of the devices listed include components working with electric power (e.g., pumps for gas supply or heat-transfer fluid circulation, electronic regulating and monitoring

devices). As a rule, the power needed for these devices is derived from the fuel cell plant itself. This leads to a certain decrease in the power level available to consumers. In most cases these needs are not very significant. In certain cases, such as when starting up a cold plant, heating using an external power supply may be required.

1.3 TYPES OF FUEL CELLS

Different attributes can be used to distinguish fuel cells:

1. *Reactant type.* As a fuel (a reducing agent), fuel cells can use hydrogen, methanol, methane, carbon monoxide (CO), and other organic substances, as well as some inorganic reducing agents [e.g., hydrogen sulfide (H_2S), hydrazine (N_2H_4)]. As the oxidizing agent, fuel cells can use pure oxygen, air oxygen, hydrogen peroxide (H_2O_2), and chlorine. Versions with other, exotic reactants have also been proposed.
2. *Electrolyte type.* Apart from the common liquid electrolytes (i.e., aqueous solutions of acids, alkalies, and salts; molten salts), fuel cells often use solid electrolytes (i.e., ionically conducting organic polymers, inorganic oxide compounds). Solid electrolytes reduce the danger of leakage of liquids from the cell (which may lead to corrosive interactions with the construction materials and also to shorts, owing to contact between electrolyte portions in different cells of a battery). Solid electrolytes also serve as separators, keeping reactants from reaching the wrong electrode space.
3. *Working temperature.* One distinguishes low-temperature fuel cells, those having a working temperature of no more than 120 to 150°C; intermediate-temperature fuel cells, 150 to 250°C; and high-temperature fuel cells, over 650°C. Low-temperature fuel cells include membrane-type fuel cells as well as most alkaline fuel cells. Intermediate-temperature fuel cells are those with phosphoric acid electrolyte as well as alkaline cells of the Bacon type. High-temperature fuel cells include fuel cells with molten carbonate (working temperature 600 to 700°C) and solid-oxide fuel cells (working temperature above 900°C). In recent years, interim-temperature fuel cells with a working temperature in the range 200 to 650°C have been introduced. These include certain varieties of solid-oxide fuel cells developed more recently. The temperature ranges are stated conditionally.

1.4 LAYOUT OF A REAL FUEL CELL: THE HYDROGEN–OXYGEN FUEL CELL WITH LIQUID ELECTROLYTE

At present, most fuel cells use either pure oxygen or air oxygen as the oxidizing agent. The most common reducing agents are either pure hydrogen or technical hydrogen produced by steam reforming or with the water gas shift reaction from coal, natural gas, petroleum products, or other organic compounds. As an example of a real fuel cell we consider the special features of a hydrogen–oxygen fuel cell with an aqueous acid electrolyte. Special features of other types of fuel cells are described in later sections.

1.4.1 Gas Electrodes

In a hydrogen–oxygen fuel cell with liquid electrolyte, the reactants are gases. Under these conditions, porous gas-diffusion electrodes are used in the cells. These electrodes (Figure 1.5) are in contact with a gas compartment (on their back side) and with the electrolyte (on their front side, facing the other electrode). A porous electrode offers a far higher true working surface area and thus a much lower true current density (current per unit surface area of the electrode). Such an electrode consists of a metal- or carbon-based screen or plate serving as the body or frame, a current collector, and support for active layers containing a highly dispersed catalyst for the electrode reaction. The pores of this layer are filled in part with the liquid electrolyte and in part with the reactant gas. The reaction itself occurs at the walls of these pores along the three-phase boundaries between the solid catalyst, the gaseous reactant, and the liquid electrolyte.

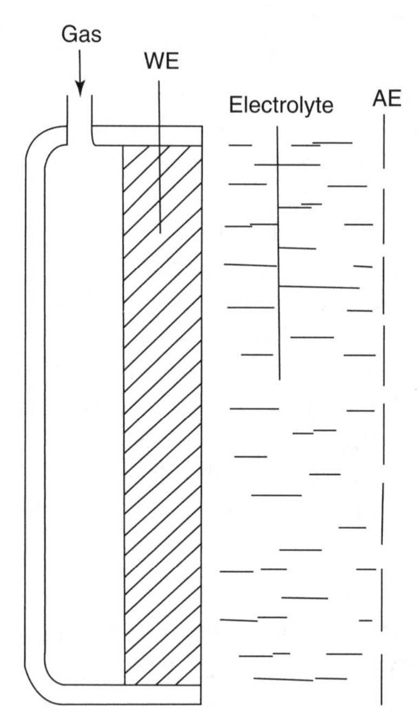

FIGURE 1.5 Schematic of a gas-diffusion electrode. WE, working electrode; AE, auxiliary electrode.

For efficient operation of the electrode, it is important to secure a uniform distribution of reaction sites throughout the porous electrode. With pores that have hydrophilic walls, walls well wetted by the aqueous electrolyte solution, the risk of flooding the electrode—or of complete displacement of gas from the pore space—exists. There are two possibilities for preventing this flooding of the electrode:

1. The electrode is made partly hydrophobic by adding water-repelling material. Here it is important to maintain an optimum degree of hydrophobicity. When there is an excess of hydrophobic material, the aqueous solution will be displaced from the pore space.
2. The porous electrode is left hydrophilic, but from the side of the gas compartment the gas is supplied with a certain excess pressure so that the liquid electrolyte is displaced in part from the pore space. To prevent gas bubbles from breaking through the porous electrode (and reaching the counterelectrode), the front side of the electrode that is in contact with the electrolyte is covered with a hydrophilic blocking layer having fine pores with a capillary pressure too high to be overcome by the gas, so that the electrolyte cannot be displaced from this layer. Here it is important to select an excess gas pressure that is sufficient to partially fill the active layer with gas, but insufficient to overcome ("break through") the blocking layer.

1.4.2 Electrochemical Reactions

A reaction of the type (1.3) occurs at the (negative) hydrogen electrode, or anode:

$$2H_2 \rightarrow 4H^+ + 4e^- \tag{1.7}$$

while a reaction of type (1.4) occurs at the (positive) oxygen electrode, or cathode*:

$$O_2 + 4H^+ + 4e^- \rightarrow 2H_2O \tag{1.8}$$

The hydrogen ions being formed in the electrolyte layer next to the anode in reaction (1.7) are transferred through the electrolyte toward the cathode, where they undergo reaction (1.8). In this way a closed electrical circuit is obtained. In the electrolyte, a (positive) electrical current flows from the anode to the

* Sometimes the opposite definition is encountered, where the anode is the positive pole of a galvanic cell and the cathode is the negative pole. This definition is valid for electrolyzers but not for fuel cells and other electrochemical power sources, the direction of current in the latter being the opposite of that in electrolyzers.

cathode; in the external circuit it flows in the opposite direction, from the cathode terminal to the anode terminal. The overall chemical reaction producing the current is

$$2H_2 + O_2 \rightarrow 2H_2O \tag{1.9}$$

which means that by reaction of 2 mol of hydrogen and 1 mol of oxygen (at atmospheric pressure and a temperature of 25°C, 1 mol of gas takes up a volume of 24.2 L), 2 mol of water (36 g) is formed as the final reaction product.

The thermal energy Q_{react} (or reaction enthalpy $-\Delta H$) set free in reaction (1.9) when this occurs as a direct chemical reaction amounts to 285.8 kJ/mol. The Gibbs free energy $-\Delta G$ of the reaction amounts to 237.1 kJ/mol. This value corresponds to the maximum electrical energy W_e^{max} that could theoretically be gained from the reaction when following the electrochemical mechanism. This means that the maximum attainable thermodynamic efficiency η_{therm} of energy conversion in this reaction is 83%.

For practical purposes it is convenient to state these energy values in electron volts (1 eV $= n \cdot 96.43$ kJ/mol, where n is the number of electrons taking part in the reaction per mole of reactant, in this case per mole of hydrogen). In these units, the enthalpy of this reaction (with $n = 2$ per mole) is 1.482 eV and the Gibbs free energy is 1.229 eV. In the following, the heat of reaction expressed in electron volts is denoted as q_{react}.

1.4.3 Electrode Potentials

At each electrode in contact with an electrolyte, a defined value of electrode potential E is set up. It can only be measured relative to the potential of another electrode. By convention, in electrochemistry the potential of any given electrode is referred to the potential of the *standard hydrogen electrode* (SHE), which in turn, by convention, is taken as zero. A practical realization of the SHE is that of an electrode made of platinized platinum dipping into an acid solution whose mean ionic activity of the hydrogen ions is unity, washed by gaseous hydrogen at a pressure of 1 bar.

In our example, the potential $E_{\mathrm{h.e.}}$ of the hydrogen electrode, to which, according to reaction (1.7), electrons are transferred from the hydrogen molecule, is more negative than the potential $E_{\mathrm{o.e.}}$ of the oxygen electrode, which, according to reaction (1.8), gives off electrons to an oxygen molecule.

The potentials of electrodes can be equilibrium or reversible, or non-equilibrium or irreversible. An electrode's equilibrium potential (denoted E^0 below) reflects the thermodynamic properties of the electrode reaction occurring at it (thermodynamic potential). The hydrogen electrode is an example of an electrode at which the equilibrium potential is established. When supplying hydrogen to the gas-diffusion electrode mentioned above, a value of electrode potential $E^0_{\mathrm{h.e.}}$ is established at it (when it is in contact with the appropriate electrolyte) that corresponds to the thermodynamic parameters of reaction (1.7). On the SHE scale, this value is close to zero (depending on the pH value of the solution, it differs insignificantly from the potential of the SHE itself).

An example of an electrode having a nonequilibrium value of potential is the oxygen electrode. The thermodynamic value of potential $E^0_{\text{o.e.}}$ of an oxygen electrode at which reaction (1.8) takes place is 1.229 V (relative to the SHE). When supplying oxygen to a gas-diffusion electrode, the potential actually established at it is 0.8 to 1.0 V, that is, 0.3 to 0.4 V less (less positive) than the thermodynamic value.

The degree to which electrode potentials are nonequilibrium values depends on the relative rates of the underlying electrode reactions. Under comparable conditions, the rate of reaction (1.8), cathodic oxygen reduction, is 10 orders of magnitude lower than that of reaction (1.7), anodic hydrogen oxidation.

In electrochemistry, reaction rates usually are characterized by values of the exchange current density i^0, in units of mA/cm^2, representing (equal values of) current density of the forward and reverse reactions at the equilibrium potential when the net reaction rate or current is zero.

The reaction rates themselves depend strongly on the conditions under which the reactions are conducted. Cathodic oxygen reduction, more particularly, which at temperatures below 150°C is far from equilibrium, comes closer to the equilibrium state as the temperature is raised.

The reasons that the real value of the electrode potential of the oxygen electrode is far from the thermodynamic value, and why cathodic oxygen reduction is so slow at low temperatures, are not clear so far, despite the large number of studies that have been undertaken to examine it.

1.4.4 Voltage of an Individual Fuel Cell

As stated earlier, the electrode potential of the oxygen electrode is more positive than that of the hydrogen electrode, the potential difference existing between them being the *voltage U* of the fuel cell:

$$U = E_{\text{o.e.}} - E_{\text{h.e.}} \tag{1.10}$$

When the two electrodes are linked by an external electrical circuit, electrons flow from the hydrogen to the oxygen electrode through the circuit, which is equivalent to (positive) electrical current flowing in the opposite direction. The fuel cell operates in a *discharge mode*, in the sense of reactions (1.7) and (1.8) taking place continuously as long as reactants are supplied.*

The thermodynamic value of voltage (i.e., the difference between the thermodynamic values of the electrode potentials) has been termed the cell's

* The term *discharge* ought to be seen as being related to a *consumption* of the reactants, which in a fuel cell are extraneous to the electrodes but in an ordinary battery are the electrodes themselves.

electromotive force (EMF), which in the following is designated as $\mathscr{E}^0(\mathscr{E}^0 = E^0_{\text{o.e.}} - E^0_{\text{h.e.}})$. The EMF of the hydrogen–oxygen fuel cell (in units of volts) corresponds numerically to the Gibbs free energy of the current-producing reaction (1.9) (in units of electron volts) [i.e., $\mathscr{E}^0(= W_e) = 1.229\,\text{V}$].

The practical value of the voltage of an idle cell is called the *open-circuit voltage* (OCV) U_0 of this cell. For a hydrogen–oxygen fuel cell, the OCV is lower than $\mathscr{E}^0$, owing to the lack of equilibrium of the oxygen electrode. Depending as well on technical factors, it is 0.85 to 1.05 V.

The working voltage of an operating fuel cell U_i is even lower because of the internal ohmic resistance of the cell and the shift of potential of the electrodes occurring when current flows, also called *electrode polarization*, and caused by slowness or lack of reversibility of the electrode reactions. The effects of polarization can be made smaller by the use of suitable catalysts applied to the electrode surface that accelerate the electrode reactions.

The voltage of a working cell will be lower the higher the current I that is drawn (the higher the current density $i = I/S$ at the electrode's working surface area S). The current–voltage relation is a cell characteristic, as shown in Figure 1.6. Sometimes this relation can be expressed by the simplified linear equation

$$U_i = U_0 - IR_{\text{app}} \qquad (1.11)$$

where the apparent internal resistance R_{app} is conditionally regarded as constant. This is a rather rough approximation, since R_{app} includes not only the cell's internal ohmic resistance but also components associated with polarization of the electrodes. These components are a complex function of current density and other factors. Often, the U_i versus I relation is S-shaped. Sometimes it is more convenient to describe the relation in the coordinates of U_i versus ln I. At moderately high values of the current, the voltage of an individual hydrogen–oxygen fuel cell, U_i, is about 0.7 V.

1.5 BASIC PARAMETERS OF FUEL CELLS

1.5.1 Operating Voltage

Fuel cell systems differ in the nature of the components selected, and thus in the nature of the current-producing chemical reaction. Each reaction is associated with a particular value of enthalpy and Gibbs free energy of the reaction, and thus also with a particular value of the heat of reaction Q_{reac} and of the thermodynamic EMF $\mathscr{E}^0$. Very important parameters of each fuel cell are its open-circuit voltage (OCV) U_0 and its discharge or operating voltage U_i as observed under given conditions (at a given discharge current). It had been shown in Section 1.4.1 that the OCV is lower than the EMF if the potential of at least one of the electrodes is a nonequilibrium potential. The difference between $\mathscr{E}$ and U_i depends on the nature of the reaction. Because of the cell's internal resistance and of electrode polarization during current flow, the discharge or operating voltage U_i is lower than the OCV, U_0. In different systems the influence of polarization of the electrodes is different; hence, the difference between U_0 and U_i also depends on the nature of the electrode reaction.

1.5.2 Discharge Current and Discharge Power

The discharge current of a fuel cell at any given voltage U_i across an external load with the resistance R_{ext} is determined by Ohm's law:

$$I = \frac{U_i}{R_{ext}} \tag{1.12}$$

Since U_i in turn depends on the current, and writing Eq. (1.11) for the current, the expression for the current becomes

$$I = \frac{U_0}{R_{app} + R_{ext}} \tag{1.13}$$

During discharge of a fuel cell the power $P = U_i I$ is delivered, or using Eqs. (1.12) and (1.13), we obtain

$$P = \frac{U_i^2 R_{app}}{(R_{app} + R_{ext})^2} \tag{1.14}$$

With increasing current (decreasing R_{ext}), the voltage decreases; hence, the power–current relation goes through a maximum (Figure 1.6, curve 2).

Neither the discharge current nor the power output are sole characteristics of a fuel cell, since both are determined by the external resistance (load) selected by the user. However, the maximum admissible discharge current I_{adm} and associated maximum power P_{adm} constitute important characteristics of all cell types. These performance characteristics place a critical lower bound U_{crit} on cell voltage; certain considerations (such as overheating) make it undesirable to operate at discharge currents above I_{adm} or cell voltages below U_{crit}. To a certain extent the choice of values for I_{adm} and U_{crit} is arbitrary. Thus, in short-duration (pulse) discharge, higher currents can be sustained than in long-term discharge.

For sustainable thermal conditions in an operating fuel cell, it will often be necessary for the discharge current not to fall below a certain lower admissible limit $I_{min,adm}$. The range of admissible values of the discharge current and the

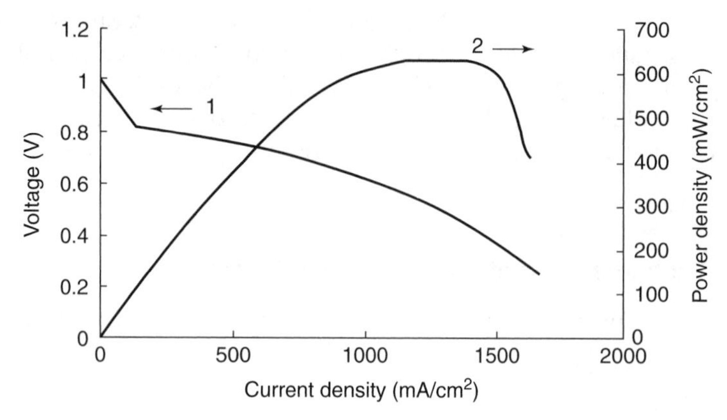

FIGURE 1.6 Typical current–voltage curve, and discharge power as a function of current load.

ability of a cell to work with different loads are important characteristics of each fuel cell.

1.5.3 Operating Efficiency of a Fuel Cell

The operating efficiency of a fuel cell is its efficiency in transforming a fuel's chemical energy to electrical energy, or the ratio between the electrical energy produced and the chemical energy of oxidation of a fuel supplied, $\eta = W_e/Q_{\text{react}}$. As a rule, the overall efficiency of fuel cells η_{total} is less than unity (less than 100%). A number of factors influence the overall efficiency.

Theoretical (Thermodynamic) Efficiency η_{therm}

The theoretical (thermodynamic efficiency was defined above by Eq. (1.6))

Voltage Efficiency η_V

The value of the voltage efficiency is given by

$$\eta_v = \frac{U_i}{\mathscr{E}^0}, \tag{1.15}$$

where U_i is the real operating voltage of the fuel cell during discharge at a current density i and $\mathscr{E}^0$ is the value of the EMF for the given fuel cell type, i.e., the highest thermodynamically possible value of the cell voltage. For hydrogen–oxygen fuel cells the value of $\mathscr{E}^0$ at a temperature of 25°C is 1.229 V.

Efficiency of Reactant Utilization: The Coulombic Efficiency η_{Coul} (Often Called Faradaic Efficiency)

Usually, not all of the mass or volume of the reactants supplied to a fuel cell stack is used for the current-producing reaction or production of electric charges (coulombs). External reasons for incomplete utilization include trivial leakage from different points in the stack. Intrinsic reasons include (1) diffusion of a reactant through the electrolyte (possibly a membrane) from "its own" to the opposite electrode, where it undergoes direct chemical reaction with the other reactant; (2) use of a reactant for certain auxiliary purposes, such as the circulation of (excess) oxygen serving to remove water vapor from parts of a membrane fuel cell and its subsequent venting to the ambient air; and (3) incomplete oxidation of individual organic fuel types: for example, an oxidation of part of methanol fuel to formic acid rather than to CO_2.

Design Efficiency η_{design}

Often, part of the electrical energy generated in a fuel cell is consumed for the (internal) needs of auxiliary equipment such as pumps supplying reactants and removing products, and devices for monitoring and controlling. The leakage of reactants mentioned above as a possibility also depends on design quality. If the fuel cells making up an electric power plant work with a secondary fuel derived on site from a primary fuel (such as with hydrogen made by steam reforming), the efficiency of such processing must also be taken into account.

Overall Efficiency η_{total}

The overall efficiency of the power plant will depend on all of the following factors:

$$\eta_{\text{total}} = \eta_{\text{therm}}\eta_{\text{volt}}\eta_{\text{Coul}}\eta_{\text{design}} \tag{1.16}$$

The overall efficiency is a very important parameter for fuel cell–based power plants, both the centralized plants of high capacity and the medium or small-capacity plants set up in large numbers in a distributed fashion. The basic goal of these setups is that of reducing the specific consumption of primary fuels for power generation.

1.5.4 Heat Generation

The amount of thermal energy liberated during operation of a fuel cell bears a direct relation to the value of the discharge operating voltage. When passing an electrical charge of λ_e coulombs, the total heat of reaction is given by $\lambda_e q_{\text{react}}$ joules (where the heat of reaction q_{react} is expressed in electron volts). The electrical energy produced is given by $\lambda_e U_i$ joules. The thermal energy produced will then be (in units of joules)

$$Q_{\text{exh}} = (q_{\text{react}} - U_i)\lambda_e \tag{1.17}$$

This includes both the latent heat Q_{lat} and all types of energy loss Q_{loss} incurred because of the efficiencies mentioned above being less than unity.

For hydrogen–oxygen fuel cells, $q_{\text{react}} = 1.48$ eV. With a discharge voltage of $U_i = 0.75$ V, heat generation amounts to $0.73\lambda_e$ joules, which is close to the value of electrical energy produced. Also, η_{volt} can be seen to be about 0.6 at this discharge voltage.

1.5.5 Ways of Comparing Fuel Cell Parameters

Often, a need arises to compare electrical and other characteristics of fuel cells that differ in their nature or size, or to compare fuel cell–based power generators with others. This is most readily achieved when using reduced or normalized parameters.

A convenient measure for the relative rates of current-producing reactions of fuel cells of a given type but differing in size is by using the current density, that is, the current per unit surface area S of the electrodes: $i = I/S$ (the units: mA/cm^2). The power density $p_s = P/S$ (the units: mW/cm^2) is a convenient measure of the relative efficiency of different varieties of fuel cells.

For users of fuel cells, important performance figures are the values of power density referred to unit mass M: $p_m = P/M$ (the units: W/kg) or unit volume V: $p_v = P/V$ (the units: W/L), and also the energy densities per unit mass (in Wh/kg) or unit volume (in Wh/L), both including the reactant supply. The power density is usually reported merely by referring to the mass or volume of the fuel cell battery itself but not to those of the power plant as a whole, since the mass and volume of reactants, including their storage containers, depend on the projected operating time of the plant. The energy density is usually reported for the power plant as a whole.

For stationary fuel cell–based power plants, the most important parameter is the energy conversion efficiency, inasmuch as this will define the fuel consumption per unit of electric power generated. For portable and other mobile power plants, the most important parameters are the power density and the energy density, inasmuch as they reflect the mass and volume of the mobile plant.

1.5.6 Lifetime

Theoretically, a fuel cell should work indefinitely, that is, as long as reactants are supplied and the reaction products and heat generated are duly removed. In practice, however, the operating efficiency of a fuel cell decreases somewhat in the long run. This is seen from a gradual decrease in the discharge or operating voltage occurring in time at any given value of the discharge or operating current. The rate of decrease depends on many factors: the type of current load (i.e., constant, variable, pulsed), observation of all operating rules, conditions of storage between assembly and use, and so on. It is usually stated in μW/h. If for a cell operated under constant load, the lifetime may be stated in hours, a better criterion for the lifetime of cells operated under a variable load is the total of energy generated, in Wh, while the rate of decrease of the voltage would then be given in μV/Wh.

The major reason for this efficiency drop is a drop in activity of the catalysts used to accelerate the electrode reactions. This activity drop may be due to:

- Spontaneous recrystallization of the highly disperse catalyst, its gradual dissolution in the electrolyte, or deposition of contaminants (inhibitors or catalytic poisons) on its surface
- A drop in ionic conductivity of the electrolyte: for example, of the polymer membrane in proton-exchange membrane and direct methanol fuel cells and that is caused by its gradual oxidative destruction
- The corrosion of different structural parts of fuel cells, leading to partial destruction and/or the formation of corrosion products that lower the activity of the electrodes, particularly in high-temperature fuel cells
- A loss of sealing of the cells: for example, because of aging of packings, so that it becomes possible for reactants to reach the "wrong" electrode

The rate of drop of fuel cell efficiency depends strongly on the mode and conditions of use. Periodic interruptions and temperature changes of idle cells from their operating temperature to ambient temperature and back when reconnected may have ill effects, and sometimes the documentation mentions an admissible number of load or temperature cycles. On relatively rare occasions, a fuel cell may suddenly fail, its voltage falling to almost zero. This type of failure is usually caused by an internal short that could occur when electrolyte leaks out through defective packing or when metal dendrites form and grow between electrodes.

It should be pointed out that since fuel cell problems are relatively new, few statistical data are available from which to judge the expected lifetime of different types of fuel cells under different operating conditions. The largest research effort goes into finding reasons for the gradual efficiency drop of fuel cells and finding possibilities to make it less important.

1.5.7 Special Operating Features

Transient Response

A fuel cell power plant is usually operated with variable loads, including the periodic connection and disconnection of different power consumers. This leads to periodic changes in the load resistance R_{ext} and the discharge or operating current. Any such act gives rise to a transient state where one parameter (e.g., current) changes and other parameters (e.g., heat removal) have to accommodate to the new conditions. For normal operation of fuel cell power plants, it is important that the time spent under transient operating conditions be as short as possible.

Startup

Problems often arise at the startup of a new cell stack after its manufacture and storage, or in repeated startup after a long idle period. Usually, the operating temperature of a fuel cell stack is higher than ambient or warehouse. If external heating is not possible, it may be possible, as pointed out in Section 1.2.3, to begin heating the battery on its own with a small discharge current and to raise its temperature gradually. An important criterion for a power plant is the time from switching on to full power.

The Effects of Climate

Any power plant should be operative over a wide range of temperatures and humidities of the surroundings. In most countries the temperature bracket needed reaches from −20 to +50°C. For countries with a cold climate, such as Russia and Canada, operation should be guaranteed down to −40°C.

Reliability and Convenient Manipulation

Power plants on the basis of fuel cell batteries constitute rather complex setups, including different operating, monitoring, and regulating units. The uninterrupted operation of these power plants depends largely on the smooth work of all these units. Their work should be governed by a single controlling unit or "brain." The work of operators running the plant should be minimized and reduced to that of "pushing buttons." The plant should also be sufficiently foolproof, in order not to react overly strongly to operator faults. Mobile plants for portable devices or transport applications should be compact and mechanically sturdy.

REFERENCE

Ostwald W., *Z. Elektrochem.*, **1**, 122 (1894).

CHAPTER 2

THE LONG HISTORY OF FUEL CELLS

In this chapter we reflect major points in the development of fuel cells, but for reasons of space limitations we cannot describe in greater detail the contributions of the many research workers who have worked in this field.

2.1 THE PERIOD PRIOR TO 1894

In 1791, the Italian physiologist Luigi Galvani (1737–1798) demonstrated in remarkable experiments that muscle contraction similar to that produced by discharge of a Leyden jar will occur when two different metals touch the exposed nerve of a frog. This phenomenon was in part interpreted correctly in 1792 by the Italian physicist Alessandro Volta (1745–1827), who showed that this galvanic effect originates from the contacts established between these metals and between them and the muscle tissue. In March 1800, Volta reported a device designed on the basis of this phenomenon of metal contact that could produce "inexhaustible electric charge." Now known as the *Volta pile*, this was the first example of an electrochemical device: an electrochemical power source (i.e., *a battery*).

The Volta pile (certainly not *in*exhaustible!) was of extraordinatory significance for developments in the science of both electricity and electrochemistry, since a new phenomenon, a continuous electric current, hitherto not

Fuel Cells: Problems and Solutions, By Vladimir S. Bagotsky

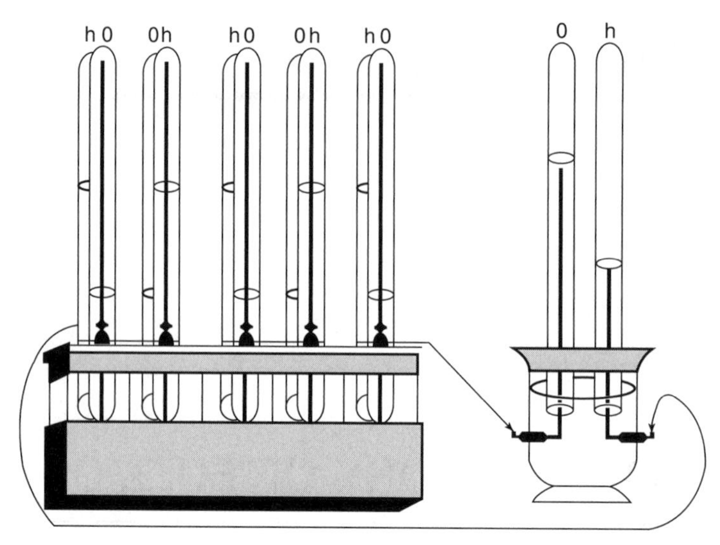

FIGURE 2.1 Grove's gas battery (1839). (From Davtyan, 1947.)

known, could now be realized. Soon various properties and effects of the electric current were discovered, including many electrochemical processes. In May 1801, William Nicholson and Sir Anthony Carlisle in London electrolyzed water, producing hydrogen and oxygen.

In the 1830s the British chemist Sir William Robert Grove (1811–1896), a trained lawyer and judge and an amateur natural scientist, conducted a series of experiments on water electrolysis. His device consisted of two platinum electrodes dipping into water acidified with sulfuric acid (Figure 2.1).*

He assumed that if by the passage of current, water can be decomposed to hydrogen and oxygen, there should be a possibility for the opposite reaction to occur. He saw, in fact, that after disconnecting the current, the electrodes at which hydrogen and oxygen had been evolved as gases were *polarized*; that is, a certain potential difference was preserved between them. When in this state they were linked by an external circuit, current was found to flow in this circuit. Grove called this invention a *gas voltaic battery*. His results were published in February 1839 in the *Philosophical Magazine*. This date is regarded as that of creation of the first prototype of a fuel cell.

Grove himself did not regard his gas battery as a practical means for producing electrical energy (instead, he developed a battery involving zinc and

* Christian Friedrich Schönbein, a German–Swiss chemist at Basel University, actually had discovered the principle in 1838 and published the finding one month earlier than Grove (Schönbein, 1839). Grove, who was doing similar experiments, then built the device often seen as an illustration in fuel cell histories and gave the correct explanation.

nitric acid). He pointed out, though, that such cells could be improved by increasing the contact area between the gases, adsorbents, and electrolyte.

Fifty years later, Ludwig Mond and Carl Langer (1889) conducted relatively successful experiments concerning the generation of electric current with hydrogen–oxygen cells. They used electrodes of platinized platinum (which have a large specific surface area). To prevent the flooding of the catalyst's pore space with electrolyte that would prevent access of gas to the catalyst, they immobilized the sulfuric acid serving as the electrolyte in a porous ceramic matrix. Their battery gave currents of 2 to 3 A at a voltage of 0.73 V; the electrodes had a surface area of 700 m^2. This battery was widely discussed in the German electrochemical literature. Its practical applications were greatly limited by its high price, also by lack of reproducibility of the results and a rapid decline in performance. "Nobody knew at that time what to do with an invention such as this fuel cell. No practical or commercial application was to be found for more than a century. Totally useless in its time" (Wand, 2006, pt. 1).

In 1894, the German physical chemist Wilhelm Ostwald (Figure 2.2) came forward in the *Zeitschrift für Elektrochemie* with the proposal to build devices for direct oxidation of natural types of fuel with air oxygen by an electrochemical mechanism without heat production (called the *cold combustion* of fuels). He

FIGURE 2.2 Friedrich Wilhelm Ostwald (1853–1932); Nobel prize, 1909.

wrote: "In the future, the production of electrical energy will be electrochemical, and not subject to the limitations of the second law of thermodynamics. The conversion efficiency thus will be higher than in heat engines." This paper of Ostwald was basic and marked the beginning of huge research in the field of fuel cells.

2.2 THE PERIOD FROM 1894 TO 1960

Ostwald had considered the thermodynamic aspects of fuel cells but had entirely disregarded their kinetic aspects: that is, the question of whether electrochemical reactions involving natural types of fuel are feasible, and how efficient they would be. Even the first experimental studies performed after publication of Ostwald's paper showed that it is very difficult to build devices for the direct electrochemical oxidation of natural kinds of fuel.

In 1896, William Jacques claimed to have developed a "coal battery" built with negative coal and positive iron electrodes immersed in molten alkali NaOH through which air was blown. In its time, this battery drew great attention and for some time was regarded as the only solution to the problem of building fuel cells. Subsequently it was shown, however (by Fritz Haber in 1896), that the anodic process taking place in Jacques' cell was oxidation of hydrogen evolved during the contact of iron with water in the electrolyte rather than the oxidation of coal.

Over the period from 1912 to 1939, a great deal of research into the electrochemical oxidation of coal and coal gasification products was done in Zürich, Switzerland and Brunswick, Germany by the Swiss scientist Emil Baur (1873–1944) and his co-workers. It was the basic assumption of these workers that the process will succeed only at those temperatures at which coal usually burns sufficiently rapidly. In their first studies, therefore (Baur et al., 1910, 1912, 1921), high-temperature molten electrolytes such as a mixture of sodium and potassium carbonates or molten caustic soda were used. Nickel, iron, and sometimes platinum were used as anodes for the oxidation of gases. Molten silver was the cathode for reduction of oxygen. With time, large difficulties were encountered due to the corrosivity of the molten electrolytes and to the instability of these fuel cells at high temperatures. Subsequent work was done with high-temperature solid electrolytes (Baur and Tobler, 1933; Baur and Preis, 1937, 1938). The basis of these electrolytes was the Nernst rod, consisting of 15% yttrium oxide and 85% zirconium oxide. Apart from its high cost, this electrolyte had another large defect, that of gradually losing conductivity during current flow.

After numerous studies, a cell for the electrochemical oxidation of coal was built (Figure 2.3). In this cell a tubular crucible consisting of a mixture of clay, cerium dioxide, and tungsten trioxide served as the electrolyte. The crucible was filled with powdered coke and placed into a clay vessel filled with annealed iron oxide (Fe_3O_4). At temperatures above 1000°C, the cell exhibited a voltage of 0.7 V and developed a power density (per unit of useful volume) of 1.33 W/dm^3.

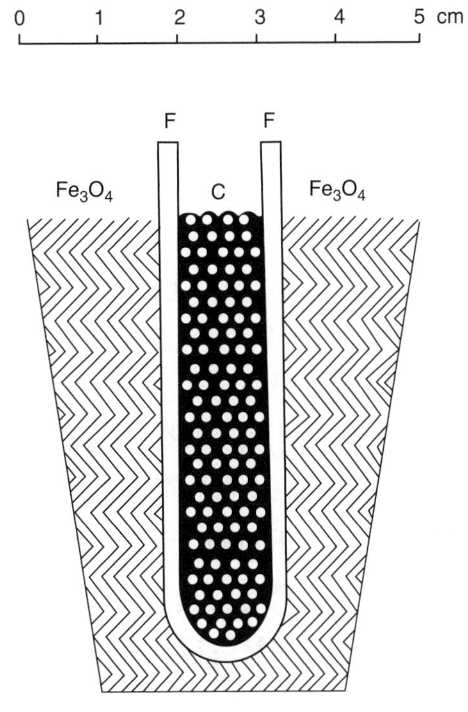

FIGURE 2.3 Coal battery of Baur and Tobler (1937). (From Davtyan, 1947.)

However, even this cell had a limited lifetime, owing to a gradual loss of conductivity, to mechanical destruction of the electrolyte, and to the effect of salt impurities in the coal. It is interesting to note that Walter Schottky, a prominent specialist in the field of solid electrolytes, found fuel cells with solid electrolytes unpromising, and favored those with salt melts (Schottky, 1936).

With the aim of producing a more highly conducting, mechanically and chemically stable electrolyte, the Armenian scientist Oganes Davtyan, who worked at the Moscow G. M. Krzizhanovsky Power Institute of the Russian Academy of Sciences in the 1930s, used a mixture of Urals' monazite sand (containing 3 to 4% ThO_2 and up to 15 to 20% rare-earth elements), tungsten trioxide (WO_3), calcium oxide, quartz, and clay. The cell developed by him was used to oxidize carbon monoxide (CO) with air oxygen at a temperature of 700°C; at a voltage of 0.79 V and a current density of 20 mA/cm^2 it could be operated for several tens of hours. Drastic temperature changes sometimes provoked cracking of the electrolyte. Davtyan also performed numerous studies on low-temperature hydrogen–oxygen cells. The results of his research were presented, together with a review of earlier work, in his book *The Problem of Direct Conversion of the Chemical Energy of Fuels into Electrical Energy*, published in 1947. This book is the first monograph in the world dedicated to fuel cells (Figure 2.4).

АКАДЕМИЯ НАУК СОЮЗА ССР

ЭНЕРГЕТИЧЕСКИЙ ИНСТИТУТ ИМ. Г. М. КРЖИЖАНОВСКОГО

О. К. ДАВТЯН

ПРОБЛЕМА НЕПОСРЕДСТВЕННОГО ПРЕВРАЩЕНИЯ ХИМИЧЕСКОЙ ЭНЕРГИИ ТОПЛИВА В ЭЛЕКТРИЧЕСКУЮ

ИЗДАТЕЛЬСТВО АКАДЕМИИ НАУК СССР

МОСКВА • 1947 • ЛЕНИНГРАД

FIGURE 2.4 The first monograph on fuel cells: O. K. Davtyan's book *The Problem of Direct Conversion of the Chemical Energy of Fuels into Electrical Energy* (1947).

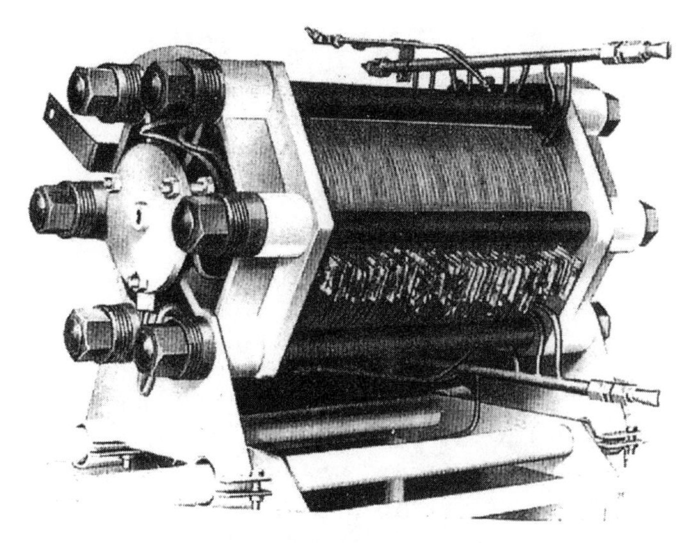

FIGURE 2.5 Bacon's battery of hydrogen–oxygen fuel cells, 1960 (From *Varta Fachbuchreihe*, Band 6, with permission of VDI Verlag.)

In 1932, Francis Thomas Bacon (1904–1992), an engineering professor at Cambridge University in England, began work on modifying the earlier Mond–Langer battery. Instead of highly corrosive acidic solutions, Bacon used an alkaline (KOH) electrolyte. He used the porous electrodes of the gas-diffusion type that had been suggested as early as 1923 by A. Schmid. To prevent gases from breaking through these electrodes, Bacon coated them on the side of the electrolyte with a gas-impermeable barrier layer with fine pores. The electrodes were made of nickel powder treated with hot lithium hydroxide solution to raise their corrosion resistance (lithiated nickel). His first alkaline fuel cell, the *Bacon cell*, was patented in 1959. In 1960, Bacon gave a public demonstration of a fuel cell battery producing a power of 5 to 6 kW. It contained a concentrated alkali solution (37 to 50% KOH). High temperatures (200°C and more) and high gas pressures (20 to 40 bar) were used to accelerate the electrode processes and attain high current densities (0.2 to 0.4 A/cm^2) (Figure 2.5). Because of the high gas pressure, the cell was very massive and heavy.

2.3 THE PERIOD FROM 1960 TO THE 1990s

Bacon's battery demonstration attracted great attention from the scientific and technical community, and in many countries research and design work in this field started on a large scale (the first "fuel cell boom"). Considerable interest was aroused in scientific circles by the work of Grubb and Niedrach

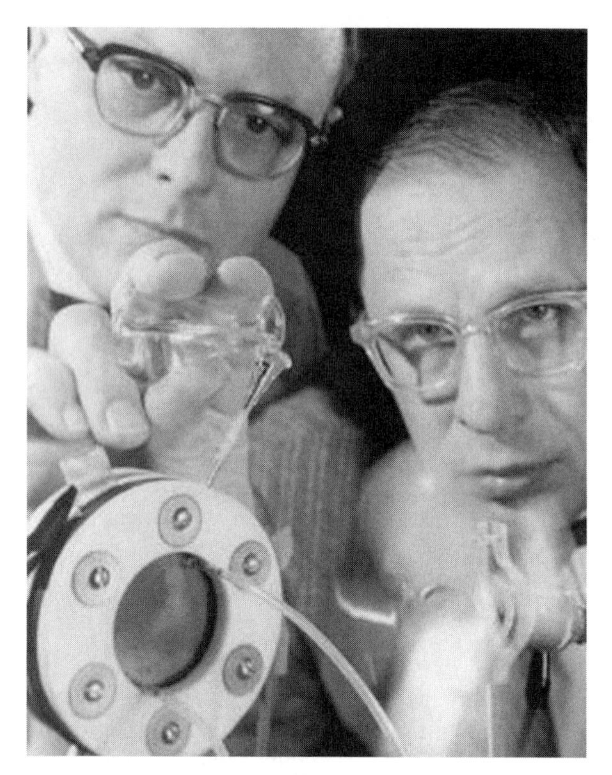

FIGURE 2.6 GE's Thomas Grubb (left) and Leonard Niedrach run a fan with a hydrocarbon-fed PEMFC, April 1963. (From Smithsonian Institution, neg. EMP059029, from the Science Service Historical Images Collection, courtesy of General Electric.)

(1960) and Grubb and Michalske (1964), who were able to show for the first time that hydrocarbons such as methane, ethane, and ethylene can be oxidized electrochemically using electrodes made of platinum metals at temperatures below 150°C (Figure 2.6). By comparison, the chemical process occurring at the same catalysts in the gas phase exhibits marked rates only at temperatures higher than 250°C. This work showed that despite the lack of success during the first half of the twentieth century, there is a basic chance of finding solutions for the direct conversion of the chemical energy of natural fuels to electrical energy.

Rising interest in fuel cells can to a considerable extent be attributed to the worldwide oil crisis that started to be felt in the late 1960s and came to a peak in 1973–1974. Great hopes were placed on fuel cells as a means to raise the efficiency of utilization of natural fuel resources. Another important factor stimulating fuel cell development was the space race between the United States and the USSR that began in the early 1960s and the beginning of the Cold War between these countries. In these two developments, fuel cells played a key role, which led to the military paying the cost of developments beneficial for later nonmilitary use. During the period examined in this section, almost all basic types of fuel cells now known evolved.

Alkaline Fuel Cells

In the early 1960s, the aircraft and engine manufacturer Pratt & Whitney obtained a license to use Bacon's patents. With the aim of simplifying the initial design and lowering the weight of the Bacon cell, very high alkali concentrations (85% KOH) were introduced so that the gas pressure could be lowered significantly. These P&W fuel cells were used in the *Apollo* program of space flights to the Moon.

Other workers gradually went to less concentrated alkali (30 to 40% KOH) than that used in Bacon's and P&W's cells. For the space shuttle program, United Technology Corporation (UTC Power) developed a battery of alkaline fuel cells (AFCs) in which 35% KOH immobilized in an asbestos matrix was used as the electrolyte. The electrodes contained a relatively large amount of platinum catalyst, so that at a temperature of 250°C it was possible to work at very high current densities, up to $1\ A/cm^2$.

In the laboratories of the agricultural equipment manufacturer Allis-Chalmers, a new version of a fuel cell with immobilized alkaline electrolyte solution was developed (Wynveen and Kirkland, 1962). The company reequipped one of its tractors to electric traction using an electric motor powered by four batteries consisting of 252 alkaline fuel cells each. The traction was strong enough for a load of 3000 lb. This tractor was exhibited at various agricultural fairs in the United States.

As early as the 1950s, the Austrian battery specialist Karl Kordesch built a rather efficient zinc air battery with a new type of carbon–air electrode (Kordesch and Marko, 1951). In later work at the U.S. company Union Carbide, Kordesch developed a fuel cell with an alkaline electrolyte using multilayer carbon electrodes with a small amount of platinum on the hydrogen side and with cobalt oxide on the oxygen side. He put a battery of such fuel cells into his car and was the first person to regularly use an electric car with fuel cells (on city roads in Cleveland and Parma, Ohio; Kordesch, 1963).

Toward the end of the 1950s a considerable contribution to the development of alkaline fuel cells was made by the German physicist Eduard Justi and his coworkers (1954). They made electrodes with nonplatinum catalysts called *Raney-type skeleton metals*: nickel for the hydrogen side and silver for the oxygen side. The catalysts were incorporated into a matrix of carbonyl nickel. These electrodes were called *Doppel-Skelett* (double skeleton) *electrodes* (Justi and Winsel, 1962).

Fuel Cells with a Proton-Exchange Membrane

Toward the end of the 1950s, several groups of scientists and engineers headed by Thomas Grubb and Leonard Niedrach (1963) at General Electric worked on

FIGURE 2.7 PEMFC in the *Gemini 7* spacecraft, 1965. (From Smithsonian Institution, neg. EMP059020, from the Science Service Historical Images Collection, courtesy of NASA.)

the development of fuel cells using a solid ion-exchange (proton-exchange) membrane as the electrolyte. After 1960, this *Grubb–Niedrach fuel cell* was commercialized and used for electric power generation in the *Gemini* spacecraft during its first flights in the early 1960s (Figure 2.7). When the *Gemini* flights ended, work on this type of fuel cell stopped almost entirely. For its subsequent space flights, NASA relied on alkaline fuel cells, known at that time as being more efficient. Work on proton-exchange membrane fuel cells (PEMFCs) was not taken up again until the very late 1980s.

High-Temperature Molten Carbonate Fuel Cells

It is the great advantage of high-temperature fuel cells that not only hydrogen but also hydrocarbons and carbon monoxide, a coal-processing product, can be used directly as a fuel. High-temperature fuel cells are intended primarily for use by large power plants, in the megawatt range. It is also important that the heat Q_{exh} evolved during the reaction is high-potential heat, that is, heat liberated at a high temperature, which can be used to generate additional electric energy in heat engines. In this way the original plant efficiency of 40 to 50% can be raised by another 10 to 20%.

In high-temperature molten carbonate fuel cells (MCFCs) a molten mixture of the carbonates of sodium, potassium, and lithium comprises the electrolyte. The working temperature of these cells is around 650°C. Such an electrolyte had been used by Emil Baur and his co-workers prior to 1920.

In the period 1958 to 1960, a large program of studies on these fuel cells was conducted at the University of Amsterdam in the Netherlands (Broers and Ketelaar, 1961). A little later, such studies were initiated at the Institute of High-Temperature Electrochemistry in Ekaterinburg, Russia (Stepanov, 1972, 1973, 1974).

Starting in 1960, work toward building real models of MCFCs was done at the Institute of Gas Technology, Chicago, and in a number of companies in the United States, Europe, and Japan. The interest in MCFCs gained significant impetus with the realization that apart from hydrogen, one could use methane and carbon monoxide as fuels. In the fuel cell itself, these fuels undergo an "internal conversion" to hydrogen, making prior reforming in an external plant unnecessary (Baker and Dharia, 1980). In various countries, tens of power plants operating with such cells and having a power output of tens to hundreds of kilowatts were built. The largest plants were demonstration plants of 2 MW in Santa Clara, California and in Japan.

High-Temperature Solid-Oxide Fuel Cells

Right after the work of Davtyan referred to earlier, vigorous studies into solid-oxide fuel cells (SOFCs) with electrolytes on the basis of zirconium dioxide doped with oxides of yttrium and other metals were begun in many places, particularly at the Institute of High-Temperature Electrochemistry already mentioned (Pal'guev and Volchenkova, 1958; Chebotin et al., 1971). The working temperature in these cells is in the range 800 to 1000°C. In 1962, important R&D work in this direction was begun at Westinghouse (Weissbart and Ruka, 1962). Analogous work was undertaken in many research and manufacturing establishments in the United States, Europe, and Japan. Operating models with powers of tens of kilowatts were built.

Medium-Temperature Phosphoric Acid Fuel Cells

The first information on medium-temperature phosphoric acid fuel cells (PAFCs) was that of Elmore and Tanner (1961). Highly concentrated phosphoric acid (85 to 95%) is used as the electrolyte in these cells. The working temperature is in the range of 180 to 200°C. These fuel cells quickly aroused great interest and found wide distribution. On the basis of such cells, numerous power plants up to 250 kW were built and used as an autonomous power supply for individual operating units such as hotels and hospitals. Also, in the United States, Japan, and other countries, megawatt plants were built on this basis and used to supply power to entire city districts. Compared to the low-temperature hydrogen–oxygen fuel cells, the phosphoric acid cells have the advantage of being able to work with less carefully purified technical hydrogen.

Fuel Cells Using Other Types of Fuel

Hydrogen is an electrochemically very active reducing agent (fuel). Thus, in hydrogen–oxygen fuel cells, relatively high values of current density and specific power per unit weight have been achieved at acceptable values of the discharge or operating voltage. However, hydrogen as a fuel is very complicated in its handling, storage, and transport. This is a problem primarily for relatively small, low-power plants of the portable or mobile type. For such plants, liquid fuels are much more realistic.

In the 1960s, several versions of alkaline fuel cells (AFCs) using liquid hydrazine as a fuel were built, including those for a motorcycle driven by Karl Kordesch. Hydrazine is also very active electrochemically and yields fuel cells with high-performance indicators. However, apart from its high cost, hydrazine has another notable defect: It is highly toxic. Therefore, the use of hydrazine fuel cells has been limited to a few special areas (chiefly for the military).

Methanol is a much more realistic fuel for fuel cells. The specific energy content of methanol when completely oxidized to CO_2 electrochemically is 0.84 Ah/g. For fuel cells with methanol as a fuel, acidic electrolyte solutions must be used. Alkaline solutions are inappropriate, since the alkali combines with CO_2 produced in the fuel cell to form insoluble carbonates. The first laboratory models of methanol–air fuel cells were built in the early 1960s (Grimes et al., 1961). Since large amounts of expensive platinum catalysts were used in these fuel cells, work in this direction soon ended and was not taken up again for many years.

Most types and models of fuel cells that had become known by the 1960s were described in monographs by E. Justi and A. Winsel (1962) and by W. Vielstich (1965), as well as in handbooks on fuel cells edited by G. J. Young (1960, 1963) and by W. Mitchell (1963).

During the next two decades, some decline of interest in fuel cells can be noted, and fewer studies appeared in this field. Technical and design improvements were introduced into models of the AFC, MCFC, and SOFC systems, and some large power plants were built. The basic structure of fuel cells themselves (composition of electrodes and electrolyte) and also the specific performance figures (per unit surface area of the electrodes) changed little during this time.

To a certain extent, the decline of interest in fuel cells was due to difficulties in the commercial realization of earlier achievements. Despite the demonstration that low-temperature electrochemical oxidation of hydrocarbons is basically possible, reaction rates realized in practice were too low, and the amounts of platinum–metal catalyst required to achieve them were so large that the economic prospects of fuel cells using these reactions were very poor. Platinum catalysts were used in most of the fuel cell built, despite the fact that in many studies it had been shown that nonplatinum catalysts could be useful for hydrogen and oxygen electrodes. For economic reasons, the number of potentially interested users decreased gradually. The financial support of work on fuel cells decreased correspondingly.

Work of an academic character concerned with the processes occurring in the different versions of fuel cells continued through the entire period mentioned. It was the aim of this work to find ways to raise the rate and efficiency of the electrochemical reactions taking place at the electrodes of fuel cells, mainly by raising their catalytic activity. These studies examined reactions at both the oxygen (air) electrode and at the hydrogen electrode (see the monograph of M. Breiter, 1969). In many countries, the special features of electrochemical oxidation of methanol at platinum catalysts were also studied. All these studies, as well as the work of Grubb and Niedrach mentioned earlier, concerning the possibility of low-temperature hydrocarbon oxidation led to the development of a new branch of electrochemical science, *electrocatalysis*.

2.4 THE PERIOD AFTER THE 1990s

Around the mid-1980s, apart from work toward building powerful plants for more efficient grid power generation, a second application of fuel cell work was discerned: that of building autonomous power sources of intermediate or small capacity. Such power sources are intended for applications where grid energy is inaccessible, such as in means of transport, in portable devices, and in remote uses. The term *fuel cell* began to lose its original meaning as an electrochemical power source operated by natural fuel and acquired the meaning of an electrochemical power source that in contrast to ordinary batteries works continuously as long as the reactants, a reducing agent (the fuel), and an oxidizer are supplied. The first application of this type of work was in the hydrogen–oxygen fuel cells used to power the equipment of the *Apollo* and *Gemini* manned spacecraft.

During the 1960s, Du Pont began development of a new polymeric ion-exchange membrane marketed as Nafion®. This had a dramatic effect on the further development of fuel cells, particularly for applications of the second type. It was soon obvious that this membrane could help to strongly enhance the characteristics and lifetime of relatively small fuel cells. At the same time, success was achieved toward substantially lower platinum catalyst outlays in such fuel cells. These developments aroused the interest of potential fuel cell users.

Beginning in the mid-1980s, published work on membrane fuel cells increased rapidly. This was driven by several factors, one of which was air pollution caused by the rising automobile population in large cities worldwide. This situation induced legislation in some places, such as in California, that required the introduction of a certain number of zero-pollution vehicles. Major carmakers worldwide began serious efforts to develop different versions of electric cars and associated power plants based on storage batteries and fuel cells.

Another factor contributing to the rising interest in fuel cells was a strong increase in the selection and number of different types of portable electronic devices (e.g., mobile phones, portable PCs). A need developed to extend drastically the time of continued operation of these devices and to replace ordinary batteries by power sources of higher capacity.

All these factors led to a very strong increase in the number of studies performed in this area, and one may safely speak of the start of a second boom in fuel cell R&D. The greatest progress was made in the area of membrane fuel cells. Today's membrane fuel cells differ widely from the prototypes of the 1960s in their design and characteristics (see Chapter 3). Improved membrane fuel cells of medium-size power output are widely used and produced on a large commercial scale.

Progress in the area of membrane hydrogen–oxygen fuel cells led to the development of fuel cells of a new type, direct methanol fuel cells (DMFCs), in which methanol is oxidized directly. The more convenient liquid fuel replaces hydrogen, a fuel that is inconvenient to handle, store, and transport.

At present, methanol is regarded as a very promising fuel for future electric cars. Two ways of using methanol are its direct oxidation in DMFCs and its prior conversion to technical hydrogen and the use of this hydrogen in hydrogen–oxygen PEMFCs. Each has shortcomings and problems. The specific performance figures for DMFCs are still too low for their use in an electric car. Power plants with prior conversion need complex and bulky additional equipment. Additional difficulties arise, since technical hydrogen always contains traces of carbon monoxide, which strongly affects the platinum catalysts used in PEMFCs.

The current development of DMFCs is based largely on results obtained in the development of PEMFCs. Similar problems must be solved, particularly those associated with the proton-exchange membrane and with catalysts for the electrodes. In addition, each type has its own problems.

Apart from the large volume of research and design work on PEMFCs and DMFCs, many studies on improved high-temperature fuel cells, SOFCs and MCFCs, have been conducted since 1990. A marked rise in the number of power plants based on MCFCs was seen between 2003 and 2005 (*Fuel Cell Today*, March 2007).

The volume of work concerned with alkaline fuel cells has declined considerably since the late 1980s. As to PAFCs, recent literature has offered almost no indications for research in this area.

REFERENCES

Bacon F. T., *Ind. Eng. Chem.*, **52**, 301 (1960); in: G. J. Young (ed.), *Fuel Cells*, Vol. 1, Reinhold, New York, 1960, p. 51.

Baker D. S., D. J. Dharia, U.S. patent 4,182,795 (1980).

Baur E., H. Preis, *Z. Elektrochem.*, **43**, 727 (1937); **44**, 695 (1938).

Baur E., J. Tobler, *Z. Elektrochem.*, **39**, 169 (1933).

Baur E., et al., *Z. Elektrochem.*, **16**, 286, 304 (1910); **18**, 1002 (1912); **22**, 409 (1916); **27**, 409 (1921).

Broers G. H. J., J. A. A. Ketelaar, in: G. J. Young (ed.), *Fuel Cells*, Vol. 1, Reinhold, New York, 1961, p. 78.

Chebotin V. N., M. V. Glumov, S. F. Palguev, A. D. Neuimin, *Elektrokhimiya*, **7**, 196 (1971).

Elmore G., H. A. Tanner, *J. Electrochem., Soc.*, **108**, 669 (1961).

Grimes P. G., B. Fiedler, J. Adams, *Proc. 15th Annual Power Sources Conference*, Atlantic City, NJ, 1961, p. 29.

Grove W., *Phil. Mag.*, **14**, 447 (1839).

Grubb W. T., U.S. patent 2,913,511 (1959).

Grubb W. T., *J. Electrochem. Soc.*, **111**, 1086 (1964).

Grubb W. T., C. J. Michalske, *J. Electrochem. Soc.*, **111**, 1015 (1964).

Grubb W. T., L. W. Niedrach, *J. Electrochem. Soc.*, **107**, 131 (1960).

Grubb W. T., L. W. Niedrach, *J. Electrochem. Soc.*, **110**, 1086 (1963); in: W. Mitchell (ed.), *Fuel Cells*, Academic Press, New York, 1963, p. 253.

Haber F., Experimental investigations on the decomposition and combustion of hydrocarbons [in German], habilitation thesis, Karlsruhe, Germany (1896).

Jacques W., *Electrical Rev.*, **38**, 896 (1896).

Justi E., W. Scheibe, A. Winsel, German patent 1,019,361 (1954).

Kordesch K., *Proc. IEEE*, **51** (5), 806 (1963); in: W. Mitchell (ed.), *Fuel Cells*, Academic Press, New York, 1963, p. 329.

Kordesch K., A. Marko, *Oesterr. Chem. Ztg.*, **52**, 125 (1951).

Mond L., C. Langer, *Proc. Roy. Soc.*, London, **46**, 296 (1889).

Ostwald W., *Z. Elektrochem.*, **1**, 122 (1894).

Pal'guev S. F., Z. S. Volchenkova, *Proc. Institute of High-Temperature Electrochemistry*, No. 2, 1958, p. 183.

Schmid A., *Die Diffusions-Elektrode*, Enke, Stuttgart, Germany, 1923.

Schönbein C. F., *Phil. Mag. London* (III), **44**, 43 (1839).

Schottky W., *Wiss.Veroeff. Siemens-Werken*, **14**, 1 (1936).

Stepanov G. K., *Proc. Institute of High-Temperature Electrochemistry*, No. 18, 1972, p. 129; No. 20, 1973, p. 95; No. 21, 1974, p. 88.

Weissbart J., R. Ruka, *J. Electrochem. Soc.*, **109**, 723 (1962).

Wynveen R. A., T. G. Kirkland, *Proc. 16th Annual Power Sources Conference*, Atlantic City, NJ, 1962, p. 24.

Monographs (1894–1990)

Baker B. S. (ed.), *Hydrocarbon Fuel Cell Technology*, Academic Press, New York, 1965.

Bockris J. O'M., S. Srinivasan, *Fuel Cells: Their Electrochemistry*, McGraw-Hill, New York, 1969.

Breiter M. W., *Electrochemical Processes in Fuel Cells*, Springer-Verlag, New York, 1969.

Davtyan O. K., *The Problem of Direct Conversion of the Chemical Energy of Fuels into Electrical Energy* [in Russian], Publishing House of the USSR Academy of Sciences, Moscow, 1947.

Justi E. W., A. Winsel, *Fuel Cells: Kalte Verbrennung*, Steiner, Wiesbaden, Germany, 1962.

Kordesch K. (ed.), *Brennstoffbatterien*, Springer-Verlag, Berlin, 1984.

Kordesch K. V., 25 years of fuel cell development (1951–1976), *J. Electrochem. Soc.*, 125, 77C (1978).

Liebhafsky H. A., E. J. Cairns, *Fuel Cells and Fuel Cell Batteries*, Wiley, New York, 1968.

Mitchell W. (ed.), *Fuel Cells*, Academic Press, New York, 1963.

Vielstich W., *Brennstoffelemente*, Verlag Chemie, Weinheim, Germany, 1965.

Young G. J. (ed.), *Fuel Cells*, Reinhold, New York, Vol. 1, 1960; Vol. 2, 1963.

Recent Reviews

Gottesfeld S., Fuel cell techno: personal milestones, 1984–2006, *J. Power Sources*, **171**, 37 (2007).

Perry M. L., T. E. Fuller, A historical perspective of fuel cell technology in the 20th century, *J. Electrochem. Soc.*, **149**, 859 (2002).

Wand G., Fuel cell history, Parts 1 and 2 *Fuel Cell Today*, Apr. 2006.

PART II

MAJOR TYPES OF FUEL CELLS

CHAPTER 3

PROTON-EXCHANGE MEMBRANE FUEL CELLS

Although not obvious from the designation, this group includes only fuel cells that use hydrogen and oxygen as the reactants. Methanol–oxygen fuel cells that use a proton-conducting membrane are known as direct methanol fuel cells (see Chapter 4).

In proton-exchange membrane fuel cells (PEMFCs),* the polymeric proton-exchange membrane serves as a solid electrolyte. The membrane's conductivity comes about because in the presence of water, it swells, a process leading to dissociation of the acidic functional groups and formation of protons free to move about throughout the membrane.

The electrochemical reactions occurring at the electrodes of PEMFCs as well as the overall current-producing reaction are the same as in the hydrogen–oxygen fuel cells with liquid acidic electrolyte discussed in Section 1.4. [reactions (1.7) and (1.8)].

It will become obvious from the following outline that an important fact to remember when looking at PEMFCs is the hydration of protons in the aqueous medium. The protons always exist in the form of the species $H^+ \cdot nH_2O$. To

*The acronym PEMFC also stands for the equivalent term, polymer electrolyte membrane fuel cell.

Fuel Cells: Problems and Solutions, By Vladimir S. Bagotsky

reflect this, instead of (1.7) and (1.8), the electrode reactions can be written as

$$2H_2 + 4nH_2O \rightarrow 4H^+ \cdot nH_2O + 4e^- \quad (3.1a)$$

at the hydrogen anode and

$$O_2 + 4H^+ \cdot nH_2O + 4e^- \rightarrow (n+2)H_2O \quad (3.2a)$$

at the oxygen cathode.

During discharge of the cell, when current flows, the hydrated protons migrate in the membrane from the anode toward the cathode, each proton dragging n water molecules along in the same direction.

3.1 HISTORY OF THE PEMFC

A first version of a PEMFC battery that had a power of 1 kW was built in the early 1960s by General Electric for the *Gemini* spacecraft. A sulfated polystyrene ion-exchange membrane was used as the electrolyte in these cells. The electrodes contained about 4 mg/cm^2 of platinum catalyst. Because of the marked ohmic resistance of the membrane, the current density was below 100 mA/cm^2, with a voltage of about 0.6 V for an individual element. This corresponds to a specific power of the cells of about 60 mW/cm^2. Due to insufficient chemical stability of the membrane used, the total lifetime of the battery was below 2000 hours. The high cost of such a battery excluded uses in fields other than space flight.

After a dormant period lasting almost three decades, striking improvements in PEMFC properties were achieved after 1990. The specific power of current PEMFCs went up to 600 to 800 mW/cm^2 while less than 0.4 mg/cm^2 of platinum catalysts are used, and the lifetime is now several tens of thousands of hours. On the strength of these achievements, P. Costamagna and S. Srinivasan were able to title their comprehensive analytical review of 2001: "Quantum jumps in PEMFC science and technology from the 1960s to the year 2000." This breakthrough, as well as the need to develop zero-emission vehicles after enactment of the corresponding laws in California in 1993, led to a brisk pace of scientific and engineering work on PEMFCs.

Basically, a combination of three factors was responsible for this important progress:

1. Introduction of Nafion proton-exchange membranes
2. The development of new methods, yielding a drastic increase in the efficiency of platinum catalyst utilization in the electrodes
3. The development of unique membrane–electrode assemblies

3.1.1 Nafion Membranes

Between 1960 and 1980, the U.S. chemical company Du Pont de Nemours began delivery of ion-exchange membranes of a new type trade-named Nafion. They are made of a perfluorinated sulfonic acid polymer (PSAP) consisting of a continuous skeleton of $–(CF_2)_n–$ groups to which a certain number of hydrophilic segments containing sulfonic acid groups $–SO_3H$ are attached. The hydrophobic skeleton is responsible for the very high chemical stability, greatly exceeding that of membranes known previously. When the membrane is wetted appropriately, the sulfonic acid groups dissociate and at a temperature of 80°C provide a relatively high protonic membrane conductivity of about 0.1 S/cm, three to four times higher than that of the ion-exchange membranes that had been used previously.

Nafion-type membranes had been developed initially for the needs of the chlorine industry. Chlorine cells with such membranes were widely introduced. Apart from chlorine, they yield as the second major electrolysis product alkali that is very pure. Quite soon after the advent of these membranes, work on fuel cells including them was begun (Grot, 1985). Fuel cells with such membranes had lifetimes two to three orders of magnitude longer.

At high current densities, the ohmic resistance of the membrane still affects the properties of the fuel cell. For improved performance, thinner membranes of Nafion 112 (50 μm) were introduced in place of the standard Nafion 115 membranes (100 μm). Even thinner membranes have been developed (Nafion 111, 25 μm), but cells with such membranes failed occasionally because of small membrane defects that allowed the gases to mix. Tests showed that apart from membrane thickness, the polymer's equivalent weight will also influence the cell parameters. Thus, cells with Nafion 1035 membranes (87 μm thick, equivalent weight 960) were superior to cells with Nafion 112 (equivalent weight 1100) (Gamburzev and Appelby, 2002). Very reliable membranes were supplied by W.L. Gore & Associates. They were made by applying dissolved Nafion onto a thin Teflon screen. The U.S. company Dow Chemical and the Japanese company Asahi Chemicals synthesized and supplied analogous membranes on the basis of PSAP having a relatively high concentration of sulfonic acid groups, leading to a certain decrease in resistance of the membrane and improved characteristics of fuel cells built with them.

Membranes of the Nafion type are very sensitive to heavy-metal contaminants. Thus, chromium ions in a concentration of 500 ppm produce an eightfold conductivity drop. High demands must therefore be satisfied with respect to the composition and purity of materials used for the electrodes and cell structure (Peled et al., 2000).

Fuel cells with Nafion-type membranes are usually operated at a temperature of 80 to 90°C. A sufficient water content of the membranes must be secured by continued moistening, which is done by saturating the reactant gases supplied to the cell with water vapor by passing them through water having a temperature somewhat higher (by 5 to 10°C) than the cell's operating temperature. The optimum moistening conditions depend on membrane thickness and on the current drawn from the fuel cell (Susai et al., 2001).

A general disadvantage of all membranes of this type is the complicated manufacturing process and associated high cost. A Nafion membrane costs about $700 per square meter, or $120 per kilowatt of designated power (at an average power density of $600\,mW/cm^2$).

3.1.2 More Efficient Utilization of Platinum Catalysts

Platinum is the basic catalyst for electrochemical reactions at hydrogen and oxygen electrodes. In the first models of membrane fuel cells, the dispersed platinum catalyst was in a pure form, so that large amounts of platinum were used. A considerably higher utilization efficiency and much lower metal input were attained when platinum deposited onto highly dispersed carbon carriers was used. The best results were attained with Vulcan XC-72 furnace carbon black as a support, possibly because of the small amounts of sulfur contained in it, which give rise to special surface properties (Swider and Rolison, 1996). In various reports, the amount of platinum in the platinized carbon black is between 10 and 40% by mass, sometimes even more. At low platinum contents, its specific surface area on this support can be as high as $100\,m^2/g$ (with a platinum crystallite size of 0.3 nm), while for unsupported dispersed platinum the specific surface area is usually not larger than $15\ m^2/g$ (with platinum crystallites of 1 to 2 nm). At low platinum contents in the carbon black, thicker catalyst layers must be used, which interferes with reactant transport to the reaction sites. The degree of dispersion of the platinum and its catalytic activity depend not only on the amount deposited on the support but also on the method of chemical or electrochemical deposition of platinum onto the support. Many studies into the correlation between the method of platinum deposition and catalyst properties have been performed in recent decades and continue today (Thompson et al., 2001; Escudero et al., 2002).

3.1.3 Membrane–Electrode Assemblies

In PEMFCs, as in many other types of fuel cells, gas-diffusion electrodes are used. These consist of a porous, hydrophobic *gas-diffusion layer* (GDL) and a *catalytically active layer*. The diffusion layers (often called *backing layers*) usually consist of a mixture of carbon black (acetylene black or other) and about 35% by mass of polytetrafluoroethylene (PTFE) applied to a conducting base (most often a thin graphitized cloth). The GDL yields a uniform supply of reactant gas to all segments of the active layer. The porosity of the diffusion layer is 45 to 60%. This high porosity is particularly important when air oxygen rather than pure oxygen is used as the oxidizing agent.

The electrode reaction proceeds within the active layer along a highly developed catalyst–electrolyte–gas three-phase boundary. The active layer is supported either by a special support (i.e., carbon cloth or carbon paper, made hydrophobic) or by the membrane itself.

An appreciable decrease in relative platinum input without any departure from the high electric parameters was attained when adding to the catalytically active layer a certain amount (30 to 40% by mass) of proton-conducting ionomer. To this end, an ink of platinized carbon black and PSAP ionomer (or a solution of low-equivalent-weight Nafion in alcohol) is homogenized by sonication and applied to the diffusion layer, the solvent is evaporated, and the whole is treated at a temperature of about 100°C. The ionomer added to the active layer leads to a considerable increase in the contact area between catalyst and electrolyte (which in the case of solid electrolytes, is actually quite small).

For further improvement in the catalyst–membrane contact, the membrane is sandwiched between the electrodes, with the active layers facing the membrane and subjected to hot pressing at a temperature of 130 to 155°C, pressures up to 100 bar, and pressing times of 1 to 5 minutes. This operation yields unique *membrane–electrode assemblies* (MEAs).

The technology just described, of applying catalysts to the diffusion layers and then hot-pressing all components, results in a *catalyst-coated diffusion layer* (CCDL). Another technology, which results in a *catalyst-coated membrane* (CCM), involves all components of the catalytic layer being evaporated as a thin layer directly onto the membrane.

The basic development work for MEAs of the present type was done in the United States by Gottesfeld and Zawodzinski (1997) at Los Alamos National Laboratory and by Srinivasan et al. (1991) at the Center for Electrochemical Systems and Hydrogen Research, Texas A&M University.

Through all these developments, toward the mid-1990s it was possible without sacrificing performance to lower the use of platinum metals to 0.4 mg/cm^2 (one order of magnitude less than in the first models), and by 2001 even further, to 0.05 mg/cm^2 for the hydrogen electrode and to 0.1 mg/cm^2 for the oxygen electrode.

By the late 1990s, the U.S. company E-Tek started commercial production of several versions of complete MEAs used by other companies to manufacture different types of fuel cell batteries.

3.2 STANDARD PEMFC VERSION FROM THE 1990s

Membrane-type hydrogen–oxygen fuel cells underwent numerous changes in design and manufacturing technology in the course of their development. In the present section we describe the structure of a PEMFC version that had become something of a standard in the mid-1990s, and in the following will be called the standard version of the 1990s.

3.2.1 Individual PEMFC

Figure 3.1 is a schematic drawing of the structure of an individual cell in a modern PEMFC battery. Each cell is separated from a neighboring cell by an electronically conducting *bipolar plate*. This plate is in contact on one side with the positive electrode of a given cell and on the other side with the negative electrode of the neighboring cell. Such plates thus function as interconnections for cells connected in series in a battery. The plate has channels cut out on either side, which via manifolds are supplied with the reactant gases (hydrogen and oxygen). The gases are carried to the electrodes by these channels. Such channels serve to eliminate water vapor as a reaction product from the oxygen side. For a uniform gas supply, the channels are laid out in spirals or serpentines, yielding a ramified flow field (Figure 3.2).

Between any two bipolar plates providing reactant access, a multilayer MEA is set up that consists of a positive electrode and a negative electrode pressed

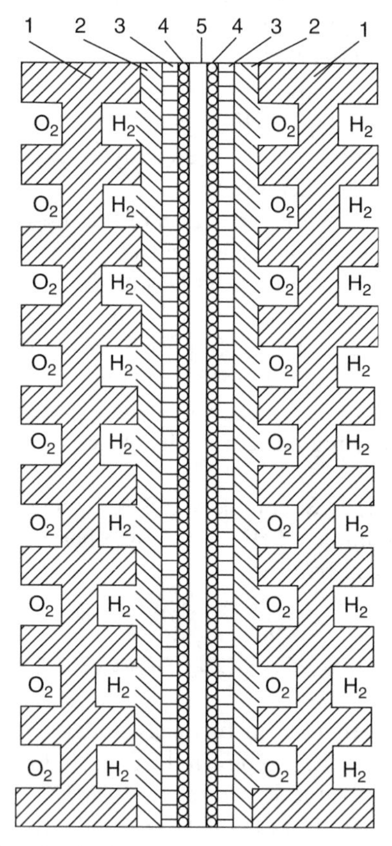

FIGURE 3.1 Schematic of a single PEMFC: 1, bipolar plates; 2, current collectors; 3, gas-diffusion layers; 4, catalytic layers; 5, membrane.

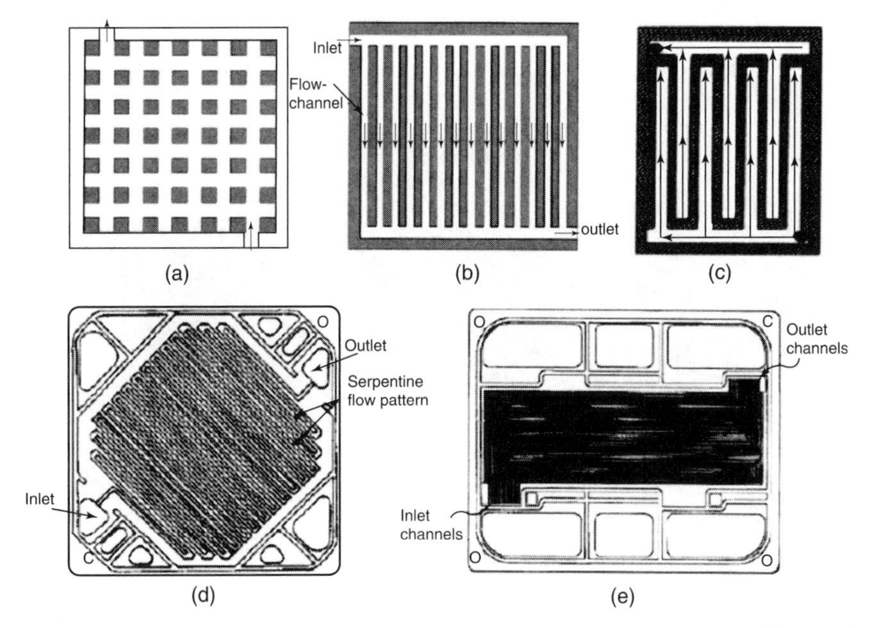

FIGURE 3.2 Schematic of flow field designs. (From Zhang et al., 2006, with permission from Elsevier.)

into the two sides of a proton-conducting membrane. For reasons discussed in Section 3.4, the catalyst for the hydrogen electrode in a PEMFC is a mixed platinum–ruthenium catalyst that applied to carbon black. The overall thickness of a modern MEA is about 0.5 to 0.6 mm (of which approximately 0.1 mm each is for the membrane, the two gas-diffusion layers, and the two active layers). The bipolar plates are about 1.5 mm thick, with the channels on both sides having a depth of about 0.5 mm.

Most companies producing PEMFCs use electrodes that have a surface area of 200 to 300 cm^2. Smaller electrodes (i.e., 10 cm^2) are used only in certain cases for batteries of less than 100 W.

3.2.2 PEMFC Stack (Battery)

The discharge voltage of an individual cell is very low, between 0.65 and 0.75 V, depending on the current density. For a battery of a given voltage, therefore, a certain number of cells are connected in series. For a 30-V battery, for example, about 40 individual cells must be put together. Since PEMFCs have a flat configuration, a battery of filter-press design is built. The required number of cells are clamped as a block between two end plates, also serving as the terminals for drawing the current. The end plates must be sufficiently massive

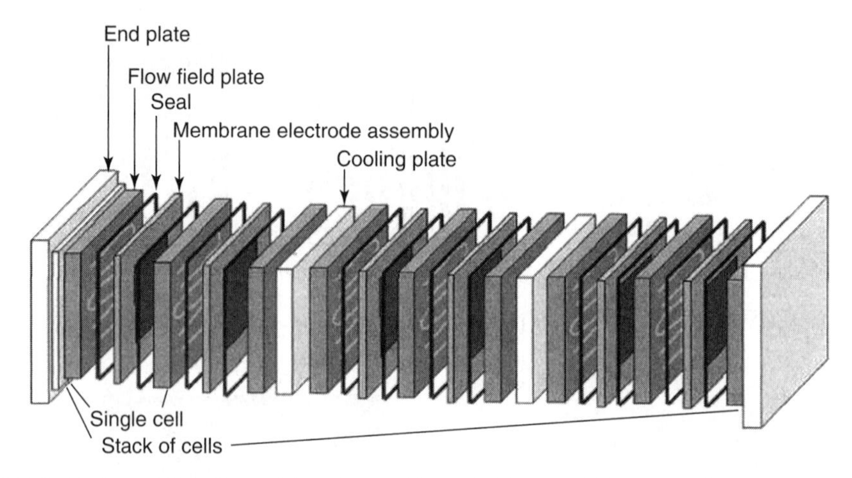

FIGURE 3.3 PEMFC stack hardware. (From Mehta and Cooper, 2003, with permission from Elsevier.)

not to become deformed when clamping the cells together, and provide a compressing force that is uniform over the entire cell surface.

As a rule, the membrane in an MEA extends about 2 cm beyond the active layers of the electrodes on all sides of the assembly. This part of the membrane serves as a peripheral seal of the individual cells clamped together into a battery stack. The seal is needed to prevent mixing of the reactant gases, hydrogen and oxygen. Additional O-rings may be provided when needed.

In addition to the individual cells and their bipolar plates, special heat-exchanger plates must be included in the battery stack. Cooling fluid is circulated through these plates to eliminate the heat produced during battery operation. At least one such plate must be provided for any two cells when the battery is to be operated at high current densities. These plates could also be used to warm the battery for a cold startup. Figure 3.3 shows the layout of PEMFC hardware.

3.2.3 Operating Conditions of PEMFCs

Modern PEMFCs are intended basically for high-energy densities at the electrodes (up to 0.6 W/cm^2). For this reason, and also because of the compact design, the maximum values of specific power per unit volume and weight are higher for these power sources than for all other batteries of conventional type. Often, these fuel cells are also used for operation at lower energy densities.

The working temperature of PEMFCs is about 80°C. The reactant gases are supplied to the battery at pressures of 2 to 5 bar. This relatively high pressure is needed because, to prevent drying out of the membrane, gases entering the cells must be presaturated with water vapor so that the resulting partial pressure of the reactant gas in the gas–vapor mixture is lower. For work at high current densities, however, a sufficiently high partial pressure must be maintained. This is attained by raising the overall pressure of the mixture.

3.2.4 Current–Voltage Curves of PEMFCs

The thermodynamic EMF of a hydrogen–oxygen fuel cell at a temperature of 25°C is given by $\mathscr{E} = 1.229\,V$. The open-circuit voltage of hydrogen–oxygen PEMFCs has values between 0.95 and 1.02 V, depending on the temperature and gas pressures.

Current–voltage curves plotted as voltage against current density that have been obtained at an overall gas pressure of 5 bar and at temperatures between 50 and 95°C are shown in Figure 3.4 for pure oxygen and air oxygen as the cathodic reactant. It can be seen from the plots that at high current densities, the voltage is much lower when air oxygen without excess pressure is used instead of pure oxygen, particularly at lower temperatures.

Hydrogen–oxygen PEMFCs containing the new membrane–electrode assemblies are convenient insofar as they operate highly efficiently (with relatively low reactant consumption) over a wide range of discharge current densities. In many applications, current densities of 0.8 to 1.0 A/cm^2 can be drawn at cell voltages U_i of about 0.75 V, corresponding to power densities of 0.6 to 0.75 W/cm^2. Under these conditions, the total voltage loss of $\mathscr{E} - U_i$ amounts to

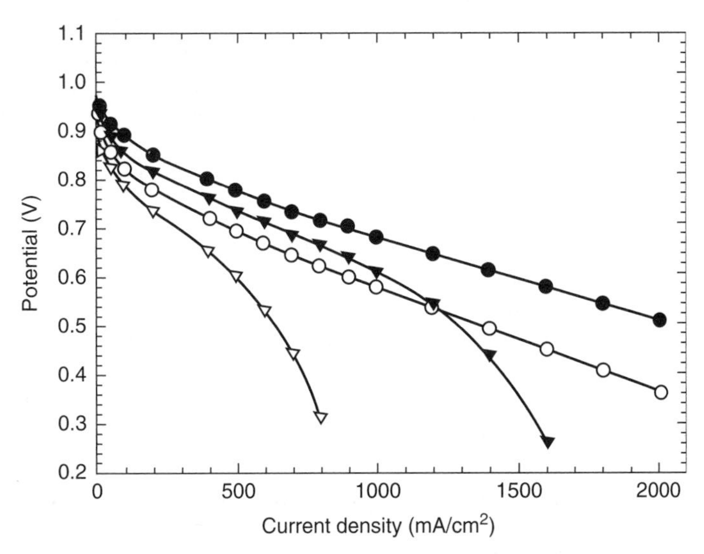

FIGURE 3.4 Effect of temperature and reactant gas on the performance of a PEM fuel cell, Pt loading of electrodes 0.45 mg/cm^2, Dow membrane, 50°C, (●) 5 atm O_2, (O) 1 atm O_2, (▽) 95°C, air, (▲) 50°C, air. (From Srinivasan et al., 1991, with permission from Elsevier.)

about 0.48 V. Of this loss, about 0.4 V is due to polarization of the oxygen electrode, about 20 mV to polarization of the hydrogen electrode, and the remainder to ohmic resistances within the MEA.

3.3 SPECIAL FEATURES OF PEMFC OPERATION

3.3.1 Water Management

In PEMFC operation, water is formed as a reaction product at the positive (oxygen) electrode. It can be seen from reaction (3.2a) that water reaches the side of the membrane that is in contact with the positive electrode, also by being dragged along in the form of hydrated protons. As a result of reaction (3.1a), on the other hand, water leaves the side of the membrane that is in contact with the hydrogen electrode. In part, this unilateral membrane water transport is compensated by the backdiffusion of water due to the concentration gradient that develops, but it is not completely compensated. Asymmetry of the hydraulic pressure on both sides of the membrane may also contribute to membrane water transport. All this leads to excess water at the oxygen electrode during cell operation, while the water content of the membrane next to the hydrogen electrode decreases.

Both of these effects have negative consequences for fuel cell operation. Excess water at the positive electrode may lead to flooding of the pores of the active layers through which oxygen reaches the catalyst. This will affect the performance of this electrode and may in the end lead to complete cessation of oxygen access to the active layer, and thus to a cessation of fuel cell operation. Dehydration of the membrane next to the hydrogen electrode, on the other hand, raises the ohmic resistance and leads to a lower discharge voltage of the cell.

To avoid these situations, the hydrogen that is supplied is usually saturated with water vapor, and the oxygen is circulated. Passing next to the electrode, the oxygen becomes saturated with water vapor. It then reaches a chamber with a lower temperature, where the water vapor condenses, and dry oxygen is returned to the electrodes. Sometimes, the periodic release of excess oxygen saturated with water vapor is practiced. The liquid water that is produced may be used for other purposes, such as drinking water.

All these processes must be controlled quite carefully. If water withdrawal is too fast, there is a risk of water loss from the swollen membrane, which not only leads to a drastic rise in resistance but also to fragilization. Cracks may then develop across which the gases may mix, yielding an explosive mixture, with all the catastrophic consequences that ensue.

A large problem in PEMFC operation is a possible partial condensation of water vapor when temperature gradients are present in the fuel cell, and a dual-phase water system develops. The liquid water forming within the MEA or in the channels of the bipolar plates interferes with the access of reactant gases to

Oszcipok et al., 2006). In this dried state, a battery will withstand tens of freeze–thaw cycles without marked performance loss (Hou et al., 2006).

3.4 PLATINUM CATALYST POISONING BY TRACES OF CO IN THE HYDROGEN

For fuel cell operation, technical hydrogen obtained by the conversion of primary fuels such as methanol or petroleum products is generally used rather than pure hydrogen obtained by electrolysis. The technical hydrogen always contains carbon monoxide (CO) and a number of other impurities, even after an initial purification. In the first experiments, conducted in the mid-1980s, it was seen that traces of CO in hydrogen used for the operation of fuel cells with phosphoric acid electrolyte lead to a marked increase in polarization of the hydrogen electrode (Dhor et al., 1987). A similar increase in polarization, but more pronounced, is seen in PEMFCs, which have a lower working temperature (80°C) than PAFCs (about 200°C). An increase in polarization is noticeable in the presence of traces of CO of less than 10 ppm. With 25 ppm of CO and a current density of 600 mA/cm^2, polarization of the electrode increases by 0.2 to 0.3 V, implying a loss of about 30 to 40% of electrical power (Gottesfeld and Pafford, 1988). A more thorough elimination of CO from hydrogen is difficult to attain.

This polarization is due to the fact that the platinum catalyst is a good adsorbent for CO. Upon CO adsorption, the fraction of the surface that is available for the adsorption of hydrogen and its subsequent electrochemical oxidation decreases drastically (the catalyst is poisoned by the catalytic poison CO). At least 10% of the surface of platinum must remain accessible for hydrogen if polarization of the hydrogen electrode is to be kept within reasonable limits (10 to 20 mV), so that the degree of surface coverage by adsorbed CO should not be higher than 0.8 to 0.9. There are various ways to fight catalyst poisoning.

3.4.1 Oxidation of CO by Current Pulses and Oxidizing Agents

Several methods of selective oxidation of CO impurities in hydrogen have been suggested. One of them (Carrette et al., 2002) involved the application of periodic current pulses of alternating sign to the hydrogen electrode. The potential of the anode then shifts periodically in the positive direction, which causes oxidation of the adsorbed CO species and their desorption from the catalyst surface. This method is actually effective over short times, but CO adsorption repeats after cessation of the pulse.

Another suggestion (Wilson and Gottesfeld, 1992) was to inject some quantity of oxygen periodically into the contaminated hydrogen. This method is effective at low CO contents. Higher CO contents demand more considerable amounts of oxygen, causing an explosion hazard. Moreover, the oxygen oxidizes not only

CO but also some hydrogen, thus lowering the Faradaic efficiency of the hydrogen gas. An analogous effect is achieved when injecting hydrogen peroxide into the hydrogen (Divisek et al., 1998; Barz and Schmidt, 2001).

3.4.2 The Use of Platinum–Ruthenium Catalysts

The most reliable and promising way of fighting poisoning of the platinum catalyst by CO impurities in the hydrogen is by modifying the catalyst itself (e.g., by adding alloying elements). When studying the anodic oxidation of methanol at platinum catalysts in the late 1960s, it was found that the reaction is much faster at mixed platinum–ruthenium (Pt–Ru) catalysts than at pure platinum catalysts (see Section 4.3). It could be shown in a number of studies (Watanabe et al., 1987; Schmidt et al., 2002) that in PEMFCs, such Pt–Ru catalysts are appreciably less sensitive than pure platinum toward CO poisoning. The reasons are not exactly clear. Possibly the energy of adsorption of CO decreases, owing to changes in the crystal lattice structure of the alloy relative to pure platinum (or changes in its electronic state). It is also possible that excess oxygen adsorbed on ruthenium sites oxidizes the adsorbed CO species, eliminating them from the platinum surface. Many studies have been performed in recent years to elucidate the operating mechanism of Pt–Ru catalysts for hydrogen oxidation in the presence of CO. It could be shown in particular (Ioroi et al., 2002) that the effectiveness of the Pt–Ru catalyst depends on the moisture content of the technical hydrogen supplied to the anode.

At present, anodes with Pt–Ru catalyst (with about 50% Ru) are used in the majority of PEMFCs designed to work with technical hydrogen. It follows from experiments with pure hydrogen (free of CO traces), however, that the catalytic activity of the mixed Pt–Ru catalyst is somewhat lower than that of pure platinum (Roth et al., 2001). In this connection, it is worth noting the suggestion of Chinese workers to use electrodes having a special structure with two catalytically active layers (Shim et al., 2001; Wan et al., 2006). The first layer, closer to the diffusion layer from which the gas is supplied, contains the Pt–Ru catalyst at which CO is oxidized. It thus acts like a filter not passing the CO. The hydrogen that has been freed of CO reaches the following layer, containing the more active, pure platinum catalyst.

3.4.3 Higher Working Temperatures

Still another way of fighting hydrogen catalyst poisoning by CO impurities is that of raising the operating temperature. When the operating temperature of PEMFCs is raised from the current range of 80 to 90°C to a level of 120 to 130°C, the adsorption equilibrium between hydrogen and CO jointly adsorbing on platinum shifts in favor of hydrogen adsorption. This raises the highest admissible threshold concentration of CO. The effect can be seen in fuel cells with phosphoric acid electrolyte (PAFCs), which work at temperatures of about 180 to 200°C and admit CO concentrations in hydrogen as high as 100 ppm, despite the fact that platinum catalysts are used. However, raising the operating temperature of PEMFCs beyond 100°C brings a number of difficulties. Work addressing these problems is described in Section 3.7.

3.5 COMMERCIAL ACTIVITIES IN RELATION TO PEMFCs

At present, PEMFC batteries and power plants based on such batteries are produced on a commercial scale by a number of companies in many countries. As a rule, the standard battery version of the 1990s is used in these batteries, although in certain cases different ways of eliminating water and regulating the water balance (water management) have been adopted.

Ballard Power Systems (Burnaby, British Columbia, Canada) started research in PEMFC manufacturing technology in 1989, and between 1992 and 1994 delivered a few prototypes of power units in different sizes. The first commercial unit, which had 1.2 kW of power, was made in 2001. At present, this company produces different types of power units between 4 and 21 kW for different applications: electric cars, power backup, and plants for heat and power cogeneration [combined heat and power (CHP) systems]. In these plants, water is eliminated by supplying to the anodes hydrogen with a low water vapor content, so that in the membrane, water is transferred by diffusion from the cathode to the anode. In this way, oxygen circulation along the cathode surface can be lowered. In 2006, Ballard reported delivery of the new power unit Mark 1020 ACS, for the first time using air cooling and a simplified membrane humidification system (www.ballard.com).

UTC Power (South Windsor, Connectient), a United Technologies Company, produces power units with PEMFCs for different military and civil applications. In 2002, regular electric bus service using fuel cell batteries developed by this company was started. The Pure Cell model 200M Power Solution power plant delivers 200 kW of electric power and about 900 Btu/h (about 950 kJ/h) of thermal power (www.utcpower.com).

Plug Power (Latham, New York), founded in 1997, has delivered since 2000 emergency power plants on the basis of PEMFC batteries providing uninterruptible power supply for hospitals and other vitally important objects in cases of sudden loss of grid power (www.plugpower.com).

At present, in addition to the United States, PEMFCs and power plants based on them have been developed in many other countries, including China, France, Germany, South Korea, and the United Kingdom. Most of the power plants delivered in 2006 (about 60%) were for power supply to portable equipment. A secondary use (about 26%) was as small stationary power plants for an uninterruptible power supply.

Approximately 75% of the work on PEMFCs is conducted in industrial organizations, the remaining 25% in academic and government organizations. This proves that the initial research and engineering stage has been completed, and commercial development of these fuel cells is under way (see the review by Crawley, 2006). A detailed analysis of the many ways used to make PEMFCs may be found in the review of Mehta and Cooper (2003).

3.6 FUTURE DEVELOPMENT OF PEMFCs

At this time in 2008, PEMFC batteries and power plants built on the basis of such batteries have attained a high degree of perfection. They work reliably, exhibit rather good electrical characteristics, and are convenient to handle. Such plants have found practical applications in many areas, such as to secure an uninterruptible power supply in the case of grid breakdowns to strategically important entities (e.g., hospitals, command stations, water works, gas suppliers, telecommunication centers). They are also used for combined power and heat supplies in individual residence and office buildings. Regular bus service with electric traction provided by such power plants is in operation in a number of places. Yet these uses are not on a mass scale, and the commercial success deriving from the production of such power plants is still very limited.

A wider use of fuel cells of this type can only be expected when they have conquered two new areas of application: light electric vehicles and portable electronic equipment. For success in this direction, a number of important and rather complex problems must first be solved:

- A longer lifetime for the power plants (Section 3.6.1) and better stability of the catalysts (Chapter 12) and membranes (Chapter 13) associated with this longer lifetime
- A lower cost of production, both for the PEMFCs as such and for the entire power plant (Section 3.6.2), and the development of catalysts without platinum (Chapter 12) and of cheaper membranes (Chapter 13) associated with this lower cost
- A higher tolerance of PEMFCs for CO impurities in the hydrogen, particularly by building versions of these fuel cells operating at higher temperatures (Section 3.7)
- The development of new installations for hydrogen production admitting a wider selection of primary fuels (Chapter 11)

Since the power supply for a variety of portable devices is one of the more important future applications of PEMFCs, great efforts are made at present to reduce the dimensions and weight and even to miniaturize both the fuel cell battery and all ancillary equipment needed for a power plant. This aspect is discussed in more detail in Chapter 14.

3.6.1 Longer Lifetime

There are a few different aspects to the concept of *lifetime*. They cover values of the following parameters of a power plant with fuel cells that have either been attained or are guaranteed or expected:

- The time required for smooth uninterrupted operation in a given operating mode
- The number of (admissible) on–off cycles
- The number of (admissible) temperature cycles between ambient and operating temperature and back

By definition, fuel cells should work without interruption as long as reactants are supplied and reaction products eliminated. Actually, almost all varieties of fuel cells exhibit some time-dependent decline of their characteristics during long-term discharge. When operating at constant current, for instance, the voltage will gradually decrease. The rate of voltage decrease of an individual cell is stated in μV/h or, where currents vary as a function of time, in μV/Ah.

The gradual decline of the indices of performance of a fuel cell power plant has many causes, related both to the work of the fuel cells themselves and to the work of all ancillary equipment needed for their function (e.g., installations for reactant supply and for the elimination of reaction products and heat, systems for controlling the entire plant and for monitoring the parameters). In the present section we examine only the reasons for degradation of PEMFC fuel cells and batteries based on them.

Few data exist in the literature as to the lifetime of fuel cells. In 2006, Cleghorn et al. reported data for a cell that had been operated continuously for 26,300 hours, which is about three years. During this time, the rate of voltage decrease was 4 to 6 μW/h on an average. The test was terminated because of failure of the membrane. In view of other tests where individual cells or batteries had been operated for a few thousand hours, the result above can be regarded as very promising, since it shows that fundamentally, a PEMFC will be able to be operated for several years. For general use of this type of fuel cell, however, not enough lifetime data are available.

Figures for the time of *smooth* operation of PEMFCs (and other fuel cells used in the same applications) are given variously as 2000 to 3000 hours for power plants in portable devices, as up to 3000 hours over a period of five to six years for power plants in electric cars, and as five to 10 years for stationary power plants. Much time will, of course, be required to collect statistical data for the potential lifetime of different types of fuel cells. Research efforts are therefore concentrated on finding reasons for the gradual decline of performance indicators and for premature failure of fuel cells. In recent years, many studies have been conducted in this area.

The decline in PEMFC parameters may be reversible or irreversible. When reversible, better performance can be reestablished by the operators by, for example, changing over to a lower operating power. When irreversible, there is no way to bring them back to the original level.

A reversible voltage drop will in most cases be caused by an upset water balance within the MEA, leading to flooding of the catalytically active layer of the cathode and/or some dehydration of the membrane layer next to the anode. Flooding of the cathodic layers leads to hindrance of oxygen transport to the catalytically active sites and decrease in its concentration (partial pressure) at these sites, an effect described as oxygen *transport difficulties* or oxygen *concentration polarization*. Membrane dehydration causes higher ohmic resistance of the membrane. When changing to lower power, the original optimum distribution of water within the MEA is often reestablished.

Irreversible changes have many causes, one of which is recrystallization of the very highly disperse platinum catalyst on its carbon support. At the positive potentials of the oxygen electrode, platinum undergoes perceptible dissolution, producing platinum ions. This has two consequences: (1) part of these ions settle in the membrane, so that the total amount of catalyst in the catalytic layer decreases; and (2) most of them diffuse to nearby platinum grains, deposit on them, and cause them to grow larger. This leads to a smaller true surface area of the catalyst and higher true current density (higher fraction of the discharge current per unit of true surface area). Polarization of the electrode increases in response to the higher true current density, and the discharge voltage of the fuel cell as a whole decreases. At the potentials existing at the hydrogen electrode, the rate of platinum dissolution is much lower, yet because of solid-phase reactions, the highly disperse platinum present in the catalytically active layer of this electrode may also become coarser by recrystallization, the true surface area thus becoming smaller.

Another reason for higher polarization of the oxygen electrode is corrosion (oxidation) of the carbon material serving as a support for the platinum catalyst. This causes loss of contact of the catalyst with the support (Cai et al., 2006) (i.e., a de facto exclusion of part of the catalyst). The loss of Pt–C bonding also favors recrystallization of the platinum catalyst.

Nafion membranes are chemically highly stable, yet under the conditions existing in an operating fuel cell, slow degradation is seen. In part, this is due to oxygen-containing free radicals forming in side reactions at the oxygen electrode. These radicals are strongly oxidizing and may attack the membrane, the attending degradation leading to higher membrane resistance and sometimes even brittleness and mechanical defects.

A shortened lifetime may be caused not only by changes in the MEA (its catalytically active layers or the membrane) but also by problems arising in other components of the fuel cell. Bipolar graphite plates have some porosity, and a perceptible permeation of gases through them gives rise not only to irreproducible reactant losses but also to certain irreversible changes. Bipolar metal plates are subject to corrosion, and heavy-metal ions produced by corrosion markedly depress the activity of the catalysts when depositing on them. The sealing materials may also be the reason for degradation of the cells: It has, in fact, long been known that polymer oxidation and decomposition products affect the active catalysts. A marked decrease in catalyst activity was noticed when seals made of organic silicon material were used and traces of silicon were deposited on the catalyst surface (Ahn et al., 2006).

The influence on PEMFC lifetime of a variety of contaminants in reactants and in the cells themselves, as well as the mechanisms of this influence, have been examined in a review by Cheng et al. (2007). In the three-year PEMFC test described above, one of the main reasons for the voltage drop was the reversible difficulty in oxygen transport. In addition, an irreversible loss of the true surface area of the cathodic catalyst of about 60% of the original value was also observed. Since silicone organic seals were used, moreover, silicon was detected in the catalyst.

The rate and character of PEMFC degradation are a function of the operating conditions. Liu and Case (2006) studied the work of two identical MEAs. One worked at constant current, the other at currents varying according to a certain program. In the latter, the formation of pinholes was observed in the membrane after 500 hours, resulting in a drastic rise of hydrogen crossover. In the former, increased crossover was not seen, and its voltage drop was reversible and due to transport limitations.

Nonuniformities in the distribution of current-producing reactions over the electrode surface have a large effect on the rate of degradation of the performance indicators of fuel cells. Such nonuniformities may have a technological origin (such as fluctuations in the thickness of individual MEA components) or arise during fuel cell operation. The latter include local formation and accumulation of water drops in the channels of the bipolar plates, causing a nonuniform supply of reactant gases across the electrode surface. Patterson and Darling (2006) described a situation where the gas compartment of the hydrogen electrode in a hydrogen–oxygen fuel cell was found to be filled, in part with hydrogen and in part with air, when the cell had been turned on and off repeatedly. This gives rise to some redistribution of potential and to a strong shift of potential of the oxygen electrode in the positive direction, which in turn leads to enhanced corrosion of both the platinum catalyst and its carbon support.

A rather dangerous situation arises when individual cells of a multicell battery deviate. Such nonuniformity is due most often to problems in reactant supply. Two systems of gas supply exist: parallel and series. In parallel supply, the gas reaches each cell through a narrow channel coming from a common manifold. The pressure in these channels is the same for all elements where they leave the manifold, but because of differences in gas flow resistance, the amounts of gas (or the pressure) reaching each cell may differ. In series supply, gas is fed to a first individual cell, flows through it and continues to the next cell, and so on. In each cell in series, the amount of gas needed for the reaction is consumed, leaving a lower pressure of reactant gas for the next cell. The cells thus operate at different working pressures of the gas, which constitutes a nonuniformity.

Most often, use of a parallel gas supply is practiced. In fact, for the oxygen that should be circulated for the purposes of water elimination, this is the only possibility. If in a battery working with an external load, one of the cells is in short supply as related to one of the gases, the corresponding electrode reaction will cease, but the closed circuit of the load and all series-connected cells implies that current is forced through the affected cell. Assuming that the hydrogen supply had stopped, then at the hydrogen electrode, anodic hydrogen oxidation is no longer possible and anodic oxygen evolution will instead be caused by the current. When the oxygen supply has stopped, then analogously, oxygen reduction at the oxygen electrode will be replaced by hydrogen evolution. De facto, the deviating cell is forced into a mode of electrolysis by the normal cells ahead of it and behind it. As a result, instead of contributing its voltage (of, e.g., 0.65 V) to the battery voltage, this cell draws voltage equivalent to a "negative contribution" of about −2.0 V. This polarity inversion produces not only a marked voltage drop of the battery but also irreversible changes in the MEA of the affected cell. More serious consequences include the formation of explosive gas in the system.

It follows from what was said above that to raise the lifetime of PEMFCs a number of problems must be solved, the most important being (1) the development of catalysts less prone to recrystallize and of supports for the catalysts that are more corrosion resistant (Chapter 12); (2) the development of more highly stable membranes (Chapter 13); (3) optimized conditions of water management, to avert temporary flooding of the oxygen electrode; and (4) finding new materials for the bipolar plates, the seals, and other structural elements of the fuel cell.

The issues affecting the life and long-term performance of PEMFCs are discussed in a recent review (Schmittinger and Vahidi, 2008) where it is shown that the main causes of short life and performance degradation are poor water management, fuel and oxidant starvation, corrosion, and chemical reactions of cell components. Problems related to long-term performance and durability are also discussed in reviews by Yu and Ye (2007) and Wu (2008).

3.6.2 Cost Estimation for PEMFCs

At this time a generally accepted way of estimating the cost of items such as fuel cells just now going into industrial mass production does not exist. The sales price of fuel cell power plants produced today is usually not made public and is decided by an arrangement between seller and buyer. The sales price of a new item usually depends not only on the manufacturing cost itself but also on additional charges for research, development, and special design features. It is only after the start of mass production that the manufacturing price becomes determining for the price of a new item.

Few data on this topic can be found in the literature on PEMFCs. In addition, these data are often contradictory. First data became known in 2002, that is, about five years after establishment of the "standard version of the 1990s."

In 2002, Schmidt et al. estimated the cost of the first PEMFCs as $20,000/kW (here and below, all prices are in U.S. dollars). Of this figure, 90% was for labor (the cells were hand-made), and only 10% was for materials. In mass production, for instance when making 1 million 50-kW power plants per year, labor cost could fall to $10/kW. In the opinion of Schmidt et al., it would be necessary to develop new membranes costing no more than $20/m^2, and to lower the platinum content of the electrodes to 0.25 mg/cm^2, in order to bring the cost of material down from $2000/kW to a desirable figure of $30/kW or at least to a temporarily acceptable value of $100 to $200/kW (also, platinum should be recovered and recycled). A very important and difficult problem is that of making cheaper corrosion-resistant metallic bipolar plates. All ancillary devices must also become much cheaper.

Also in 2002, Isa Bar-On et al. provided a more optimistic estimate of $200/kW as the manufacturing cost of a fuel cell and thought it necessary to bring the price down to $20 to $50/kW by 2004. They also stressed the need for an important price cut in the manufacture of bipolar plates.

A detailed cost analysis for a PEMFC power plant of 5 kW was provided in 2006 by Kamarudin et al. According to their data, the total cost of such a plant will be about $1200, of which $500 is for the actual fuel cell stack and $700 is for the ancillary equipment (pumps, heat exchangers, etc.). The cost of the fuel cell stack is derived from the components as $55/kW for the membranes, $52/kW for the platinum, $128/kW for the electrodes, and $148/kW for the bipolar plates.

By comparison, the cost of an internal combustion engine of the same power is between $500 and $1500/kW according to data of different authors, so that in this respect, a power plant with fuel cells hardly yields to a combustion engine.

In addition to costs, one must take into account the availability of raw materials needed for fuel cell production. Assuming a total platinum content in both electrodes of 0.8 mg/cm^2 and an optimistic value of 1 W/cm^2 for the specific power, one will need 0.8 g of platinum per kilowatt. With a price of platinum of $30/g, this gives $24/kW. Therefore, an electric car with a power of 50 kW would have a price tag of $1200 only for the platinum in the fuel cells. It must, of course, be taken into account here that with a production volume of 1 million electric cars per year, 40 tons of platinum metal were needed, representing about 20% of current world production. An even more difficult situation would arise in the mass production of fuel cells with mixed Pt–Ru catalysts. World reserves of ruthenium are very limited and would not permit mass production on such a scale.

For this reason it will be a prime task in further fuel cell development to search broadly for ways to lower the platinum metal content of the catalytic layer and to find new nonplatinum catalysts (Chapter 12).

3.6.3 Bipolar Plates

Bipolar plates are a very important component of fuel cell batteries. They largely determine their efficiency and possible lifetime. The bipolar plates take up considerably more than half of the total battery volume (sometimes as much as 80%) and have the corresponding share of the battery's weight. The cost of the bipolar plates is up to 15% of a battery's total cost.

Bipolar plates have a number of functions in fuel cell batteries. Mechanically, they are the backbone of the individual fuel cells and of the battery as a whole. They provide the electrical contact between the individual cells in the battery and channel the reactant supply to the entire working surface area of the electrodes.

To provide all these functions the plates must meet a number of requirements. They must (1) be sufficiently sturdy, (2) be electronically conducting, (3) have a low surface resistance in contact with other conductors, (4) be impermeable to gases, and (5) be corrosion resistant under the operating conditions of fuel cells.

Materials that can be considered in the manufacture of bipolar plates are graphite, various carbon–polymer composites, and various metals. Graphite and the composites are sufficiently highly conducting and have a low surface resistance but are not sufficiently strong; they are brittle and withstand shock and vibration poorly. Machining the gas channels in graphite plates is laborious and expensive. Composites are more favorable in this respect, since the channels may be made by pressing and stamping. Both the composites and the graphite are porous to some extent, and permeable to hydrogen. In view of their mechanical strength and impermeability to gases, metals such as stainless steel, titanium, and aluminum are very attractive. In metals, the channels are readily made by the separate stamping of metal sheets for each side of the bipolar plate and subsequent welding of the two halves.

An important defect of metallic bipolar plates is their inadequate corrosion resistance under the operating conditions of fuel cells (high temperature and humidity; contact with the acidic, proton-conducting membrane; oxygen atmosphere). Corrosion of the plates not only detracts from their mechanical properties but also gives rise to undesirable corrosion products, that is, heavy metal ions which when depositing on the catalysts, strongly depress their activity. The corrosion processes also give rise to superficial oxide films on the metal parts, and these cause contact resistance of the surfaces. For a lower contact resistance, metallic bipolar plates sometimes have a surface layer of a more stable metal. Thus, in the first PEMFC developed by General Electric for the *Gemini* spacecraft, the bipolar plates consisted of niobium and tantalum coated with a thin layer of gold. A metal part could also be coated with a layer of carbide or nitride.

Metals as the material for bipolar plates have the additional advantage that thinner, lighter plates can be made. Future developments in this area will probably yield new ways to protect the metal surfaces from corrosive attack by the medium and from formation of the superficial oxide films that lead to contact resistance. All the problems associated with bipolar plates are discussed in detail in a review by Tawfik et al. (2007).

3.7 ELEVATED-TEMPERATURE PEMFCs

Fuel cells of this variety are sometimes called high-temperature or midtemperature PEMFCs, but it is preferable to use the designation elevated-temperature PEMCFs (ET-PEMFCs), since in their application to solid-oxide fuel cells, *high-* and *midtemperature* refer to different temperature ranges (see Chapter 8).

The primary interest in studies of PEMFCs working at temperatures of 120 to 130°C rather than the usual 80°C was that of attaining a higher tolerance by the platinum catalysts for the carbon monoxide (CO) present as an impurity in the hydrogen supply. In fact, if as mentioned in Section 3.4.4, at 80°C the admissible CO concentration in hydrogen is no higher than 10 ppm, at 130°C this value goes up to 1000 ppm, making prior purification of the hydrogen much simpler.

Apart from CO tolerance, which is highly important, the higher working temperature of PEMFCs also has a favorable effect on other processes occurring in these cells:

1. The electrode processes are greatly accelerated, the polarization of the electrodes decreases accordingly, and the discharge voltage of the cell rises. This is seen particularly distinctly at the oxygen electrode, where as mentioned earlier, the energy lost through lack of equilibrium and through polarization of the electrode is particularly high.
2. Cooling of the battery is greatly simplified and utilization of the heat set free at higher temperature becomes possible (e.g., for the steam reforming of methanol or for indoor heating). Another fact to consider is the inability of radiators of modern cars to cope with dissipating heat that is set free at a temperature of only 80°C.
3. The elimination of water from the operating fuel cell is greatly facilitated. At a temperature of 80°C, part of the water accumulates in liquid form in the pores of the diffusion layer and in the gas supply channels, hampering gas transport to the catalyst. In winter, this residual water may even freeze when the battery has been turned off and damage the MEA. In ordinary PEMFCs, a careful purge with dry gas is needed to eliminate these problems. At temperatures higher than 100°C, they no longer exist.

In view of all these advantages, most of the research into PEMFCs concentrates on the elevated-temperature variant. At higher temperatures, the thermodynamic EMF value of a hydrogen–oxygen fuel cell is somewhat lower, but it can be seen from data reported by Zhang et al. (2006a) that the cell's OCV is practically unaffected.

Temperature (°C)	EMF, $\mathscr{E}$(V)	OCV, U_0(V)
25	1.229	1.05
80	1.19	1.05
120	1.14	1.05

The cell's operating voltage U_i increases considerably with increasing temperature because of the factors mentioned in point 1 above.

A number of challenges also arise at higher operating temperatures of PEMFCs:

1. At temperatures of 120 to 130°C, the vapor pressure of water is about 2.5 bar. The total pressure in the system must then be at least 3 to 4 bar, so that the partial pressures of the reactant gases will retain acceptable values of about 0.5 bar.
2. At the higher temperature, the rate of hydrogen diffusion through the membrane (crossover) increases. At 80°C, no more than 3% of the hydrogen is lost by this crossover, which is equivalent to a current loss of about 3 mA/cm^2. These losses rise markedly with temperature. The hydrogen diffusing to the oxygen electrode is oxidized there by a chemical mechanism. Then free oxygen radicals of $OH^\bullet$ or $OOH^\bullet$ type may form and attack the membrane, accelerating its degradation.
3. Degradation processes of fuel–cell components are accelerated at higher temperatures:
   a. Catalyst recrystallization and the associated decrease in working surface area of the catalyst
   b. Superficial oxidation of the catalyst's carbon support and the associated loss of contact between the support and the catalyst
   c. Oxidative attack of the membrane (see above)
   d. Superficial oxidation and corrosion of the bipolar plates
   e. Attack of various structural materials (seals, etc.)

   The data available up to now as to the effect of higher temperatures on all these phenomena of degradation and destruction are very limited. Research into these problems needs considerable time. Various accelerated degradation tests that are conducted often lead to ambiguous results.
4. The main problem in ET-PEMFC operation is degradation of the membrane at the higher temperature. Marked water loss raises the ohmic resistance of the membrane, causes brittleness, and may give rise to crack formation. For this reason, most PEMFC research at present addresses the question of how to maintain the membrane in good working condition in an elevated-temperature PEMFC.

Various approaches to this problem are discussed next.

Nafion-Type Membranes with Hydrophilic Additives

When inorganic compounds such as SiO_2, TiO_2, $Zr(HPO_4)$, or heteropoly acids of the type of phosphotungstic acid (PWA; $H_3PW_{12}O_{40}$) or silicotungstic acid (SiWA; $H_4SiW_{12}O_{40}$) are introduced into a membrane, they will combine with the water present in the membrane and make the water's escape at elevated temperatures more difficult. Such additives may be introduced both during membrane manufacture and by posttreatment of finished membranes.

Solvents Other Than Water

The water present in the membrane makes possible the dissociation of the sulfonic acid groups, providing protonic conductivity. The water could be replaced by other solvents that have the same effect but a higher boiling point, and hence would be less easily lost from the membrane. Phosphoric acid and imidazoles [e.g., poly(2,5)benzimidazole (PBI)] could be used in this capacity. Sometimes they are used together with added heteropoly acids as mentioned above, which by themselves are proton conducting.

Membranes Made from Other Polymers

It has been suggested that instead of membranes made from perfluorinated sulfonic acids (PFSAs), membranes be used made from other, heat-resistant polymers containing the sulfonic acid group. One could think here of sulfonated polyimides (SPIs) and sulfonated polyether ether ketones (SPEEKs). Membranes made from such polymers have a sufficiently high protonic conductivity at elevated temperatures but are less sensitive to lower humidities and water loss.

Membranes that display protonic conductivity in the complete absence of water are of interest as well. A material of this type is cesium hydrogen phosphate (CsH_2PO_4) described by Norby (2001). At temperatures above 230°C, where a "superprotonic state" is attained, the protonic conductivity of this material is as high as 10^{-2} S/cm. Uda and Haile (2005) have reported results obtained when working with a test cell that includes such a membrane (in a thickness of about 35 μm) at a temperature of 240°C.

In their review of 2007, Shao et al. called attention to the fact that in classical versions of PEMFCs, Nafion membranes were used while adding a polymer of the same type as a proton-conducting additive to the catalytically active layers. If for the membrane one now uses materials other than for the proton-conducting additive, complications may arise during humidity changes in the system that lead to differences in the degrees of swelling or compression and to a loss of contact between individual MEA components. It had taken about five years from the time when new samples of a Nafion membrane had become available until an optimum structure of the MEA and an optimum technology for its manufacture was established. With the advent of new membrane materials, existing experience can only be used in part, and considerable effort must be expended to arrive at optimization.

From all that has been said above, it can be concluded that PEMFCs working at elevated temperatures are highly promising. Many difficulties must still be overcome to develop models that will function in a stable and reliable manner and for extended periods of time. At present, about 90% of all publications on fuel cells are concerned precisely with attempts to overcome these difficulties. Most of the publications deal with research into new varieties of membrane materials, and some of them are discussed in Chapter 13. One may also consult reviews on ET-PEMFCs (Zhang et al., 2006b; Shao et al., 2007).

REFERENCES

Ahluwalia R. K., X. Wang, *J. Power Sources*, **162**, 502 (2006).

Ahn S.-Y., S.-J. Shin, H. Y. Ha, et al., *J. Power Sources*, **106**, 295 (2006).

Bar-On I., R. Kirchain, R. Roth, *J. Power Sources*, **109**, 71 (2002).

Barz D. P. J., V. M. Schmidt, *Phys. Chem. Chem. Phys.*, **3**, 330 (2001).

Cai M., M. S. Ruthkosky, B. Merzougui, et al., *J. Power Sources*, **160**, 977 (2006).

Carrette L. P. L., K. A. Friedrich, M. Huber, U. Stimming, *Phys. Chem. Chem. Phys.*, **3**, 320 (2002).

Cleghorn S. J. C., D. K. Mayfield, D. A. Moore, et al., *J. Power Sources*, **158**, 446 (2006).

Dhor H. P, L. G. Christner, A. K. Kush, *J. Electrochem. Soc.*, **134**, 3021 (1987).

Divisek J., V. Peinecke, V. M. Schmidt, U. Stimming, *Electrochim. Acta*, **43**, 3811 (1998).

Escudero M. J., E. Hontanon, S. Schwartz, et al., *J. Power Sources*, **106**, 206 (2002).

Gamburzev S., A. J. Appleby, *J. Power Sources*, **107**, 5 (2002).

Gottesfeld S., J. Pafford, *J. Electrochem. Soc*, **135**, 3651 (1988).

Grot W. G., *Chem. Ind.*, **19**, 647 (1985).

Guo Q., Z. Qi, *J. Power Sources*, **160**, 1269 (2006).

Hou J., H. Yu, S. Zhang, et al., *J. Power Sources*, **162**, 513 (2006).

Ioroi T., K. Yasuda, Y. Miyazaki, *Phys. Chem. Chem. Phys.*, **4**, 2337 (2002).

Kamarudin S. R., W. R. W. Daud, A. Md. Som, et al., *J. Power Sources*, **157**, 641 (2006).

Kazim A., *J. Power Sources*. **143**, 9 (2005).

Liu D., S. Case, *J. Power Sources*, **162**, 521 (2006).

Norby T., *Nature* (*London*), **410**, 877 (2001).

Oszcipok M., M. Zedda, D. Riemann, D. Geckeler, *J. Power Sources*, **154**, 404 (2006).

Patterson, T. W., R. M. Darling, *Electrochem. Solid-State Lett.*, **9**, A183 (2006).

Peled E., T. Dudevan, A. Aharon, A. Melman, *Electrochem. Solid-State Lett.*, **3**, 525 (2000).

Perry M. L., S. Kotso, *Intelec 2004 Proc.*, 2004, pp. 210–217.

Roth C., M. Goetz, H. Fuess, *J. Appl. Electrochem.*, **31**, 793 (2001).

Schmidt V. M., P. Brockerhoff, B. Hohlein, et al., *J. Power Sources*, **46**, 299 (1994).

Schmidt H., P. Buchner, A. Datz, et al., *J. Power Sources*, **105**, 243 (2002).

Shim J., C.-R. Lee, H.-K. Lee, J.-S. Lee, E. J. Cairns, *J. Power Sources*, **102**, 72 (2001).

Susai T., M. Kaneko, K. Nakato, et al., *Int. J. Hydrogen Energy*, **26**, 631 (2001).

Swider K. E., D. R. Rolison, *J. Electrochem. Soc.*, **143**, 813 (1996).

Thompson S. D., L. R. Jordan, M. Forsyth, *Electrochim. Acta*, **84**, 167 (2001).

Uda T., S. M. Haile, *Electrochem. Solid-State Lett.*, **8**, A245 (2005).

Wan C.-H., Q.-H. Zhuang, C.-H. Lin, M.-T. Lin, C. Shih, *J. Power Sources*, **162**, 41 (2006).

Watanabe M., M. Uchida, S. Motoo, *J. Electroanlyt. Chem.*, **146**, 395 (1987).

Wilson M. S., S. Gottesfeld, *J. Appl. Electrochem.*, **22**, 1 (1992).

Zhang J., Y. Tang, C. Song, et al., *J. Power Sources*, **163**, 532 (2006a).

Reviews

Cheng X., Z. Shi, N. Glass, et al., A review of PEM fuel cell contamination: Impacts, mechanisms, and mitigation, *J. Power Sources*, **165**, 739 (2007).

Costamagna P., S. Srivinasan, Quantum jumps in PEMFC science and technology from the 1960s to the year 2000: I. Fundamental scientific aspects; II. Engineering, technology development and application aspects, *J. Power Sources*, **102**, 242, 253 (2001).

Crawley G., Alkaline fuel cells (AFC), *Fuel Cell Today*, Mar. 2006, Art. 1087.

Gottesfeld S., T. A. Zawodzinski, Polymer electrolyte fuel cells, in: R. C. Alkire, H. Gerischer, D. M. Kolb, C. W. Tobias (eds.), *Advances in Electrochemical Science and Engineering*, Vol. 5, Wiley-VCH, Weinheim, Germany, 1997, p. 195.

Li H., Y. Tang, Z. Wang, et al., A review of water flooding issues in the proton exchange membrane fuel cell, *J, Power Sources*, **178**, 103 (2008).

Mehta V., J. S. Cooper, Review and analysis of PEM fuel cell design and manufacturing, *J. Power Sources*, **114**, 32 (2003).

Schmittinger W., A. Vahidi, A review of the main parameters influencing long-term performance and durability of PEM fuel cells, *J. Power Sources*, **180**, 1 (2008).

Shao Y., G. Yin, Z. Wang, Y. Gao, Proton exchange membrane fuel cell, from low temperature to high temperature: material challenges, *J. Power Sources*, **167**, 235 (2007).

Shao Y., G. Yin, Y. Gao, Understanding and approaches for the durability issues of Pt-based catalysts for PEM fuel cells, *J. Power Sources*, **171**, 558 (2007).

Srinivasan S., O. A. Velev, A. Parthasarathy, D. J. Manko, A. J. Appleby, High energy efficiency and high power density proton exchange membrane fuel cells, *J. Power Sources*, **36**, 299 (1991).

Tawfik H., Y. Hung, D. Mahajan, Metal bipolar plates for PEM fuel cell, *J. Power Sources*, **163**, 755 (2007).

Wu J., Yuan X.Z., J. Martin et al., A review of PEM fuel cell durability: Degradation mechanisms and mitigation strategies, *J. Power Sources*, **184**, 104 (2008).

Yang C., P. Costamagna, S. Srinivasan, et al., Approaches and technical challenges to high temperature operation of proton exchange membrane fuel cells, *J. Power Sources*, **103**, 1 (2001).

Yousfi-Steiner N., P. Moçotéguy, D. Hissel, et al., A review of PEM voltage degradation associated with water management: impacts, influent factors and characterization, *J. Power Sources*, **183**, 260 (2008).

Yu X., S. Ye, Recent advances in activity and durability enhancement of Pt/C catalytic cathode in PEMFC, *J. Power Sources*, pt. 1, **172**, 133 (2007); pt. 2, **172**, 145 (2007).

Zhang J., Z. Xie, J. Zhang, Y. Tang, et al., High-temperature PEM fuel cells, *J. Power Sources*, **160**, 872 (2006b).

CHAPTER 4

DIRECT LIQUID FUEL CELLS

Fuel cells using different types of liquid fuel (reducing agent) and an electrolyte in the form of a proton-conducting polymer membrane are examined in this chapter. Historically, the first fuel cells of this type were those using methanol, which are described in Part A. In Part B, cells using other liquid reducing agents are considered. Fuel cells with alkaline electrolyte using different types of liquid fuel are examined in Chapter 6.

PART A: DIRECT METHANOL FUEL CELLS

4.1 METHANOL AS A FUEL FOR FUEL CELLS

Methanol is a highly promising type of fuel for fuel cells. It is considerably more convenient and less dangerous than gaseous hydrogen. These advantages of methanol are particularly important for applications in mobile power plants, and more particularly, small fuel cells of low power intended to power portable electronic devices (e.g., personal computers, mobile phones, and the like). In contrast to petroleum products and other organic types of fuel, methanol has a rather high electrochemical activity. Its specific energy content of about 6 kWh/kg, although lower than that of gasoline (about 10 kWh/kg), is still quite satisfactory. For this reason, even its use in fuel cells for large power plants in electric vehicles is widely discussed today.

Fuel Cells: Problems and Solutions, By Vladimir S. Bagotsky

Two possibilities exist for the use of methanol in the fuel supply for fuel cells: (1) its prior catalytic or oxidizing conversion to technical hydrogen, and (2) its direct anodic oxidation at the electrodes in the fuel cell. The former possibility implies that additional unwieldy equipment for the conversion of methanol to technical hydrogen and for subsequent purification of this hydrogen is needed. The second possibility is more attractive but involves certain difficulties related to the relatively slow anodic oxidation of methanol even at highly active platinum electrodes. In the face of these difficulties, much attention is given at present, to the development of such direct oxidation methanol fuel cells.

4.2 CURRENT-PRODUCING REACTIONS AND THERMODYNAMIC PARAMETERS

The electrode reactions taking place at the electrodes of DMFCs, the overall current-producing reactions and the corresponding thermodynamic values of equilibrium electrode potentials E^0 and EMF $\mathcal{E}^0$ of the fuel cell are as follows:

$$\text{Anode:} \quad CH_3OH + H_2O \rightarrow CO_2 + 6H^+ + 6e^- \qquad E^0 = 0.02\,\text{V} \tag{4.1}$$

$$\text{Cathode:} \quad \tfrac{3}{2}O_2 + 6H^+ + 6e^- \rightarrow 3H_2O \qquad E^0 = 1.23\,\text{V} \tag{4.2}$$

$$\text{Overall:} \quad CH_3OH + \tfrac{3}{2}O_2 \rightarrow CO_2 + 2H_2O \qquad \mathcal{E}^0 = 1.21\,\text{V} \tag{4.3}$$

The thermodynamic parameters of reaction (4.3) are:

Reaction enthalpy (or heat Q_{react}):

$$\Delta H = -726\ \text{kJ/mol} = 1.25\,\text{eV}$$

Gibbs reaction energy (or maximum work W_e):

$$\Delta G = -702\,\text{kJ/mol} = 1.21\,\text{eV}$$

4.3 ANODIC OXIDATION OF METHANOL

First studies of the special features of the kinetics and mechanism of anodic methanol oxidation at platinum electrodes, which began in the early 1960s (Frumkin and Podlovchenko, 1963; Brummer and Makrides, 1964; Petry et al., 1965; Khazova et al., 1966), constituted the "first boom of work in fuel cells."

Following that, this reaction was the subject of countless studies by many groups in different countries. In summary, one can say of all this work that by now the mechanism of this reaction has been established rather reliably (for reviews, see Bagotsky et al., 1977; Iwasita and Vielstich, 1990; Kauranen et al., 1996), while conflicting views persist on certain detailed aspects. Work on these questions is continuing even now (Léger, 2001).

The major reaction product is carbon dioxide, but in certain cases, transient production of small amounts of other oxidation products, such as formaldehyde and formic acid, is seen. Six electrons are given off in the complete oxidation of methanol to CO_2, so that the specific capacity of methanol is close to 0.84 Ah/g.

Methanol oxidation is a reaction with several consecutive stages. At the first stage, the methanol molecules undergo dehydrogenation:

$$CH_3OH \rightarrow COH_{ads} + 3H_{ads} \tag{4.4}$$

to chemisorbed species COH_{ads}.

At the next stage these species are oxidized by way of chemical interaction with oxygen-containing species OH_{ads} adsorbed on neighboring sites of the platinum surface:

$$COH_{ads} + 3OH_{ads} \rightarrow CO_2 + 2H_2O \tag{4.5}$$

Ionization of the adsorbed hydrogen atoms and the anodic formation of OH_{ads} species from water molecules are the steps actually producing the current:

$$H_{ads} \rightarrow H^+ + e^- \tag{4.6}$$

$$H_2O \rightarrow OH_{ads} + H^+ + e^- \tag{4.7}$$

Under certain conditions, as when the cell is temporarily at open circuit, when the formation of species OH_{ads} is not possible, species COH_{ads} "age" and change to species CO_{ads} that are hard to oxidize and capable of inhibiting further methanol oxidation.

The thermodynamic potential of the methanol electrode is +0.02 V, a value that is rather close to the hydrogen electrode potential. The steady-state potential of a platinum electrode in methanol solution is about +0.3 V. The working potential of steady-state methanol oxidation depends on the current density and varies within the range 0.35 to 0.65 V. This means that the working voltage of a methanol–oxygen cell will have values within the range 0.4 to −0.7 V.

Interest in anodic methanol oxidation received a marked boost when it was shown (Entina and Petry, 1968; Binder et al., 1972) that the platinum–ruthenium system has a significantly higher catalytic activity for this reaction than pure platinum (whereas pure ruthenium has no activity at all for this reaction). This synergy effect was the subject of many studies, which continue to this day. Some authors (e.g., Brankovic et al., 2001; Waszczuk et al., 2001) tend to believe that this effect is due to changes in the electronic structure of platinum when alloyed with ruthenium. The concept of a bifunctional mechanism, according to which the organic species adsorb primarily on platinum sites while ruthenium sites facilitate adsorption of the OH_{ads} species needed for oxidation of the organic species, is widely accepted (Arico et al., 1996; Kauranen et al., 1996; Watanabe et al., 2002).

Agreement as to the optimum composition of Pt–Ru alloys does not exist. Like many other workers, Chu and Gilman (1996) assume that at a temperature of 60°C, an alloy with 50 at% ruthenium represents the optimum. Gasteiger et al. (1994) demonstrated that the optimum ruthenium content depends on temperature and drops to 10 at% at 25°C.

4.4 MILESTONES IN DMFC DEVELOPMENT

Right from the beginning of research into anodic methanol oxidation in the 1960s, repeated attempts were made to build test models of methanol–oxygen or methanol–air fuel cells. In the first of these attempts, an alkaline electrolyte solution was used (Justi and Winsel, 1955; Murray and Grimes, 1963). Due to undesirable carbonate formation in this electrolyte, for most later studies of anodic methanol oxidation, solutions of methanol in aqueous sulfuric acid were used, and the same solutions were used when building the first models of methanol fuel cells, particularly by Shell in England (Hampson et al., 1979; Glazebrook, 1982) and Hitachi in Japan (Tamura and Tsukui, 1984).

Owing to the relatively low rate of methanol oxidation at platinum, considerable amounts of platinum catalyst had to be used in these models (up to $10\,mg/cm^2$). Still, the specific power attained was quite small (about $20\,mW/cm^2$). For this reason, interest gradually subsided. Just a few papers have since been published in the scientific and technical literature on this topic.

A turning point in this area was reached in the mid-1990s after the considerable success achieved in the development of hydrogen–oxygen fuel cells with proton-conducting membranes, PEMFCs, and when it had become possible to transfer these achievements to other types of fuel cells.

At present, most work toward building methanol fuel cells relies on technical and design principles developed previously for PEMFCs. In both types of fuel cells, it is common to use Pt–Ru catalysts at the anode and a catalyst of pure platinum at the cathode. In methanol fuel cells, the membrane commonly used is of the same type as in hydrogen–oxygen fuel cells. The basic differences between these versions are considered in Section 4.7.

4.5 MEMBRANE PENETRATION BY METHANOL (METHANOL CROSSOVER)

A basic problem associated with the operation of methanol fuel cells is the gradual penetration of methanol to the oxygen electrode by diffusion through the membrane. This has two undesirable consequences: (1) part of the methanol is unproductively lost; and (2) the potential of the oxygen electrode moves to more negative values (a "mixed oxygen–methanol potential" becoming established at it; i.e., the working voltage of the cell decreases).

The rate of methanol penetration, i_{cross}, is often given in electrical units of current density equivalent to oxidation of the amount of methanol involved. Many factors influence this rate, including the nature of the membrane, its thickness, and the temperature. The rate increases with increasing methanol concentration in the aqueous–methanolic solution. The crossover rate decreases with increasing working current density. Ren et al. (2000) report the following relation between the values of i_{cross} and i_{load} at a temperature of 80°C and two methanol concentrations, 0.5 M and 1 M:

i_{load}, mA/cm^2	i_{cross}, mA/cm^2 (c = 1 M)	i_{cross}, mA/cm^2 (c = 0.5 M)
0	100	60
100	85	40
300	60	20
600	40	(0)

It can be seen from these data that the methanol concentration should be below 0.5 M when working at low current densities, to avoid large losses of methanol (i.e., realize a high efficiency of methanol utilization). From the same point of view, somewhat higher concentrations could be admitted when working at high current densities. However, to realize really high current densities (i.e., raise the reaction rate of methanol), an important increase in methanol concentration is needed. This leads to antagonism between the conditions needed for working at high current densities (higher methanol concentrations) and the conditions needed for working with a high efficiency of methanol utilization (lower methanol concentrations). In recent years, a large amount of research was done to overcome this antagonism.

Negative consequences of methanol crossover were first detected for Nafion-type membranes. Three possibilities exist to avoid them:

1. Finding ways to treat the membrane that will lower its permeability for methanol, or finding new membrane materials where permeation is not at all observed. Some success has been achieved in lowering the permeability, but the results (see Chapter 13) so far are not sufficient to solve this problem.
2. Replacing methanol in the solution with another substance readily undergoing electrochemical oxidation but not penetrating the membrane. Work in this direction is discussed in Section 4.11.

3. Attempting to act on the crossover rate of methanol by improving MEA design and the design of the cell as a whole.

The effect mentioned above, of a crossover rate decreasing with increasing working current density, is due to the drastic decrease in methanol concentration in the anode's catalytic layer that occurs when methanol undergoes rapid consumption at high current densities. This, in turn, leads to a decrease in methanol concentration in the membrane's surface layer adjacent to the catalytic layer. As methanol diffusion into the membrane starts precisely in this surface layer, the diffusion rate also decreases.

Some workers have attempted to attain the same effect by artificially delaying access of the methanol solution to the MEA. In 2006, Abdelkareem and Nakagawa suggested putting a barrier between the bipolar plate and the anodic part of the MEA that should slow down access of the concentrated methanol solution to the electrode boundary. A porous plate of carbon material is used as such a barrier. Between this plate and the MEA, the CO_2 evolved in the reaction accumulates and leads to an additional delay in access to the methanol solution. The rate of access depends on the thickness and porosity of the plate and on its distance from the anode. A similar suggestion was made in 2006 by Pan, who used four layers of Nafion as the barrier. It should be mentioned that as early as 1999, Scott et al. had pointed out that the CO_2 present in the pores of the anodic diffusion layer will lower the maximum current density that can be attained, because of slow access of the methanol solution to the MEA.

In 1997 and 1999 U.S. patents of Wilkinson et al. it was suggested that a multilayer electrode be used as the anode in DMFCs. The first layer consists of catalyst applied to carbon paper. Most of the methanol is oxidized in this layer. The second catalytic layer, applied directly to the membrane, already has a diluted methanol solution. Other ways of limiting methanol crossover are discussed in the next section when we discuss improvements in water management in fuel cells.

At present, as a practical way of overcoming excessive methanol penetration through the membrane, most DMFC types use aqueous solutions that have a methanol concentration of no more than 1 to 2 M. Some success was attained in overcoming the second of the negative consequences of methanol crossover: poorer performance of the oxygen electrode with platinum catalyst. It was shown by a number of workers that if a dispersed alloy of platinum with a metal of the iron group, rather than pure platinum, is used as the catalyst for the oxygen electrode (Yuan et al., 2006), or if platinum deposited on a carbon base that contains FeTPP (iron tetramethoxyphenylporphyrin) is used (Wang et al., 2004), the influence of methanol on the cathodic catalyst decreases while the catalyst's activity for the oxygen reaction increases.

Apart from the undesirable consequences mentioned above, methanol crossover also has some useful influence (small, to be true) on DMFC operation. Methanol that has diffused to the cathode will be chemically oxidized by oxygen to CO_2 (i.e., without generating current) under the influence of the platinum catalyst. This reaction produces additional heat, and this heat may serve to accelerate startup of a cold fuel battery.

4.6 VARIETIES OF DMFCs

4.6.1 Varieties in Terms of Reactant Supply

Not only must methanol be supplied to the anode space of a DMFC but also water, to secure a sufficiently high conductivity of the proton-exchange membrane. This can be done in two ways: (1) by supplying a mixture of methanol vapor and water vapor produced by evaporation of an aqueous methanol solution in a special evaporator; or (2) by direct feeding of a liquid aqueous methanol solution. The first of these possibilities has the advantage of keeping the fuel cell at a higher temperature and thus realizing better electrical performance. Shukla et al. (1995) have described such a DMFC. A temperature of 200°C was maintained in an evaporator for 2.5 M aqueous methanol solution. The fuel cell's working temperature was as high as 125°C. At a temperature of 100°C, a current density of $200\,mA/cm^2$ could be realized at a voltage of about 0.5 V (the total platinum content of both electrodes was $5\,mg/cm^2$).

The electrical parameters of varieties supplied with a liquid solution are lower than those of varieties supplied with gases, but research in DMFCs has mainly been done exactly with a liquid supply. This type of cell is much simpler to design and operate, inasmuch as neither a special evaporator nor dual temperature control (for the evaporator and for the reaction zone of the fuel cell) is needed. With the supply of a liquid methanol solution, all risk of the membrane drying out close to the anode is eliminated. The elimination of heat is also easier with cells that have a liquid solution supply. All subsequent information on DMFCs in this chapter refers to the variety supplied with a liquid water–methanol solution.

One distinguishes fuel cells with active and passive reactant (fuel and/or oxygen) supply. In a *passive DMFC*, the methanol gets into the cell, either under the effect of gravitational forces (where the vessel with methanol solution is situated above the fuel cell itself) or under the effect of capillary forces through special wicks. A third method of passive methanol supply is hydraulically under the effect of pressure from the gases evolved in the reaction. In completely passive cells, the surrounding air, having direct access to the cathode's gas-diffusion layer, is used as a source of oxygen (air-breathing cells).

In an *active DMFC*, special pumps are used to supply the methanol solution. They yield a higher rate of methanol flowing into the fuel cell, and when needed, better control of this rate. Moreover, forced methanol supply secures a more uniform access of methanol to the various cells of a multicell battery. An active oxygen supply can be achieved using high-pressure cylinders across regulating valves. When air is used as the oxidizing agent, fans or precompression to a relatively high pressure can be used for a forced-air supply.

The electrical performance figures of active DMFCs are much higher than those of passive DMFCs. Yet owing to the need to use many ancillary pieces of equipment, such as pumps, valves, and controllers, power plants with active DMFCs are much more complex and in a number of cases are less reliable in their operation. Also, part of the electrical energy generated by them is used up in the ancillary equipment. Chen et al. (2007) describe a 20-W power plant with active DMFCs operating with four pumps (for water, methanol, and air, and for circulation of the aqueous methanol solution). The plant yields 28 W, of which 8 W is used for internal needs (e.g., the pumps).

As a rule, passive fuel cell systems are simpler to operate, more reliable, and cheaper than active systems. They are used primarily for power plants with low output, such as those supplying power for portable devices. Active systems are more appropriate for power plants with higher output.

4.6.2 Varieties by Application

Basically, all variants of DMFCs are built following a single principle, but depending on the intended use and method of use, different requirements may have to be met and will be reflected in special design and operating features. In contrast to plants built up from other types of fuel cells (of the PEMFC and PAFC systems and with the MCFC and SOFC high-temperature systems), on economic grounds DMFCs are not meant to build up relatively large fixed-site plants having hundreds of kilowatts of power, or plants of a few megawatts for centralized or decentralized power supplies isolated settlements or towns.

As mentioned by Ren et al. (2000), main applications of DMFCs in the near future will be as a power supply for various portable electronic devices with civil, commercial, or military uses. In the future, power plants for electric vehicles and other transport media will join these applications. All these applications have specific features:

1. Power plants for portable devices ranging in power from a few watts to 20 W, must have a low working temperature (e.g., not higher than 60°C), should work with a passive supply of air oxygen, and must be easy to handle. These power plants replace ordinary batteries, both disposable and rechargeable, and should secure long, uninterrupted operation of the portable devices powered by them. The basic criterion reflecting this requirement are the specific energy output per unit mass (Wh/kg) or unit volume (Wh/L) of the battery. For the rechargeable batteries that today have the largest power density (lithium-ion batteries), the specific energy output is about 150 Wh/kg or 250 Wh/L. In the case of power plants with fuel cells, this criterion should account as well for the weight and volume of the reactants needed for uninterrupted work during a given period of time (including their storage containers).

2. Power plants for electric vehicles having tens of kilowatts of power can have working temperatures of up to 100°C, may use precompressed air, and their specific cost (per kilowatt of power output) should be lower than for less powerful plants for portable devices. These power plants replace ordinary internal combustion engines. The basic criterion for the operating efficiency is the average or maximum available power and the reactant consumption per unit electrical energy output (per unit driving distance, the analog of miles per gallon).

In the present chapter we examine the basic questions pertaining to all types of DMFCs, regardless of their future applications. Specific features of small plants for portable devices are considered in Chapter 14.

4.7 SPECIAL OPERATING FEATURES OF DMFCs

The peripheral equipment needed for DMFCs is largely analogous to that of PEMFCs. The mechanical basis of cells and batteries on the whole consists of the bipolar plates between which the membrane–electrode sandwiches (MEAs) are arranged. For the venting of heat, cooling plates with a circulating heat-transfer agent are set up in a particular order between individual cells in batteries.

The major distinguishing features of DMFCs relative to PEMFCs are:

- The use of an aqueous methanol solution as a reactant for the anode, rather than hydrogen gas.
- The liberation of a reaction product [carbon dioxide gas (CO_2)], not at the cathode but at the anode.
- A relatively slow anodic reaction (oxidation of methanol), leading to considerable polarization of the negative electrode and thus to a lower working voltage of the fuel cell.
- A supply of a large amount of water to the MEA, inasmuch as the feed consists of an aqueous solution; hence, water elimination must be just as intense.

Various provisions have been suggested to account for these special features.

4.7.1 Reactant Supply and Product Elimination

Considerable change is needed in the anodic part of MEAs to accommodate the first two points given above. Instead of the porous gas-diffusion layer that in PEMFCs ensures a uniform distribution of hydrogen across the surface, a gas–liquid diffusion layer (GLDL) is needed here that contains a set of hydrophilic as well as a set of hydrophobic pores. Through the hydrophilic pores, this layer must secure unobstructed access of the aqueous methanol solution to the reaction zone and its uniform distribution. Through the hydrophobic pores, this layer must secure the unobstructed elimination of CO_2 as the gaseous reaction product from the reaction zone. Analogous changes must be made in the catalytically active anode layer of the MEA, where the gas is actually formed, and must be removed in the direction of the GLDL. By introducing carefully determined amounts of Nafion and PTFE into the GLDL and into the catalytically active layer, the necessary hydrophilic/hydrophobic micropore ratio is achieved.

It was pointed out by Oedegaard et al. (2004) that to secure free flow of both the methanol solution and the gas, the pore distribution in the GLDL must be optimized not only in terms of the degree of hydrophobicity but also in terms of dimensions. Layers with pore diameters in the range 200 to 300 μm are more efficient than the standard GDL, with a pore size of about 50 μm used in PEMFCs. Such layers may be prepared from fine metal felt materials made hydrophobic.

4.7.2 Water Management in DMFCs

The operation of DMFCs is based on principles that at first seem absurd: water, which together with CO_2 is the major product of the current-producing reaction [see reaction (4.1)], is supplied to the fuel cell in large amounts as an aqueous methanol solution. Certainly, this water is useful for moistening the membrane and possibly for the elimination of the heat of reaction, but when considering the electrochemical reaction, this water is merely inert ballast supplied to and withdrawn from the cell. The aqueous solution needs much larger and heavier containers for reactant storage than would be needed for pure methanol.

This raises the question of whether water could be recirculated. One makes a distinction between outer and inner water recirculation. In the former, water vapor arising from the anode compartments (and consisting of two components: water as a reaction product and evaporated solution water) is condensed and redirected into a special mixer. In this mixer, pure methanol is admixed as needed to maintain the required solution concentration while the reaction proceeds.

In a model system described by Ren et al. (2000), the methanol concentration in the mixer was controlled with a special concentration sensor. Chen et al. (2007) used signals from a chip sensing the fuel cell's voltage to meter in the pure methanol.

Inner water circulation is much fancier, but also more difficult to realize. Under ordinary operating conditions, a flow of water from the anode toward the cathode is observed in both DMFCs and PEMFCs. This flow has two origins: water molecules are dragged along by hydrated hydrogen ions moving from the anode to the cathode in the electric field, and water diffuses under the influence of its own concentration gradient.

To arrange an inverse flow of water from the cathode to the anode, Blum et al. (2003) suggested arranging several layers of paper treated with a mixture of PTFE and carbon black, in a total thickness of 0.5 μm, on each side of the cathode's current collector. These hydrophobic layers do not allow liquid water to pass through, and redirect the water flow, aiming for the cathode on the opposite side. Water vapor that has gone through this barrier layer is eliminated together with the flow of dry oxygen. This exhaust contained about two molecules of water per molecule of methanol that had reacted. This indicates that only that part of the water formed in the reaction quits the fuel cell, while there is no change in the overall water content within the MEA. This would eliminate any need for introducing any water together with methanol into the fuel cell. The cell that operated under these conditions was termed a *water-neutral DMFC*.

An analogous system with hydrophobic barrier layers at the oxygen electrode was described in the work of Kim et al. (2006a). These workers noticed that with this arrangement, methanol crossover decreases as well, inasmuch as a low water concentration is maintained near the anode, owing to the water being turned back.

4.7.3 CO_2 Buildup in the Bipolar Plate Channels

A further problem in DMFC operation is due to the evolution of gaseous CO_2 at the anode (Ye et al., 2005b). Then gas bubbles that can locally interfere with the flow of the aqueous methanol solution may form in the flow field on the anodic side of the bipolar plates. This leads to a nonuniform distribution of the reaction (and thus current) across the MEA surface. This effect is particularly noticeable when the solution is supplied passively (e.g., by free flow from a tank above). To overcome it, one should use an active reactant supply at flow rates several times in excess of the stoichiometric requirements (Cowart, 2005). This raises the question of how to dimension the means of pumping the solution through (and what energy they would consume). The dimensions would depend on the pressure drop within the flow field channels between solution input and outlet (Yang et al., 2005).

4.8 PRACTICAL MODELS OF DMFCs AND THEIR FEATURES

In parallel with the large amount of work done to study the mechanism and operating features of methanol fuel cells with proton-conducting membranes, operating models of such fuel cells started to appear in the mid-1990s, first as laboratory-type small single-element fuel cells and finally, in the form of multielement batteries of relatively large power.

The electrical parameters (particularly the current–voltage relationship) were found to depend on many factors. These include factors related to the technical parameters, such as thickness and the nature of the membrane, the nature of the catalyst, and its amount per unit surface area of the electrode, as well as factors related to the test and operating conditions: the temperature, methanol solution concentration, rate of flow of the solution through the cell, use of pure oxygen or of air, pressure of these gases, and so on. In published data on the development and testing of DMFC models, the factors above exhibit a wide variation. For this reason it is impossible to compare the performance parameters obtained by different workers and in different laboratories. It is merely possible, with some approximation, to select an optimum technical solution (for particular operating conditions) or to select optimum operating conditions (for particular technical solutions). To provide examples, we report below data selected from the development and testing of experimental DMFC models.

Some of the earliest work (Sarumpudi et al., 1994) was done in the Jet Propulsion Laboratory at the University of California in Pasadena. Laboratory samples of cells with $5 \times 5\,cm^2$ electrodes were examined. It was seen that for a cell with a 2 M methanol solution working at a current density of $100\,mA/cm^2$, the working voltage rose from 0.35 V to 0.55 V when the temperature was raised from 30°C to 90°C. Changing the solution concentration from 0.5 M to 2.0 M, at the same current density and a temperature of 60°C had practically no effect on the working voltage (which was about 0.5 V), but when the concentration was raised to 4.0 M, the voltage dropped by almost 0.1 V. At this concentration, the negative effect of methanol crossover was noticed distinctly.

In other early work done at Los Alamos National Laboratory (LANL) (Ren et al., 1996), test cells were studied at temperatures above 100°C with air and pure oxygen. With a cell with a Nafion 112 membrane at a temperature of 130°C and pure oxygen at a pressure of 5 bar, a current density of $670\,mA/cm^2$ was attained at a working voltage of 0.5 V, which corresponds to a power density of $400\,mW/cm^2$. With the same cell at a temperature of 110°C using air at a pressure of 3 bar, a current density $370\,mA/cm^2$ was recorded at a working voltage of 0.5 V, the maximum power density being $250\,mW/cm^2$.

In somewhat later work at the same laboratory (Ren et al., 2000), data were reported for a five-cell battery that was tested at a temperature of 100°C with 1 M methanol solution and air at a pressure of about 2 bar. At a working voltage of 0.4 V (per cell), this battery provided a current density of $500\,mA/cm^2$.

In the Jülich Forschungszentrum in Germany (Dohle et al., 2002) a battery model with 500 W of power was built. The battery consisted of 71 cells, each with a surface area of $144\,cm^2$. The battery was intended to work with air oxygen under a pressure of 1.5 to 3 bar. In the paper, mainly data on the testing of individual cells or groups of three cells are reported. Using a 1 M methanol solution and air oxygen at a pressure of 3 bar, the current density was about $320\,mA/cm^2$ at a temperature of 80°C and a working voltage of 0.3 V.

From the University of Hongkong, China, Ye and Zhao (2005a) reported building a methanol cell with a passive supply of the methanol solution by natural convection from a vessel above the cell. The CO_2 evolved from the cell was fed into this vessel, which was tightly closed off. Increasing current density led to a higher rate of gas evolution and to higher pressure in the vessel. This also produced a larger flow of methanol solution into the cell. This provided a system in which the rate of methanol supply is controlled by the load on the cell. At current densities of up to 150 mA/cm^2, this cell had approximately the same characteristics as the same cell when operated with an active supply of methanol solution via pumps. At higher current densities, natural convection proved inadequate. Difficulties were also seen at low current densities (below 50 mA/cm^2) when operation became unstable. At current densities below 5 mA/cm^2, gas evolution is too insignificant, and the system ceases to function.

At the Korea Institute for Science and Technology, Kim et al. (2006b) studied the operation of a six-cell battery that had a total power of 50 W. The battery was fed with a 2 M methanol solution. The battery was supposed to work at ambient temperature, but internal heat production raised its temperature to above 80°C. Using oxygen, a maximum power of 254 mW/cm^2 was attained; with air, this value was 85 mW/cm^2.

In all the DMFC models mentioned above, Nafion-type membranes were used. In work performed at LANL (Kim et al., 2004), data were obtained for a test cell having a membrane made of sulfonated poly(arylene) ether copolymers. According to these data, methanol crossover for this membrane is only half that for a Nafion membrane. A cell with this new membrane, working at a temperature of 80°C and in ambient air, yielded a current density of 200 mA/cm^2 when the working voltage was 0.4 V. In a cell with Nafion working under analogous conditions, the current density was just 150 mA/cm^2.

4.9 PROBLEMS TO BE SOLVED IN FUTURE DMFCs

In contrast to PEMFCs, and despite the large volume of research performed, DMFC fuel cells are still not in commercial production or in wide practical use. As we have seen from what has been reported earlier in the chapter, the true performance indicators of these fuel cells when used for different needs are difficult to assess from the experience gathered in tests of individual samples performed under a variety of conditions. For such an assessment one would need statistical data obtained from tests on a sufficiently large number of cells of any single type and accounting for all cell parameters and test conditions.

Yet even now, one potential area of application can be recognized distinctively: for relatively low-power energy sources (no more than 20 W) in electronic equipment such as notebooks, cameras and videocameras, DVD players, and some medical devices. In these areas, DMFCs taking up little space should replace the disposable or rechargeable batteries used at present. Even now, a good amount of work is performed that aims at preparing the commercial conquest of these areas by miniaturizing cells of the DMFC type and the analogous cells of the DLFC type using a variety of liquid fuels. This work is described in detail in Chapter 14.

So far, another potential field of application of DMFCs, as power sources for electric vehicles, is too remote. A large amount of research and engineering still has to be done to master this application, primarily work aimed at improving the technical and economic parameters of these cells. Work is needed more particularly to solve the problems described below.

4.9.1 Longer Lifetime

The production volume of DMFCs and long-term testing results are still too small for any reliable estimate of the lifetime of methanol fuel cells. Workers therefore look primarily at all the major reasons giving rise to the gradual performance drop and/or premature failure of these cells. The reasons considered in the corresponding section dealing with cells of the PEMFC type must be supplemented in the case of methanol fuel cells by two more, described below.

Crossover of Ruthenium Ions

At the working potential of an electrode, the Pt–Ru catalyst is rather stable, but in long runs and following the slightest shift of potential in the positive direction (as may occur following an arrest of the methanol supply), ruthenium is seen to dissolve selectively from the catalyst, transferring into the solution or membrane in the form of ruthenium ions. Different optical techniques have been used for a rather detailed examination of this phenomenon (Piela et al., 2004; Sarma et al., 2007). The dissolution of ruthenium not only leads to a gradual drop in the activity of the Pt–Ru catalyst but also, when these ions pass the membrane and reach the cathode, to an important inhibition of the cathodic reduction of oxygen. This may lead to a decrease in the working voltage of the cell by almost 0.2 V.

Aging of Methanol Adsorption Products on Platinum

During the electrochemical oxidation of methanol, products of the –COH type become adsorbed on the platinum catalyst surface. They are the product of dehydrogenation of the original methanol molecules. Under normal conditions, these species are oxidized rapidly by –OH species adsorbed on neighboring segments of the platinum or ruthenium. Under certain conditions, such as long current breaks, these particles may age, change into species of the –CO type, and be unwilling to leave the surface. They then cause an important slowdown in the continued oxidation of new quantities of methanol.

4.9.2 Greater Efficiency

The operating efficiency of DMFCs is lowered considerably by methanol crossover. This effect leads to unproductive methanol consumption and to a marked decrease in working voltage caused by the action of methanol on the potential of the oxygen electrode. So far, only two possibilities are known to lessen or eliminate this effect completely:

1. Developing ways to treat membranes of the Nafion type so as to lower their permeability for methanol, or developing new types of proton-conducting membrane materials that are totally impermeable to methanol. Some work in this direction is considered in Chapter 13.
2. Replacing the methanol in cells of the DMFC type by other liquid reducing agents not penetrating the membrane: that is, changing over from DMFCs to other DLFCs. Some varieties of this type are considered beginning in Section 4.10.

4.9.3 Lower Cost

Reasonable estimates for the possible production cost of DMFCs are almost nonexistent. Even today it can be seen, however, that apart from technical problems, their large-scale use in such an area as future electric vehicles will be possible only if two basic problems are solved:

1. Developing new membrane materials that cost at least an order of magnitude less than Nafion.
2. Developing new types of catalysts with a much lower platinum metal content, yet with an activity level higher than that of catalysts in current use. A special concern is ruthenium world production, which would barely be able to supply the needs arising in the production of many millions of electric vehicles at the current ruthenium content of the catalyst.

PART B: DIRECT LIQUID FUEL CELLS

4.10 THE PROBLEM OF REPLACING METHANOL

The basic advantage of methanol–oxygen fuel cells (DMFCs) with proton-conducting membranes over analogous hydrogen–oxygen fuel cells (PEMFCs) resides in a much more convenient liquid reactant replacing hydrogen, a gas that has drawbacks in transport, storage, and manipulation. Undoubtedly, the specific energy content per unit weight is much larger for hydrogen than for methanol, but when including all equipment needed for reactant storage (e.g., cylinders, containers for metal hydrides), the energy content per unit weight turns out to be three times higher for methanol than for hydrogen.

Yet using methanol in fuel cells is associated with certain difficulties and inconvenient features:

- The relatively low rate of electrochemical reaction.
- The need to eliminate from the system the large amount of water introduced together with methanol as an aqueous methanol solution.

- The need to eliminate from the system CO_2 formed as a gaseous reaction product.
- The considerable amount of heat evolved, owing to the relatively low values of working voltage.
- The boiling point of methanol being around 65°C, raising the temperature in order to accelerate the electrode reactions will be possible only when working under higher pressure.
- Methanol crossover through the membrane and the need to deal with all consequences of this effect.
- Methanol being toxic, certain precautions must be taken against the leakage of methanol and its vapors from the system as a whole and from its storage vessels.

Methanol is not the only possible liquid reactant for anodes in fuel cells. During the last two decades, in parallel with the very active work on methanol fuel cells, numerous studies were made to examine the possibility of using other liquid reactants with reducing properties (both organic and inorganic).

A number of factors are considered when selecting a new reducing agent: (1) the equilibrium potential of the anodic oxidation reaction of this substance, which will influence the EMF and OCV of the fuel cell; (2) the relative rate of this reaction, which will influence the polarization of the electrode and thus the working voltage; (3) the price and availability of the substance; and (4) the ecological cleanliness and harmlessness of the reactant itself and of its direct and potential secondary oxidation products. The latter question was reviewed in 2007 by Demirci.

4.11 FUEL CELLS USING ORGANIC LIQUIDS AS FUELS

4.11.1 Direct Ethanol Fuel Cells

In its chemical properties, ethanol is similar to methanol but has a considerable advantage over methanol of much lower toxicity. It must be pointed out that from an ecological viewpoint, ethanol is exceptional among all other types of fuel. In the oxidation of all organic fuels (e.g., natural gas, oil-derived products, coal), carbon dioxide is formed, which leads to the well-known global warming effect in the atmosphere of the Earth: a global temperature rise. This is equally true for chemical uses of these fuels in heat engines by "hot combustion" as for electrochemical uses in fuel cells by "cold combustion."

Ethanol can be obtained by the fermentation of various agricultural biomasses, which in turn are formed by photosynthesis involving CO_2 and solar energy. This implies that combustion, hot or cold, of ethanol will not lead to the accumulation of excess CO_2 and will not upset the overall balance of this gas in the atmosphere. Using ethanol as an energy vector in essence is a practical way of using solar energy. Ethanol is the only chemical fuel in renewable supply. In Brasil, mass production of ethanol from biomass has already started, the corresponding infrastructure was created, and large part of automotive transport is transformed so as to work with ethanol rather than with gasoline. In the European Union, ethanol is treated as a desirable fuel additive. Attention should be drawn to current discussions of the ecological (and other) implications, including those of ethanol competing with foodstuff agriculture. The overall energy content of ethanol (about 8 kWh/kg) is rather close to that of gasoline (about 10 kWh/kg).

It is quite natural that all these considerations have led to enhanced interest in the uses of ethanol in fuel cells. A lot of research went into the development of direct ethanol fuel cells (DEFCs) during the last decade. Much of the research in this direction was performed in France in the group of Claude Lamy (see the review of Lamy et al., 2002).

The reactions that occur ideally in an ethanol–oxygen fuel cell, and the associated thermodynamic parameters, are as follows:

$$\text{Anode:} \quad C_2H_5OH + 3H_2O \rightarrow 2CO_2 + 12H^+ + 12e^- \quad E^0 = 0.084\,\text{V} \tag{4.8}$$

$$\text{Cathode:} \quad 6O_2 + 12H^+ + 12e^- \rightarrow 6H_2O \quad E^0 = 1.229\,\text{V} \tag{4.9}$$

$$\text{Overall:} \quad C_2H_5OH + 3O_2 \rightarrow 2CO_2 + 3H_2O \quad \mathscr{E}^0 = 1.145\,\text{V} \tag{4.10}$$

$$\Delta G = -1325\,\text{kJ/mol} = 1.145\,\text{eV} \qquad \Delta H = -1367\,\text{kJ/mol} = 0.969\,\text{eV}$$

The structure of a fuel cell using ethanol (DEFC) is little different from that of a fuel cell using methanol (DMFC). The same proton-conducting membranes made of Nafion or its replacements are used. As the catalyst for the anodic process, here again the Pt–Ru system provides better electrical performance indicators than does pure platinum. An important difference between these two varieties of fuel cells is the higher working temperature required with ethanol. This is due to the fact that the electrochemical oxidation of ethanol occurs with greater difficulties than the analogous reaction of methanol and is markedly slower. For acceptable electrical parameters, therefore, higher temperatures, ranging up to 200°C, must be used. For this reason, the problems that arise when building DEFCs often are the same as those described in Section 3.7 for PEMFCs working at elevated temperatures.

In experimental DEFCs working at a temperature of 90°C, values of power density of 33 mW/cm^2 were realized at a voltage of 0.55 V (Lamy et al., 2004), which is a rather modest result. Arico et al. in 1998 reported that in an ethanol fuel cell with a Nafion membrane containing added silica (SiO_2), the power density attained 110 mW/cm^2 at a temperature of 145°C and a voltage of 0.35 V.

In 1995, when studying an ethanol fuel cell with a membrane made of PBI impregnated with concentrated phosphoric acid at a temperature of 170°C and a voltage of 0.30 V, Wang et al. obtained a power density of 75 mW/cm^2. These data, although obained under different conditions and hardly comparable, show that when working at elevated temperatures, ethanol fuel cells may yield electrical performance indicators comparable to those of methanol fuel cells.

On the basis of these results, it is assumed by a number of workers that it will be possible to replace methanol with ethanol in membrane-type fuel cells working with liquid fuels. There is, however, a basic point that casts doubt on this conclusion.

In the electrochemical oxidation of methanol, carbon dioxide gas is the chief reaction product. The yields of other potential products of the oxidation reaction, such as formaldehyde, formic acid, and the like, are a few percent at most. Arico et al. (1998) concluded from a chromatographic analysis of the reaction products that the chief product of electrochemical oxidation of ethanol (with a yield of about 98%) is CO_2, just as for methanol. This conclusion is inconsistent with the results obtained by other workers. Wang et al. (1995) studied the reaction products of ethanol and propanol oxidation by differential electrochemical mass spectrometry. They found that during the reaction, only 20 to 40% of the theoretical yield of CO_2 is produced, whereas acetaldehyde is formed to 60 to 80% (even traces of acetic acid are formed). Rousseau et al. (2006) used a high-performance liquid chromatograph for analysis of the products of ethanol oxidation. According to their data, about 50% aldehyde, 30% acetic acid, and only about 20% CO_2 are formed at a temperature of 90°C at a platinum catalyst. With Pt–Sn or Pt–Sn–Ru catalysts, somewhat different numbers were obtained: 15% aldehyde, 75% acid, and 10% CO_2. It follows from these data that the composition of the reaction products depends heavily on the catalyst used for the electrode reaction. Conditions have not been found that would reliably provide CO_2 yields approaching 100%. This is of basic significance.

The formation of acetic acid by ethanol oxidation is a four-electron process:

$$C_2H_5OH + H_2O \rightarrow CH_3COOH + 4H^+ + 4e^- \qquad (4.11)$$

while the formation of acetaldehyde is a two-electron process:

$$C_2H_5OH \rightarrow CH_3CHO + 2H^+ + 2e^- \qquad (4.12)$$

For the sake of comparison, note that the electrochemical oxidation of ethanol to CO_2 according to reaction (4.8) is a 12-electron process.

The smaller number of electrons involved in the reaction leads to a considerable decrease in the de facto energy content of ethanol: from 8 kWh/kg for the 12-electron process to 2.6 kWh/kg for a four-electron process and to as little as 1.3 kWh/kg for a two-electron process.

The drop in energy content is not the only negative consequence of these side reactions. In contrast to CO_2 gas, which is readily vented from the system into the atmosphere, a number of problems of removal and disposal arise when aldehyde and acetic acid are formed in the reaction.

It is surprising that in many publications on ethanol fuel cells, this aspect is not examined more closely. It is only in the conclusions of the review of Lamy et al. (2002) that the importance of this problem is stressed.*

The formation of acetaldehyde and acetic acid is due to the fact that rupture of the C–C bond in the original ethanol molecule would be required for CO_2 formation. In the chemical oxidation of ethanol at high temperatures (combustion), this bond is readily ruptured in the hot flame, and the only reaction product (in addition to water) is CO_2. In electrochemical oxidation occurring at temperatures below 200°C, this bond is very difficult to rupture, and reactions involving the sole rupture of C–H bonds occur much more readily, thus leading to the side products noted.

The very idea of building ethanol fuel cells is very attractive and promising. To this day, development of these fuel cells is still in its initial stages, a large amount of research in electrocatalysis being required to find ways of making fully adequate ethanol fuel cells. This work should lead to the development of basically new polyfunctional catalysts that would secure a high rate not only of low-temperature oxidation of organic substances through C–H bond rupture but simultaneously, reactions involving rupture of the C–C bonds. Only then will it be possible to think of a widespread application of ethanol fuel cells for future highly efficient, ecologically harmless electric vehicles.

In many publications cited in the literature, the possibility of using higher alcohols (e.g., 1-propanol, 2-propanol) and other, analogous organic compounds (e.g., ethylene glycol, acetaldehyde, dimethyl ether) in fuel cells has been studied. The electrochemical oxidation mechanism of these substances is more complex and associated with the formation of many more side products than in the case of ethanol. In all these papers the high theoretical specific energy content of these compounds is mentioned, and experimental results concerning the power density and a few other parameters are reported, but as a rule, nothing is said as to the depth of oxidation attained (the percentage of CO_2 among the reaction products).

4.11.2 Direct Formic Acid Fuel Cells

Formic acid (HCOOH) is a somewhat unusual type of "fuel" for fuel cells, and in many of its properties differs from other substances used as fuels. On the one hand, the theoretical energy content of formic acid (1.6 kWh/kg) is much lower than for all other reactants considered in this section. On the other hand, the equilibrium electrode potential for the oxidation of formic acid (−0.171 V) is more negative than that for the other organic fuels; that is, in a thermodynamic

* This aspect was also mentioned (without discussion) in a review by Antolini (2007).

sense, formic acid is a very strong organic reducing agent. The thermodynamic EMF of a formic acid–oxygen cell is 1.45 V.

Formic acid has a number of properties that make its use in fuel cells very attractive. First, this substance is ecologically absolutely harmless (the U.S. Food and Drug Administration allows it to be used as a food additive). Its only oxidation products are CO_2 and water. It is, essentially, not possible that side products or intermediates will be formed. Formic acid is a liquid. In aqueous solutions it dissociates, yielding $HCOO^-$ ions. This is of basic significance for its use in fuel cells with proton-conducting membranes. These membranes have a skeleton containing negatively charged ionic groups [in the case of Nafion, sulfonic acid groups (SO_3^-)]. Because of the electrostatic repulsion, these groups hinder (or at least strongly retard) the penetration of formate ions into the membrane. Thus, in the case of formic acid, the effect that constituted the major difficulty for the development of methanol fuel cells, crossover of the anodic reactant from the anodic region through the membrane to the cathodic region, is practically inexistent.

The electrochemical oxidation of formic acid,

$$HCOOH \rightarrow CO_2 + 2H^+ + 2e^- \tag{4.13}$$

occurs along one of two possible pathways:

1. By a step of catalytic decomposition (dehydrogenation):

$$HCOOH \rightarrow CO_2 + H_2 \text{ (or } 2M-H_{ads}) \tag{4.14a}$$

followed by hydrogen ionization:

$$H_2 \rightarrow 2H^+ + 2e^- \tag{4.14b}$$

2. By a chemisorption step, including dehydration:

$$HCOOH \rightarrow M-CO_{ads} + H_2O \tag{4.15a}$$

followed by the steps of

$$H_2O + M \rightarrow M - H_{ads} + H^+ + e^- \tag{4.15b}$$

and

$$M-CO_{ads} + M-OH_{ads} \rightarrow CO_2 + H^+ + e^- \tag{4.15c}$$

(here M is a metallic catalyst surface site).

The overall current-producing reaction in the cell is

$$HCOOH + \tfrac{1}{2}O_2 \rightarrow CO_2 + H_2O \quad (4.16)$$

The low theoretical energy content of this substance is due to the fact that only two electrons per formic acid molecule are involved in the reaction. Yet the reaction is so simple that undesirable side products cannot be formed in it.

Formic acid is used in membrane-type fuel cells as an aqueous solution. A 20 M HCOOH solution contains about 75% formic acid. Owing to the low water content, the membrane is not sufficiently moistened in such a concentrated solution, and its resistance increases (Rice et al., 2002). In solutions less concentrated than 5 M, the current densities that can be realized are low, owing to the slow reactant supply by diffusion to the catalyst surface. An optimum concentration for fuel cell operation is 10 to 15 M. In contrast to DMFCs, an increase in reactant concentration in DFAFCs does not produce complications related to reactant crossover.

At a temperature of 70°C, a power density of about 50 mW/cm^2 was attained in 12 M HCOOH at a working voltage of 0.4 V. By comparison, the power density in a typical methanol fuel cell under the same conditions is about 30 mW/cm^2.

With the experience gathered in the development of DMFCs, Pt–Ru catalysts were used for the anodic process in the early studies on DFAFCs. Ha et al. (2006) showed that much better electrical characteristics can be obtained with palladium black as the catalyst. Importantly, with this catalyst one can work at much lower temperatures. In particular, at a temperature of 30°C, power densities of 300 mW/cm^2 were obtained with a voltage of 0.46 V, and about 120 mW/cm^2 with a voltage of 0.7 V. The differences between the two catalysts probably are due to the fact that with Pt–Ru, formic acid oxidation follows the second of the mechanisms mentioned (chemisorption with dehydration), while palladium black is a highly effective catalyst for the dehydrogenation of formic acid, the first mechanism being followed on it. It must be pointed out that this effect is highly specific; in methanol oxidation, the catalytic activity of palladium is lower than that of Pt–Ru.

Considering all these special features, it will be very convenient to use formic acid as a reactant in fuel cells of small size for a power supply in portable equipment ordinarily operated at ambient temperature. Such fuel cells are described in Chapter 14.

In a review by Yu and Pickup (2008), recent advances in DFAFCs are presented, focusing mainly on anodic catalysts for the electrooxidation of formic acid. The problem of formic acid crossover through Nafion membranes is also discussed.

4.12 FUEL CELLS USING INORGANIC LIQUIDS AS FUELS

4.12.1 Direct Borohydride Fuel Cells (DBHFC)

Sodium borohydride ($NaBH_4$) is a substance with a relatively high hydrogen content (10.6 wt%). It is a strong reducing agent. In strongly alkaline aqueous solutions, sodium borohydride is stable but undergoes hydrolysis, forming $NaBO_2$ and evolving hydrogen gas in neutral and acidic solutions:

$$NaBH_4 + 2H_2O \rightarrow NaBO_2 + 4H_2 \tag{4.17}$$

Because of this feature, the compound is sometimes used as a convenient hydrogen source for small PEMFCs (Wainright et al., 2003).

The first data concerning a possible direct use of this compound as a reducing agent (fuel) in fuel cells of the DBHFC type appeared in the early 1960s (Indig and Snyder, 1962; Jasinski, 1965). In 1999, Amendola et al. reported building a fuel cell with an anion-conducting membrane using a solution with 5% $NaBH_4$ + 25% NaOH. At a temperature of 70°C, this cell yielded a power density of about 60 mW/cm^2. Work in this direction was not followed up, apparently due to the fact that so far, anion-exchange membranes stable enough in concentrated alkaline solutions are not available. In 2003, Li et al. suggested that a proton-conducting membrane be used to this end. This variant is the one generally adopted at present.

Usually, a solution containing 10 to 30 wt% $NaBH_4$ and 10 to 40% NaOH is supplied to the anode compartment of borohydride–oxygen fuel cells of the membrane type. The reactions occurring in a borohydride–oxygen cell and their thermodynamic parameters are:

$$\text{Anode:}\quad NaBH_4 + 8OH^- \rightarrow NaBO_2 + 6H_2O + 8e^- \quad E^0 = -1.25\,V \tag{4.18}$$

$$\text{Cathode:}\quad 2O_2 + 8e^- + 4H_2O \rightarrow 8OH^- \quad E^0 = +0.40\,V \tag{4.19)*}$$

$$\text{Overall:}\quad NaBH_4 + 2O_2 \rightarrow NaBO_2 + 2H_2O \quad \mathscr{E}^0 = 1.65\,V \tag{4.20}$$

$$\Delta G = -1272\,kJ/mol = 1.65\,eV \qquad \Delta G = -1360\,kJ/mol = 1.77\,eV$$

Theoretically, sodium borohydride yields eight electrons upon oxidation; that is, it is a reactant with a very high energy content (9.3 kWh/kg). As a

* Since the solution is alkaline, the electrode reactions are formulated with hydroxyl ions OH^- rather than with hydrogen ions, and the electrode potentials are referred to the alkaline variant of the SHE. This explains the difference between equations (4.19) and (4.2), which from a thermodynamic point of view, are completely equivalent.

matter of fact, fewer electrons are involved in the current-producing reaction. This is due to the fact that the electrode potential of borohydride is more negative than the potential of the hydrogen electrode in the same alkaline solution (−0.828 V). When borohydride solution is getting into the fuel cell, cathodic hydrogen evolution is possible at the anodic metal catalyst because of the negative potential, and this is attended by the coupled reaction of borohydride oxidation (which is not current-producing). This is analogous to the corrosion of an electronegative metal under the effect of "local elements" on which hydrogen evolution is facilitated. As a result, part of the borohydride is lost unproductively. At nickel catalysts for the anodic reaction, the effective number of electrons in the current-producing reaction is about four. At platinum catalysts, this number is even lower. High values of this number (6.9) were observed for gold (Amendola et al., 1999). Different intermetallic compounds, including some containing rare-earth elements, have also been suggested as anodic catalysts.

Even when taking into account the lower number of electrons, the specific energy content of borohydride remains rather high. Under the assumption of six electrons, it is close to 7 kWh/kg. This substance therefore is rather promising for the development of small fuel cells as a power supply for portable equipment.

So far it is not clear how to solve the following problem. Upon contact with NaOH solution, H^+ ions in the ion-exchange membrane are exchanged (at least in part) for Na^+ ions. During current flow, these sodium ions start to migrate from the anode toward the cathode. When reaching the cathodic zone, hydrogen ions become involved in oxygen reduction and then are eliminated as water vapor. However, the sodium ions will accumulate in the cathodic zone as NaOH, which must then be eliminated from the fuel cell separately.

An interesting proposal is that of building a borohydride fuel cell using hydrogen peroxide (H_2O_2) instead of oxygen as the oxidizing agent (Raman et al., 2004; Ponce de León et al., 2007). In such a cell, both electrodes work with liquid reactants, which simplifies the design and manipulation since the need for sealing that always arises when working with gaseous reactants is eliminated. An alkaline hydrogen peroxide solution is rather stable.

The thermodynamic value of the electrode potential of the peroxide electrode, 0.87 V, is markedly more positive than the potential of the oxygen electrode, 0.4 V. Accordingly, the EMF the cell rises to 2.11 V, which in turn leads to a higher working voltage in the fuel cell. The possibility pointed out by Raman et al. (2004) of being able to reach voltages of more than 3 V in such a cell refers to a hypothetical case in which an alkaline borohydride solution is combined in the cell with an acidic hydrogen peroxide solution. Such a cell cannot be operated for any length of time, since the large pH gradient of the solutions inevitably leads to interdiffusion and mixing of the solutions.

A problem associated with borohydride fuel cells is the fact that both reactants fundamentally may undergo catalytic decomposition, which leads to a lower utilization efficiency. Also, gaseous decomposition products may hamper access of the liquid reactants to the catalyst.

In a review by Wee (2006), the parameters of DBFCs are compared with those of DMFCs. It was shown that a DBFC system is superior to that of a DMFC system in terms of cell size and fuel solution consumption. From an economic point of view (total operating costs) a DBFC system is more favorable in specific portable applications such as miniaturized or micro power systems with short operational time spans.

4.12.2 Direct Hydrazine Fuel Cells

In the 1970s, a number of studies were published dealing with the design of hydrazine–oxygen fuel cells with an alkaline electrolyte. Several models of such batteries were actually built for portable devices and for various military objects, but owing to the high toxicity and the high price of hydrazine, this work was not developed further (see Chapter 6). Hydrazine is a strong reducing agent, both in a thermodynamic sense (the equilibrium potential of its oxidation reaction is rather negative) and from the viewpoint of an uncomplicated electrochemical reaction. The theoretical value of the EMF of a hydrazine–oxygen cell is 1.56 V.

In 2003, Yamada et al. suggested using this substance in a fuel cell with a proton-conducting (i.e., acid) membrane. Hydrazine was used as a 10% aqueous solution of hydrazine hydrate ($N_2H_4{\cdot}H_2O$). In the aqueous solution, because of its strong alkaline properties, hydrazine dissociates into the ions $N_2H_5^+$ and OH^-. The anodic oxidation of hydrazine can be written

$$N_2H_5^+ + 2OH^- \rightarrow N_2 + H^+ + e^- + 2H_2O \tag{4.21}$$

the only reaction products being nitrogen and water.

Disperse powders of platinum and Pt–Ru were used as catalysts for the anodic reaction in the experimental cell. When working at a temperature of 80°C, this DHFC had the following parameters (the parameters for a DMFC of analogous design working at 100°C are given in parentheses): open-circuit voltage (OCV), 1.2 V (0.7 V); voltages at current densities of 50, 100, and 150 mA/cm^2: 0.95 V (0.58 V), 0.57 V (0.5 V), and 0.48 V (0.46 V). It can be seen that under these conditions at current densities below 50 mA/cm^2, a hydrazine fuel cell has definite advantages with respect to voltage (owing to the high OCV). At higher current densities these advantages disappear. In the gases evolving during operation of the cell, noticeable quantities of ammonia have been detected. This is due to the fact that apart from the current-producing reaction, catalytic decomposition of hydrazine occurs:

$$N_2H_4 \rightarrow N_2 + 2H_2 \tag{4.22a}$$

and/or

$$3N_2H_4 \rightarrow N_2 + 4NH_3 \tag{4.22b}$$

The first of these reactions does not lead to a decrease in hydrazine utilization, since the hydrogen that is formed can take part in the anodic reaction. The second reaction, hydrazine decomposition, leads to a lower utilization factor.

In the opinion of the authors, the basic problem arising in fuel cells of this type is free penetration of the $N_2H_5^+$ ions through the proton-conducting membrane to the cell's cathodic zone, which leads to those consequences that had been described above with respect to methanol crossover.

REFERENCES

Abdelkareem M. A., N. Nakagawa, *J. Power Sources*, **162**, 114 (2006).

Amendola S. C., P. Onnerud, M. Kelly, et al., *J. Power Sources*, **84**, 130 (1999).

Arico A. S., P. Creti, H. Kim, et al., *J. Electrochem. Soc.*, **143**, 3950 (1996).

Arico A. S., P. Creti, P. L. Antonucci, V. Antonucci, *Electrochem. Solid-State Lett.*, **1**, 68 (1998).

Binder H., A. Koehling, G. Sandstede, in: G. Sandstede (ed.), *From Electrocatalysis to Fuel Cells*, University of Washington Press, Seattle, WA 1972, pp. 43–58.

Blum A., T. Duvdevani, M. Philosoph, N. Rudoy, E. Peled, *J. Power Sources*, **148**, A87 (2003).

Brankovic S. R., J. McBreen, R. R. Adžić, *J. Electroanal. Chem.*, **503**, 99 (2001).

Brummer S. B., A. C. Makrides, *J. Phys. Chem.*, **68**, 1448 (1964).

Chen C. Y., D. H. Liu, C. L. Huang, C. L. Chang, *J. Power Sources*, **167**, 442 (2007).

Chu D., S. Gilman, *J. Electrochem. Soc.*, **143**, 1685 (1996).

Cowart J. S., *J. Power Sources*, **143**, 30 (2005).

Dohle H., H. Schmitz, T. Bewer, J. Mergel, D. Stolten, *J. Power Sources*, **106**, 313 (2002).

Entina V. S., O. A. Petry, *Elektrokhimiya*, **4**, 111 (1968).

Frumkin A. N., B. I. Podlovchenko, *Dokl. Akad. Nauk SSR*, **150**, 349 (1963).

Gasteiger H. A., N. Markovic, P. N. Ross, E. J. Cairns, *J. Electrochem. Soc.*, **141**, 1795 (1994).

Glazebrook R. W., *J. Power Sources*, **7**, 215 (1982).

Ha S., Z. Dunbar, R. I. Masel, *J. Power Sources*, **158**, 129 (2006).

Hampson N. A., M. J.Willars, B. D. McNicol, *J. Power Sources*, **4**, 191 (1979).

Indig M. E., R. N. Snyder, *J. Electrochem. Soc.*, **109**, 1104 (1962).

Jasinski R., *Electrochem. Technol.*, **3**, 40 (1965).

Justi E. W., A. W. Winsel, Brit. patent 821,688 (1955).

Kauranen P. S., E. Skou, J. Munk, *J. Electroanal. Chem.*, **404**, 1 (1996).

Khazova O. A., Yu. B. Vassiliev, V. S. Bagotsky, *Electrokhimiya*, **2**, 267 (1966).

Kim Y. S., M. J. Sumner, W. L. Harrison, et al., *J. Electrochem. Soc.*, **151**, A2156 (2004).

Kim H. K., J. M. Oh, J. H. Kim, H. Chang, *J. Power Sources*, **162**, 497 (2006a).

Kim D., J. Lee,T.-H. Lim, et al., *J. Power Sources*, **155**, 203 (2006b).

Lamy C., S. Rousseau, E. M. Belgsir, et al., *Electrochim. Acta*, **49**, 3901 (2004).

Léger J.-M., *J. Appl. Electrochem.*, **31**, 767 (2001).

Li Z. P., B. H. Liu, K. Arai, S. Suda, *J. Electrochem. Soc.*, **150**, A868 (2003).

Murray J. N., P. G. Grimes, in: *Fuel Cells*, American Institute of Chemical Engineers, New York, 1963, p. 57.

Oedegaard A., C. Hebling, A. Schmitz, et al., *J. Power Sources*, **127**, 187 (2004).

Pan H., *Electrochem. Solid-State Lett.*, **9**, A349 (2006).

Petry O. A., B. I. Podlovchenko, A. N. Frumkin, Hira Lal, *J. Electroanal. Chem.*, **10**, 253 (1965).

Piela P., C. Eickes, E. Brosha, et al., *J. Electrochem. Soc.*, **151**, A2053 (2004).

Ponce de León C., F. C. Walsh, A. Rose, et al., *J. Power Sources*, **164**, 441 (2007).

Raman R. K., N. A. Choudbury, A. K. Shukla, *Electrochem. Solid-State Lett.*, **7**, A491 (2004).

Ren X., M. S. Wilson, S. Gottesfeld, *J. Electrochem. Soc.*, **143**, L12 (1996).

Ren X., P. Zelenay, Sh. Thomas, J. Davey, S. Gottesfeld, *J. Power Sources*, **86**, 111 (2000).

Rice C., R. I. Masel, P. Waszczuk, A. Wieckowski, N. Barnard, *J. Power Sources*, **111**, 83 (2002).

Rousseau S., C. Coutanceau, C. Lamy, J.-M. Léger, *J. Power Sources*, **158**, 18 (2006).

Sarma S. L., C.-H. Chen, G.-R. Wang, et al., *J. Power Sources*, **167**, 358 (2007).

Sarumpudi S., S. R. Narayanan, E. Vamos, H. Frank, G. Halpert, *J. Power Sources*, **47**, 377 (1994).

Scott K., W. M. Taama, P. Argyropoulos, K. Sundmacher, *J. Power Sources*, **83**, 204 (1999).

Shukla A. K., P. A. Christensen, A. Hamnett, M. P. Hogarth, *J. Power Sources*, **55**, 87 (1995).

Simões F. C., D. M. dos Anjos, F. Vigier, et al., *J. Power Sources*, **167**, 1 (2007).

Tamura K., T. Tsukui, *Hitachi Hyoron*, **66**, 49 (1984).

Wainright J. S., R. F. Savinell, C. C. Liu, M. Litt, *Electrochim. Acta*, **48**, 2869 (2003).

Wang J., S. Wasmus, R. F. Savinell, *J. Electrochem. Soc.*, **142**, 4218 (1995).

Wang X., M. Waje, Y. Yan, *J. Electrochem. Soc.*, **151**, A2183 (2004).

Waszczuk P., A. Wieckowski, P. Zelenay, et al., *J. Electroanal. Chem.*, **511**, 55 (2001).

Watanabe M., H. Uchida, T. Yajima, in: *Book of Abstracts*, 53rd Annual Meeting of ISE, Düsseldorf, Germany, 2002, p. 267.

Wilkinson D. P., M. C. Johnson, K. M. Colbow, S. A. Campbell, U.S. patents 5,672,439 (1997), 5,874,182 (1999).

Yamada K., K. Asazawa, R. Yasuda, et al., *J. Power Sources*, **115**, 236 (2003).

Yang H., T. S. Zhao, Q. Ye, *J. Power Sources*, **142**, 117 (2005).

Ye Q., T. S. Zhao, *J. Power Sources*, **147**, 196 (2005a).

Ye Q., T. S. Zhao, H. Yang, J. Prabhuran, *Electrochem. Solid State Lett.*, **8**, A52 (2005b).

Yuan W., K. Scott, H. Cheng, *J. Power Sources*, **163**, 323 (2006).

Reviews

Antolini E., Catalyst for direct ethanol fuel cells, *J. Power Sources*, **170**, 1 (2007).

Bagotsky V. S., Yu. B. Vassiliev, O. A. Khazova, Generalized scheme of chemisorption, electrooxidation and electroreduction of simple organic compounds on platinum group metals, *J. Electroanal. Chem.*, **81**, 229 (1977).

Demirci U. B., Direct liquid-feed fuel cells: thermodynamic and environmental concerns, *J. Power Sources*, **169**, 239 (2007).

Dillon R., S. Srinivasan, A. S. Arico, V. Antonucci, International activities in DMFC R&D: status of technologies and potential applications, *J. Power Sources*, **127**, 112 (2004).

S. Gottesfeld, T. A. Zawodzinski, Direct methanol oxidation fuel cells, in: R. C. Alkire et al. (eds.), *Electrochemical Science and Technology*, Vol. 5, Wiley, New York, 1988.

Iwasita T., W. Vielstich, Progress in the study of methanol oxidation by in situ, ex situ and on-line methods, in: Advances in Electrochemical Science and Engineering, H. Gerischer, C. W. Tobias (eds.), Vol. 1, VCH, New York, 1990, p. 127.

Kamarudin S. K., W. R. W. Daud, S. L. Ho, U. A. Hasran, Overview of the challenges and development of micro-direct methanol fuel cells(DMFC), *J. Power Sources*, 163, 743 (2007).

Kauranen P., E. Skou, J. Munk, Kinetics of methanol oxidation on carbon–supported Pt and Pt + Ru catalysts, *J. Electroanal. Chem.*, **404**, 1 (1996).

Lamy C., A. Lima, V. LeRhun, F. Delime, C. Coutanceau, J.-M. Léger, Recent advances in the development of direct alcohol fuel cells (DAFC), *J. Power Sources*, **105**, 283 (2002).

Ponce de León, C., F. C. Walsh, D. Pletcher, D. J. Browning, J. B. Lakeman, Direct borohydride fuel cells, *J. Power Sources*, **155**, 172 (2006).

Qian W., D. P. Wilkinson, J. Shen, H. Wang, J. Zhang, Architecture for portaiable direct liquid fuel cells, *J. Power Sources*, **154**, 202 (2006).

Wee J.-H., Which type of fuel cell is more competitive for portable application: direct methanol fuel cells or borohydride fuel cells? *J. Power Sources*, **161**, 1 (2006).

Wee J.-H., A comparison of sodium borohydride as a fuel for proton exchange fuel cells and for direct borohydride cells, *J. Power Sources*, **155**, 329 (2006).

Yu, X., P. G. Pickup, Recent advances in direct formic acid fuel cells (DFAFC), *J. Power Sources*, 182, 124 (2008).

CHAPTER 5

PHOSPHORIC ACID FUEL CELLS

5.1 EARLY WORK ON PHOSPHORIC ACID FUEL CELLS

A hydrogen–oxygen fuel cell with liquid acidic electrolyte was described in Section 1.4. For that discussion it was not necessary to specify a particular acid, since the acid anion does not take part in the electrode reactions.

In electrochemical studies, solutions of sulfuric acid are generally used in different concentrations as an acid electrolyte. Throughout the history of fuel cell development, however, very few attempts were made to build a hydrogen–oxygen fuel cell with sulfuric acid solution as an electrolyte. First models of operative fuel cells used alkaline electrolytes. This development and its very successful practical demonstration in 1960 are associated with Francis Thomas Bacon.

Developments after 1960 went in three different directions:

1. A number of variants of fuel cells with alkaline electrolyte having a simpler design and milder working conditions (temperature and pressure) than the Bacon cell were built and used in different applications. This work culminated in a number of fuel battery versions used in NASA's *Apollo* spacecraft and space shuttles (details in Chapter 6).
2. First models of hydrogen–oxygen fuel cells with an ion-exchange membrane as the electrolyte were built and used in NASA's *Gemini* spacecraft.

Fuel Cells: Problems and Solutions, By Vladimir S. Bagotsky

After a very long break, this development was taken up and led to the modern PEMFCs described in Chapter 3.

3. Attempts were made to realize the ancient dream of all electrochemists: that is, building fuel cells working directly with natural fuels. It was with this aim that work on two high-temperature fuel cells was began (actually, as a resumption of very early work): with molten carbonate electrolytes (Chapter 7) and with solid-oxide electrolytes (Chapter 8).

In parallel with the high-temperature work, attempts continued to achieve low-temperature oxidation of methane and other hydrocarbons in aqueous solutions, a possibility first pointed out by Grubb and Niedrach in 1963. Alkaline electrolytes cannot be used with such carbon-based fuels: the carbon dioxide that is a reaction product, together with water, would transform the alkaline solution to carbonate solutions not suitable as an aqueous fuel cell electrolyte.

As we noted earlier, sulfuric acid proved not to work well as an acidic electrolyte. In the early 1960s, concentrated solutions of phosphoric acid with which the working temperature could be raised to 150°C were introduced. Even at that temperature, and even in large quantities, the platinum catalysts proved not to be sufficiently active for a practical hydrocarbon fuel cell. The experience gathered in these attempts then led to successful hydrogen–oxygen phosphoric acid fuel cells (see Srinivasan, 2006, Sec. 9.5).

5.2 SPECIAL FEATURES OF AQUEOUS PHOSPHORIC ACID SOLUTIONS

Like other acids, phosphoric acid (H_3PO_4) in aqueous solutions dissociates into ions according to

$$H_3PO_4 \rightarrow H^+ + H_2PO_4^- \qquad (5.1)$$

Unlike other acids, in concentrated phosphoric acid solutions (where the water concentration is low) the hydrogen ions exist not as hydrated ions $H^+ \cdot nH_2O$ but as ions solvated by phosphoric acid molecules: $H^+ \cdot nH_3PO_4$. For this reason, the conductivity of these solutions is a complex, nonmonotonous function of their concentration. As a conduction mechanism, these ions do not move like a spherical particle in a viscous medium when an electric field is applied to the solution (the Stokes mechanism), but rather, the protons alone jump in the field from one acid molecule to another (the Grotthuss mechanism, suggested as long ago as 1806 as an explanation for the conductivity behavior in aqueous solutions).

Another special feature of phosphoric acid is its dimerization to $H_4P_2O_7$ in aqueous solutions at a concentration of about 85 wt% and higher temperatures:

$$2H_3PO_4 \rightarrow H_4P_2O_7 + H_2O \tag{5.2}$$

This change is very important for fuel cell operation. Phosphate ions ($H_2PO_4^-$) adsorb well on a platinum catalyst surface, displacing the electrochemical reactants, which leads to an appreciably slower reaction. The pyrophosphate ions, $H_3P_2O_7^-$, adsorb much less, so when they are present, the reaction goes much faster than in the presence of phosphate ions, reducing the polarization of the electrodes and raising the cell voltage at high current densities.

Another feature of concentrated phosphoric acid solutions that is very important for fuel cells is the water vapor pressure, which decreases drastically with increasing acid concentration. This feature allows the phosphoric acid solution to be immobilized in a porous solid matrix, greatly simplifying the elimination of water as a reaction product from the fuel cell's cathode space by gas (oxygen or air) circulation. It is safe to adjust this circulation to the maximum current load (maximum rate of water production) without the need to readjust it at lower loads, as because of the immobilization, there is no risk of excessive drying of the matrix. In this way, water elimination has a peculiar self-regulation. No such feature exists in sulfuric acid solution, where for water elimination, the acid itself would have to be circulated, which would cause problems of sealing and of corrosion.

5.3 CONSTRUCTION OF PAFCs

Basically, the construction of PAFCs differs little from what was said in Section 1.4 about fuel cells with liquid acidic electrolyte. In the development of PAFCs and two decades later in the development of PEMFCs (described in Chapter 3), many similar steps can be distinguished, such as the change from pure platinum catalysts to catalysts consisting of highly disperse platinum deposited on a carbon support with a gradual reduction of platinum content in the catalyst from 4 to 0.4 and then to 0.25 mg/cm^2, and the change from pure platinum to Pt–Ru catalysts. The bipolar graphite plates that have special channels for reactant supply and distribution over the entire electrode surface now used widely in PEMFC stacks were first used in PAFCs.

The concentrated phosphoric acid solution in a PAFC is absorbed into the pores of a porous matrix with fine pores and a total thickness of about 50 μm. From the outside, this matrix electrolyte behaves like a solid electrolyte (like the membrane in PEMFCs), preventing the reactant gases hydrogen and oxygen from getting to the "foreign" electrode and mixing.

In early work, Kynar poly(vinylidene fluoride), a thermoplastic material, was used to make the matrix. It was soon discovered that in concentrated phosphoric acid at high temperatures, it is not sufficiently stable chemically and produces fluorine-containing impurities tending to adsorb on the catalyst surface, lowering the catalyst's activity and with it the fuel cell's performance.

Among new materials suggested for the porous electrolyte matrix in PAFCs, we mention a mixture of silicon carbide (SiC) and PTFE (Mori et al., 1998). A suspension of the components is mixed in a ball mill for a long time, then spread onto the surfaces of the cathode and anode. This assures good contact between the electrodes and the electrolyte immobilized in the matrix.

Song et al. (2002) suggested using a mixture of fine and coarse silicon carbide particles to improve electrolyte and gas management in the matrix. Similar results can be obtained with a combination of silicon carbide and zirconium silicate particles (Neergat and Shukla, 2001).

The first hydrogen–oxygen PAFCs in the mid-1960s had 85% phosphoric acid and were operated at temperatures not higher than 100°C. Relative to the results obtained with alkaline hydrogen–oxygen fuel cells, the performance was poor. For this reason, subsequently the phosphoric acid concentration was gradually raised, first to 95% and then to 100%, and the temperature was brought up to 200°C.

During the decade between 1975 and 1985, research in this field was widespread, and large industrial organizations gradually joined the efforts.

5.4 COMMERCIAL PRODUCTION OF PAFCs

Due to the special features mentioned above, of the partial water vapor pressure decreasing with increasing phosphoric acid concentration, the elimination of product water from these fuel cells becomes markedly simpler. It was thus possible to make electrodes as large as 1 m^2. In fuel cells of the PEMFC type, electrodes of such a size would give rise to unmanageably large difficulties in water elimination and the maintenance of water balance within MEAs. Thus, the first models of large-scale power plants producing a power of hundreds of kilowatts and more, and thus having industrial importance, were built from PAFCs.

United Technologies Corporation (UTC) built a large plant for PAFC production in 1969. Together with the Japanese company Toshiba, a special enterprise, Fuel Cells International (FCI), was created for mass production of these fuel cells and for their further improvement. A little later the FCI subsidiary ONSI was set up. In these industries, mass production of PC-25 power plants with an output of 200 kW was begun. These power plants are designated for the combined on-site heat and power supply of individual residential and municipal structures such as hospitals. These power plants were operated autonomously with natural gas, and in addition to the PAFC battery, included equipment for the conversion of natural gas to hydrogen and for subsequent hydrogen purification. Such a plant had a total weight of about 16 tons and occupied 4 m^2 of floor space. The electrical efficiency was 35 to 40%. Including the thermal energy produced, the total energy conversion efficiency was as high as 85%. The thermal energy was produced in the form of hot water with a temperature around 80°C or as superheated steam at 120°C (the total thermal power was about 800 MJ/h). Unlike ordinary power plants of similar size, a PC-25 operates without producing noise or vibrations and without noticeable output of contaminants into the ambient air. Power plants of the PC-25 type have had a relatively large commercial success. In the United States, Japan, and various European countries, about 300 such plants have been installed.

5.5 DEVELOPMENT OF LARGE STATIONARY POWER PLANTS

A large effort toward large-scale PAFC-based power plants was made in Japan, in part with the involvement of UTC. Within the framework of a national program backed by the Japanese gas companies, numerous models of power plants between 25 kW and 1 MW were built. In 1983, UTC developed and set up in Japan a power plant with an electrical output of 4.8 MW. The PAFC battery in this plant had electrodes measuring 0.34 m^2. The cells in this battery worked at 190°C and developed a voltage of 0.65 V when the current density was 250 mA/cm^2. This plant was operative during the period 1983–1985 and produced a total of 5.4 GWh of electrical energy. In 1991 the same company built an even-larger PAFC-based power plant in Japan, that had an electrical output of 11 MW. This plant had 1-m^2 electrodes and a working temperature of 250°C.

More complete data for the large-scale PAFC-based power plants built or tested in Japan can be found in papers by Hojo et al. (1996) and Kasahara et al. (2000).

5.6 THE FUTURE OF PAFCs

Toward the end of the 1990s, interest in PAFCs and PAFC-based power plants gradually waned, despite the success that had been achieved, the relatively large number of intermediate-power PC-25 plants built, and the installation of several megawatt-sized power plants in a number of places. On the one hand, this had a strictly economic basis: the high cost of such plants. On the other hand, there were strictly technical problems, that is, insufficient operating reliability in the long term.

Cost of PAFC-Based Power Plants

It was reported by an official of UTC and Toshiba, R. Whitaker (1998), that the joint company, together with several other companies and a number of government agencies, had spent about $200 million on research and technology to develop and manufacture the PC-25 plants. Not surprisingly, part of this expenditure had to be recovered in the sales price. The production cost of each of the early units was slightly over $1 million, more than $5000/kW. All 144 units delivered up to the time of Whitaker's article were given away at half of this figure, that is, with a considerable subsidy provided by the company. This subsidy should have served to enlarge the future market. It should be pointed out for comparison that the cost of an alternative power plant in the same class (e.g., wind turbines) is about $500 to $700/kW. A further decrease in the production cost of fuel cell power plants would be possible by lowering the labor cost in higher production volumes. One had hoped for a "virtuous cycle" but experienced a "vicious cycle," seeing that increases in production and sales volume were not possible without lower prices, but lowering the prices without raising the production volume was equally impossible. Measures have been taken, of course, to get through this impasse, but substantial results have not yet been achieved.

Reliability and Long-Term Operation of PAFC-Based Power Plants

PAFC-based power plants of intermediate and large power output are designed to work for 50,000 hours, which is approximately five years. During this period, mandatory checking and tune-ups are performed. Many of the large number of PC-25-type plants installed have worked for the designated period and continue to work. Yet in individual units, malfunctions and a gradual performance decline were seen. According to data reported by Blomen and Mugerwa (1990), almost 95% of the sudden interruptions, particularly in large plants, were caused by a mismatch in the work and the effect of individual components, such as electronic monitoring and controlling equipment, mechanical gear, and sensors: that is, by events not related directly to the fuel cell battery. However, during prolonged operation, processes causing a gradual decline in performance also take place in the fuel cells.

In 1991, Paffett et al. reported on an autopsy and detailed examination of individual PAFCs that had worked 5000 or 16,000 hours. By various methods, such as electron microscopy and x-ray photoelectron spectroscopy, they were able to show that during 5000 hours of operation, strong corrosion (electrochemical oxidation) occurred at the carbon support of the platinum catalyst in the catalytic layer of the cathode; this corrosion led to a loss of contact between the support and the catalyst particles. In addition, the highly disperse platinum particles recrystallized and the catalyst's working surface area decreased. No such effects were seen at the anode (the hydrogen electrode), even after 16,000 hours. The mechanical integrity of the electrodes was quite satisfactory after 5000 hours, but much less so after 16,000 hours. The authors unambiguously attributed the performance loss of the cells to this carbon and platinum corrosion. These corrosion processes have to do with the rather highly positive potential of the oxygen electrode. At lower current densities (more so at open circuit and zero current), the potential shifts even further to the positive side, and the corrosion rate increases further.

Song et al. (2000) found that a reduction in the rate of gas supply to the electrodes or, worse, the complete cessation of gas supply may in individual cases lead to irreversible performance loss.

5.7 IMPORTANCE OF PAFCs FOR FUEL CELL DEVELOPMENT

First data on fuel cells based on concentrated phosphoric acid solutions date from the mid-1960s. After three decades, beginning about 1995, very few papers on phosphoric acid fuel cells have appeared in the scientific literature. Of course, many of the numerous intermediate-size and large power plants that had been built in prior years still function, but work toward their further development and improvement has practically ceased.

The period during which these fuel cells were receiving prime interest lasted only about three decades, but they played an important role in the development of fuel cells as such. For the first time a relatively large-scale industrial production of fuel cell–based power plants was initiated, and such plants were spread widely among many users. As a result, many scientific and business circles worldwide have recognized that fuel cells are an entirely real possibility for applications useful to humankind. During this period of PAFC development, the first large government schemes and projects for fuel cell research and engineering were implemented in many countries of the world (United States, Japan, Russia, and others).

During PAFC research and development, technical solutions were found which were then adopted successfully in the development of other fuel cell types. This is true in particular for the use of platinum catalysts, not in a pure form but as deposits on a carbon support (e.g., carbon black), leading to a considerable drop in the amount of platinum needed to manufacture fuel cells. During this period, the influence of traces of carbon monoxide in hydrogen on the performance of platinum catalysts was first investigated. It was shown that Pt–Ru catalysts could be used to reduce the influence of carbon monoxide. It was also shown for the first time that the performance of an oxygen electrode could be improved by using catalysts of platinum alloyed with iron-group metals (Kim et al., 1993; Watanabe et al., 1994).

REFERENCES

Blomen L. J. M. J., M. N. Mugerwa, *J. Power Sources*, **29**, 71 (1990).

Grothuss T., *Ann. Chim.*, **58**, 54 (1806).

Grubb W. T., L. W. Niedrach, *J. Electrochem. Soc.*, **110**, 1086 (1963).

Hojo N., M. Okuda, M. Nakamura, *J. Power Sources*, **61**, 73 (1996).

Kasahara K., M. Morioka, H. Yoshida, H. Shingai, *J. Power Sources*, **86**, 298 (2000).

Kim K. T., J. T. Hwang, Y. G. Kim, J. S. Chung, *J. Electrochem. Soc.*, **140**, 31 (1993).

Mori T., A. Honji, T. Kahara, Y. Hishinuma, *J. Electrochem. Soc.*, **135**, 1104 (1998).

Neergat M., A. K. Shukla, *J. Power Sources*, **102**, 317 (2001).
Paffett M. T., W. Hutchinson, J. D. Farr, et al., *J. Power Sources*, **36**, 137 (1991).
Song R.-H., C.-S. Kim, D. R. Shin, *J. Power Sources*, **86**, 289 (2000).
Song R.-H., S. Dheenadayalan, D.-R. Shin, *J. Power Sources*, **106**, 167 (2002).
Watanabe M., K. Tsurumi, N. Mizukami, et al., *J. Electrochem. Soc.*, **141**, 2659 (1994).
Whitaker R., *J. Power Sources*, **71**, 71 (1998).

Monographs

Kinoshita K., *Electrochemical Oxygen Technology*, Wiley, New York, 1992.
Srinivasan S., *Fuel Cells: From Fundamentals to Applications*, Springer Science + Business Media, New York, 2006.

CHAPTER 6

ALKALINE FUEL CELLS

The fuel cells of the preceding chapters had (acidic) proton-conducting membranes (Chapters 3 and 4) or phosphoric acid solution (Chapter 5) as an electrolyte. Due to corrosion problems, only metals of the platinum group can be used as catalysts for the electrode reactions in such fuel cells.

In fuel cells that have an alkaline electrolyte, that is, alkaline fuel cells (AFCs), catalysts may be selected advantageously from a much wider range of materials, some of them relatively inexpensive. Thus, highly disperse nickel is a good catalyst for the electrochemical oxidation of hydrogen in alkaline solutions. Under certain conditions, these catalysts will even catalyze methanol oxidation. For the electrochemical reduction of oxygen in alkaline solutions, good catalysts are highly disperse silver and gold, but again, considerably cheaper materials exist as well (various types of activated carbon, metal oxides of the spinel and perovskite type). In fuel cells (as in other electrochemical devices), solutions of potassium hydroxide (KOH) prepared in different concentrations are generally used as the alkaline electrolytes.

Alkaline electrolytes offer further advantages over acidic electrolytes. In them, oxygen reduction is considerably faster, which implies that the working potential of the oxygen electrode is more positive and a larger cell voltage can be realized. Also, apart from the advantages in catalyst selection, less severe corrosion conditions allow nickel and alloys of iron to be used as structural materials in AFCs.

Fuel Cells: Problems and Solutions, By Vladimir S. Bagotsky

Yet alkaline electrolytes also have some severe disadvantages relative to acidic electrolytes. The most important disadvantage is their reactivity with carbon dioxide (CO_2), which they absorb readily, forming carbonates. As CO_2 is always present in air, an alkaline hydrogen–air fuel cell requires the prior stripping of CO_2 from the air that reaches the oxygen electrode in contact with the electrolyte. Otherwise, KOH would be transformed to K_2CO_3, with two negative consequences: a decrease in the concentration of the free alkali needed for the electrochemical reactions, and the precipitation of crystals of potassium carbonate when its concentration has become high enough, which may mechanically upset the structure of the electrodes and catalyst. Another defect of alkaline solutions is their "creeping," or ability to permeate the smallest cracks and holes, making tight sealing of AFCs a difficult problem.

6.1 HYDROGEN–OXYGEN AFCs

In hydrogen–oxygen fuel cells with an alkaline electrolyte, the reactions at the electrodes and the overall current-generating reaction can be formulated as follows*:

$$\text{Anode:} \quad 2H_2 + 4OH^- \rightarrow 4H_2O + 4e^- \qquad E^0 = -0.828\,V \tag{6.1}$$

$$\text{Cathode:} \quad O_2 + 2H_2O + 4e^- \rightarrow 4OH^- \qquad E^0 = 0.40\,V \tag{6.2}$$

$$\text{Overall:} \quad O_2 + 2H_2 \rightarrow 2H_2O \qquad \mathscr{E}^0 = 1.228\,V \tag{6.3}$$

6.1.1 Bacon's Battery

The modern era of fuel cell development started with the alkaline hydrogen–oxygen battery, demonstrated practically by the British engineer Francis Thomas Bacon in 1960, although it had been under development since 1932 (Section 2.2). It was the first real fuel battery that developed a power of more than 1 kW and was able to operate for a relatively long period of time. The operating conditions were extreme in modern terms: The working temperature was above 200°C and the gases (hydrogen and oxygen) were compressed to over 20 bar, with the pressures even going up to 45 bar. This necessitated a very bulky and heavy design.

* The potentials of the hydrogen and oxygen electrodes have values that depend on the pH value of the solution, implying a change by −0.828 V between pH 0 (acidic solutions) and pH 14 (alkaline solutions). The EMF value for reaction (6.3), which is the difference between the two electrode potentials, is independent of solution pH.

In Bacon's battery, a KOH solution of intermediate concentration (37 to 50%) was the electrolyte. The electrodes were made of porous nickel. This worked without any problems for hydrogen electrodes, as hydrogen is readily oxidized at nickel as a catalytic anode. But problems had to be solved for the oxygen cathodes, since their surface is converted to poorly conducting nickel oxides in the presence of oxygen. Therefore, Bacon treated the positive electrodes with LiOH solution and subjected them to heat treatment. This produced lithiated nickel oxide as a corrosion-resistant, electrically conducting oxide film on the nickel surface at which oxygen is readily reduced electrochemically.

Gas-diffusion electrodes having a barrier layer for the gases were used for the first time in Bacon's battery. These electrodes were made of coarse nickel powder, giving a structure with relatively large pores. A layer of finer nickel powder yielding finer pores was applied to the electrodes on the side facing the electrolyte. When in contact with the electrolyte solution, both layers soak up the solution by capillary forces. Gases supplied from outside under a certain pressure will then displace the solution from the coarse pores, but the pressure is not high enough to displace it from the fine pores, where the capillary forces are much higher. This prevented permeation of the gases through the electrodes, which would lead to unproductive losses.

The electrodes in Bacon's battery measured $370\,cm^2$, and they were 1.8 mm thick. At current densities of 200 to $400\,mA/cm^2$, the voltage of an individual cell in the battery was 0.90 to 0.95 V, which is considerably more than that in PAFCs and in modern PEMFCs. A variety of corrosion problems were responsible for the fact that the total lifetime of the Bacon battery reached no more than a few hundred hours.

6.1.2 Batteries for the *Apollo* Spacecraft

In the early 1960s, the U.S. Pratt & Whitney company, part of United Technologies Corporation (UTC), acquired all rights to the patents of Bacon and started work on further development of alkaline hydrogen–oxygen fuel cells. The major goal was that of giving up the high gas pressures that had been used in Bacon's battery. High working temperatures had to be maintained to preserve performance. This was attained by using a very highly concentrated (85%) KOH solution, de facto molten KOH. A battery working at 250°C was built. The electrodes were made using an improved and simplified technology.

A battery weighing about 114 kg was designed for delivering a power of 1.5 kW and sustaining short-term loads of up to 2.2 kW. The lifetime of this improved battery was extended to several thousands of hours, much more than that of Bacon's original battery.

Batteries of this type were used in the *Apollo* space flights. The spacecraft included three batteries that fully satisfied all the electrical energy requirements. Successful return of a spacecraft would have been possible on merely one of the batteries, with the other two out of commission. The water generated as the reaction product by operation of the hydrogen–oxygen battery was recovered and used as the drinking water supply of the crew. During the 18 flights of *Apollo* spacecraft, batteries of this type logged a total of more than 10,000 operating hours without any important failure.

6.1.3 First AFCs with Matrix Electrolyte

In 1962, a research group at Allis-Chalmers (a U.S. manufacturer of agricultural machinery) began development of a new type of hydrogen–oxygen fuel cell with an alkaline electrolyte (Wynveen and Kirkland, 1962). The distinguishing feature of this cell was not use of a freely flowing liquid electrolyte (KOH solution or melt, as described above), but of a quasisolid electrolyte in the form of KOH solution immobilized in an asbestos matrix. Asbestos was found to be a material very well suited to this use, as it has very high chemical and thermal resistance. It is readily made into thin sheets (asbestos paper) and has a relatively large volume of fine pores which strongly retain the alkaline solution through capillary forces. Using this matrix electrolyte, one need not have a special gas barrier within the electrodes: Even at high pressures the gases will not be able to overcome the capillary pressure in the matrix and push the alkaline solution out of the matrix pores.

The Allis-Chalmers battery had a filter-press design. It contained a 5 M KOH solution in asbestos diaphragms 0.75 mm thick. The working temperature was 65°C, much lower than that in the batteries described above. At a current density of 200 mA/cm^2 and a gas pressure of 1.35 bar, the voltage per cell was 0.75 V, which is somewhat lower than in battery types working at higher temperatures.

Allis-Chalmers installed a battery in an experimental tractor from their product line. It was actually operated under field conditions for some time and was then demonstrated in many exhibitions and fairs. This was the first use of fuel cells as the power source in an electrically powered vehicle.

6.1.4 Batteries for the *Orbiter* Space Shuttle

After the successful completion of all *Apollo* missions, UTC began developing batteries for a new spacecraft, the *Orbiter* space shuttle. This had much larger energy requirements and needed a much more powerful battery. This led to a radical redesign of the fuel battery. For a more compact design, 35% KOH solution immobilized in an asbestos diaphragm was selected, as in the Allis-Chalmers version. The electrodes contained large amounts of noble-metal catalyst: 10 mg/cm^2 of Pt + Pd for the anode and 20 mg/cm^2 of Pt + Au for the cathode. This combination gave a very high specific power of more than 1 W/cm^2. The power plant installed in the spacecraft consisted of three independent blocks, each with 96 alkaline fuel cells, a weight of about 120 kg, and a power output of 12 kW (with a short-term capability of 16 kW).

6.1.5 AFCs with Skeleton-Type Catalysts

It was shown in the late 1950s (Justi and Winsel, 1959) that highly efficient hydrogen gas-diffusion electrodes could be prepared using *skeleton nickel* (Raney nickel) as the catalyst. This is made from nickel–aluminum alloy subjected to leaching with hot alkali solution (Raney, 1927). The aluminum dissolves selectively, leaving behind a very highly disperse "skeleton" of nickel metal. The powder is highly pyrophoric and cannot be used as such for making strong electrodes. Justi and Winsel suggested pressing the electrodes from a powdered mixture of carbonyl nickel and nickel–aluminum alloy. Leaching the aluminum then leaves a dual skeleton (double-skeleton electrodes). The sintered carbonyl nickel yields the supporting skeleton structure, while the Raney-type skeleton nickel produced by leaching provides the catalytic activity. Using the same principle, highly active oxygen electrodes were produced from skeleton silver.

Laboratory models of such hydrogen–oxygen fuel cells worked for a long time in 6 M KOH solution at temperatures of 30 to 35°C, yielding a current density of 30 to 50 mA/cm^2 and a voltage of 0.75 to 0.85 V; the excess pressure of the gases was about 3 bar.

The German company Siemens later modified these electrodes with skeleton metal catalysts. Small amounts of titanium were added to the anodic nickel catalysts, and nickel, bismuth, and titanium were added to the cathodic silver catalysts. Fuel cells with such electrodes and a matrix electrolyte operated at 95°C and a current density of 400 mA/cm^2 had a working voltage of 0.8 to 0.9 V.

6.1.6 AFCs with Carbon Electrodes

In 1955, the U.S. company Union Carbide Corporation (UCC) started a large development program headed by Karl Kordesch for alkaline hydrogen–oxygen fuel cells with carbon electrodes. From the outset, a model with circulating alkaline solution was selected. In first experiments, the relatively thick pitch-bonded carbon electrodes (6.4 mm) developed previously for regular zinc–air batteries were used. Later, relatively thinner multilayer carbon electrodes containing a metal support (a nickel screen or Exmet extended nickel sheet) were used. The electrodes included several hydrophobized diffusion layers (consisting of a mixture of carbon black, PTFE, and a variety of pore-forming and wet-proofing additives) and one or two catalytically active layers. In some versions, silver and platinum catalysts were added to the electrodes. The overall thickness of the electrodes was about 0.4 mm. The electrodes prepared were 17×17 cm^2 in size. They were made by a rolling technology. At a current density of 100 mA/cm^2, the voltage produced was about 0.6 V in air and with additional catalysts attained 0.67 V in oxygen, while the current density could be raised to more than 200 mA/cm^2.

Using fuel cells of this type, Kordesch built in 1967 a power plant to power a converted Austin A40, a light passenger vehicle which for a number of years he used for urban driving. The vehicle had hybrid power, including a storage battery providing boost power for acceleration and hills. Otherwise, under normal road driving conditions, the full load was satisfied by the fuel cells, which at the same time recharged the storage battery. The fuel cell power plant had an output of 6 kW at a voltage of 90 V (0.7 V per cell). There were 13 tanks with hydrogen providing fuel for producing 45 kWh of electrical energy. Air oxygen was the oxidizing agent. The total weight of the fuel cell battery was 250 kg, and the additional storage battery weighed 150 kg. This was the first demonstration of relatively long term practical use of an electric car powered by fuel cells (Kordesch et al., 1999).

After the tragical 1984 Bhopal accident in India, UCC stopped all work on this project. Karl Kordesch continued his research at the Technical University of Graz in Austria (Cifrain and Kordesch, 2004).

6.1.7 Problems in the AFC Field

Matrix Electrolyte or Circulating Liquid Electrolyte?

It can be seen from the historical excursion in previous sections that basically two solutions for electrolyte management have emerged at the different stages of development of hydrogen–oxygen AFCs. One solution has the alkaline solution (or melt) immobilized in a porous matrix consisting of asbestos or similar material with fine pores. The other solution has the liquid electrolyte being pumped around. What are the advantages and disadvantages of these two versions?

The matrix version of AFCs offers a more compact design. It lacks the electrolyte loop needed in the other case, where pumps, valves, tanks, regulators, and so on, must handle the caustic, and sometimes hot, liquid. Thus, the many problems that arise due to the need for sealing not only the cells but also the loop do not exist in the matrix version. It is another advantage of the matrix version that because the matrix holds the electrolyte solution tightly in its fine pores, the gas-diffusion electrodes do not need an additional porous barrier layer for the gases, which greatly simplifies electrode manufacturing technology and makes it much cheaper.

The version with circulating liquid electrolyte has the following advantages: (1) it becomes much simpler to maintain the water balance in the cell, since there is no chance for the electrodes to dry out, and the electrolyte flow can be used to eliminate product water; (2) the electrolyte flow can also be used to eliminate heat, so that special cooling plates or loops for a cooling fluid are not required; (3) the electrolyte flow serves to level any gradients of electrolyte concentration that could arise within an individual cell or within the battery; and (4) the electrolyte flow may serve to eliminate "foreign" matter (e.g., gas bubbles, corrosion products, insoluble carbonates) that turns up in the liquid.

In either version, cells of the bipolar type assembled to batteries of the filter-press type can be used. The version with circulating electrolyte is feasible for building monopolar ("jar"-type) cells, which are cheaper than bipolar types (as they do not require expensive bipolar plates) and offer a highly flexible series–parallel switching within the battery to accommodate specific user needs.

An important advantage of batteries with circulating electrolyte is the much simpler shutdown and restart procedure, which is achieved simply by stopping and restarting electrolyte circulation. For a cold startup, in particular, only the tank holding the electrolyte solution has to be heated, rather than the entire battery.

The principal problem that arises in the design of fuel cells with a circulating electrolyte is that of providing perfect sealing of the full loop and of a sufficiently high corrosion resistance of all parts (i.e., pipe sections, pipe junctions, pumps, valves) toward the hot alkaline solution.

The Problem of Electrolyte Carbonation

Any carbon dioxide getting into the KOH solution binds (neutralizes) alkali, yielding potassium carbonate:

$$CO_2 + 2KOH \rightarrow K_2CO_3 + H_2O \tag{6.4}$$

This is a very dangerous event in alkaline fuel cells with matrix electrolyte, since the total amount of alkali in the fine pores of the matrix is limited and there is no way to remove carbonates from the pores. Loss of alkalinity due to partial neutralization by CO_2 affects performance, and carbonate crystals affecting the integrity of the catalytic layers in the electrodes may form in the solution when the carbonate concentration becomes high enough. In almost all practical applications of matrix-type fuel cells, therefore, pure oxygen was used as the oxidizing agent, particularly so in spacecraft, rather than air, which always contains traces of CO_2.

In fuel cells that have a free liquid electrolyte, particularly a circulating electrolyte, it was found (Gouérec et al., 2004; Gülzow and Schulze, 2004) that initial fears were exaggerated. Individual AFC electrodes functioned without any important loss of performance for more than 3500 hours at a current density of 150 mA/cm^2 when supplied with oxygen containing 5% CO_2, which is 150-fold higher than in air. For the formation of a crystalline precipitate of potassium carbonate at 70°C, the local KOH concentration should be at least 16 mol/L. When there is a real need, one may use relatively simple, cheap devices consisting of scrubbers filled with an alkali solution or other chemicals in order to remove CO_2 from the air.

The Lifetime of AFCs and Reasons for Gradual Degradation

The AFC-based power plants for the *Apollo* and *Orbiter* spacecraft described above have been used for numerous flights and attained a total operating time of tens of thousands of hours. Very few systematic long-term tests of simpler power plants for terrestrial applications have been reported in the literature. It was said that in a number of tests, the time to failure was 2000 to 3000 hours (Gouérec et al., 2004).

In Kordesch's group (Tomantschger and Findlay, 1992), reasons for the degradation of alkaline fuel cells with carbon electrodes were analyzed. Degradation was found to have several causes, including gradual corrosion of the carbon materials in the oxygen electrode and a gradual loss of hydrophobicity of the electrodes opening the way for alkaline solution to penetrate to the gas side of the diffusion layer. Drops of alkali solution could sometimes be seen with the naked eye on the electrode's gas side (called "weeping" electrodes). It was reported that rising polarization of the cathode led to a cell voltage decline of 100 to 300 μV/h. Similar data have been reported by Gülzow (1996) for electrodes with silver catalyst.

6.1.8 Present State and Future Prospects of AFC Work

As with the other fuel cell types, periods of pronounced ups and downs can be distinguished in R&D for AFCs. Work in the United States, Europe, and a number of Asian countries evolved very vigorously after the 1960 demonstration of Bacon's battery, and more particularly after the first flight of the *Apollo* spacecraft with the new power plant on board.

In Germany, the storage battery maker Varta, together with other organizations, started making AFC batteries with flowing electrolyte. Raney nickel was used at the anode and Raney silver at the cathode. Other than the electrodes of Justi and Winsel, the Varta electrodes included PTFE and had a mixed hydrophobic–hydrophilic structure. In their batteries with matrix electrolyte, Siemens also used skeleton catalysts, but included additional metals (titanium, bismuth, and others).

The Belgian–Dutch consortium ELENCO, which included the Belgian Nuclear Energy Agency and a number of Belgian and Dutch companies and organizations, developed monopolar batteries containing small amounts of platinum catalysts on a carbon support.

In the Soviet Union, the power plant Photon was developed jointly for the *Buran* space shuttle by the Urals Integrated Electrochemical Plant (which had also been concerned with problems of nuclear energy) and the S. P. Korolev Rocket and Space Corporation. The Photon power plant weighed 160 kg and had a sustained power of 10 kW (with a maximum load of 15 kW). The alkaline fuel cells in this plant were of the matrix type and had hydrophobized electrodes with a porous nickel support and 20 mg/cm^2 of platinum catalysts. Because of discontinuation of the Buran project, the experience gained when developing the Photon battery was used subsequently in building power plants for test models of electric cars (Korovin, 2005).

It has been entirely justified to build spacecraft power plants from AFCs. This has also to do with the fact that in this application, cryogenic storage of the fuel and oxidant, hydrogen and oxygen, is feasible economically and technically. In addition, it was an advantage that the water produced during operation could be used as drinking water for the crew. The same situation exists in a number of other manned stationary and mobile installations, such as bathyscaphs, submarine stations, submarines, and submarine equipment of various sizes. Different fuel cell models that have been put to such use have been described in the literature (see Chapter 16).

In terrestrial applications, certain difficulties were encountered. It was a natural desire to use air oxygen rather than pure oxygen whenever this was feasible. This change is associated, however, with a marked drop in the electrical performance figures relative to those achieved in spacecraft power plants. Also, the complications associated with the presence of CO_2 in the air that were mentioned earlier come into play here. As the fuel in AFCs, only pure electrolytic hydrogen is used. It is not possible to use the much cheaper technical hydrogen produced by the reforming of hydrocarbons or other organic compounds or by carbon gasification, since it contains not only traces of carbon monoxide (poisoning the platinum catalysts) but also appreciable amounts of CO_2. Eliminating the CO_2 from hydrogen would be complicated and expensive. Both gases, hydrogen and oxygen, pose problems in storage and transport. A cryogenic storage is not feasible in many terrestrial applications, and storage of the compressed gases in tanks has a weight penalty which is admissible for stationary installations but undesirable or impossible for mobile and portable devices.

All these points, and uncertainties as to the potential lifetime, have led to a strongly reduced set of candidate applications for AFC-based power plants. The volume of research and engineering work has shrunk accordingly.

Most organizations and companies stopped work in this area between 1990 and 1995, and some of them switched to other systems when work on fuel cells with acidic, proton-conducting membranes began to take off in the mid-1990s.

Still, ideas concerning further improvements in hydrogen–oxygen AFCs and potential new practical applications have remained attractive. Quite a few papers appeared in the years approaching the twenty-first century, among them papers from Germany (Gülzow, 1996), Canada (Lin et al., 2006), and France (Gouérec et al., 2004). In Austria, Kordesch et al. (1999) have continued with analysis of further uses of AFCs in electric cars.

6.2 ALKALINE HYDRAZINE FUEL CELLS

Fuel cells with a proton-conducting membrane that use hydrazine as a fuel were mentioned in Section 4.12.2. But long before 2003, when the first reports on these cells began to appear, hydrazine–oxygen fuel cells with alkaline electrolytes had been the subject of research and engineering work. Hydrazine N_2H_4 is a strong reducing agent that can be used as a rocket fuel. It can be synthesized from ammonia and can exist in the form of stable aqueous solutions in concentrations of up to 100% hydrazine hydrate ($N_2H_4{\cdot}H_2O$ or N_2H_5OH).

Hydrazine is extremely toxic. This point, and its high price (about $3/kg), make it difficult to use widely, so it has been used only in special cases.

In a hydrazine–oxygen (or air) fuel cell with alkaline electrolyte, the following reactions take place:

$$\text{Anode:} \quad N_2H_4 + 4OH^- \rightarrow N_2 + 4e^- + 4H_2O \quad E^0 = -1.22\,V \tag{6.5}$$

$$\text{Cathode:} \quad O_2 + 4e^- + 2H_2O \rightarrow 4OH^- \quad E^0 = 0.40\,V \tag{6.6}$$

$$\text{Overall:} \quad O_2 + N_2H_4 \rightarrow N_2 + 2H_2O \quad \mathscr{E}^0 = 1.62\,V \tag{6.7}$$

$$-\Delta G^0 = 623.5\,kJ/mol = 1.62\,eV \qquad -\Delta H^0 = 534.2\,kJ/mol = 1.39\,eV$$

First news about hydrazine–oxygen (or hydrazine–air) fuel cells with alkaline electrolytes appeared in the early 1960s. Gillibrand and Lomax (1962) reported building a rather primitive hydrazine–oxygen fuel cell of the jar type without a diaphragm using four pairs of nickel electrodes (anodes and cathodes). This cell had the defect of admitting direct contact of the oxygen electrode with the hydrazine solution, which made it much less efficient. The electrolyte was 7 M KOH and contained 1.6% hydrazine. The cell produced 50 mA/cm^2 at a cell voltage of 0.65–0.70 V.

During the decade from 1962 to 1972, a large volume of work on hydrazine–air fuel cells was done in the Shell research center in the United Kingdom (Andrew et al., 1972). Highly active hydrazine electrodes were developed on supports of expanded nickel mesh receiving additional surface activation. A thin layer of nickel–aluminum alloy was formed on the surface of this screen, then leached with alkali, producing a highly disperse nickel catalyst of the skeleton type. This catalyst was activated additionally by depositing ruthenium on it. Silver catalyst was used on the oxygen (or air) electrode; it was deposited on a support of porous poly(vinyl chloride) (PVC) and contained rhodium. In cells having a thickness of about 4 mm, a flowing alkaline electrolyte containing 1 to 2% hydrazine was used (the alkali concentration was not reported). There was no special diaphragm in the cell. Apparently, the porous PVC support for the oxýgen electrode provided sufficient protection against a direct contact between the catalyst of the oxygen electrode and the hydrazine in the electrolyte. It is also possible that the silver catalyst was sufficiently inert toward hydrazine that its activity toward oxygen reduction was unaffected by any hydrazine that was present. A battery assembled from 10 cells of this type working at a temperature of 60°C gave a current density of 160 mA/cm^2 and a voltage per cell of 0.6 V. A battery consisting of 120 cells weighing 57 kg was built which had a useful electrical power output (after subtracting power consumed for its internal needs) of 10 kW.

In 1963, work on hydrazine–oxygen fuel cells was also under way at Allis-Chalmers. A 3-kW battery with a voltage of 36 V was developed and intended for golf carts. In the battery, an electrolyte of 5.5 M KOH containing 3% hydrazine was circulated. As a catalyst for the hydrazine electrode, Jasinski (1963) first suggested nickel boride (Ni_2B), which proved to be more active than palladium-activated nickel used in the original experiments. Silver was the catalyst for the oxygen electrode. The working temperature of the battery was 70°C. At a current density of 100 mA/cm^2, the average voltage of the individual cells was 0.63 V (see Vielstich, 1965, Sec. 4.3.2.2).

Karl Kordesch equipped a motorcycle with a similar battery that he developed at UCC (Figure 6.1) and ran it for 300 miles, giving an electrochemical periodical the occasion to publish a hilarious German–English poem.*

Many research efforts went into elucidating the behavior of different catalysts in the hydrazine oxidation reaction. Like the oxygen electrode, the open-circuit potential of a hydrazine electrode deviates from the thermodynamic value (−1.22 V) and in certain cases comes close to that of the hydrogen electrode (−0.828 V). This is due to the fact that thermodynamically hydrazine is unstable and may decompose catalytically to nitrogen and hydrogen:

$$N_2H_4 \rightarrow N_2 + 2H_2 \tag{6.8}$$

a reaction that will, of course, detract from the faradaic efficiency of hydrazine utilization in a fuel cell. The rate of catalytic decomposition depends on many factors: the nature of the catalyst, solution composition, temperature, and so on. When the metal catalyst for hydrazine oxidation is a good catalyst for cathodic hydrogen evolution as well, hydrazine decomposition is greatly accelerated due to an electrochemical mechanism coupling the two reactions: anodic hydrazine oxidation by reaction (6.5) and cathodic hydrogen evolution by the reaction

$$H^+ + 2e^- \rightarrow H_2 \tag{6.9}$$

This coupling of the two reactions (a kind of local-cell action) is possible because the equilibrium potential of the hydrazine electrode is markedly more negative (by almost 0.4 V) than that of the hydrogen electrode. In an operating hydrazine–air fuel cell, the potential of the hydrazine electrode moves in the positive direction when an anodic current flows, and the rate of cathodic hydrogen evolution [reaction (6.9)] decreases accordingly. In an ideal case, catalytic decomposition should cease completely in an operating fuel cell, so

* Der Kordesch mit sei Motorrad/der racet de highway krumm und grad/Downhill goes it mighty schnell,/aber uphill push like hell.

FIGURE 6.1 Karl Kordesch rides his hydrazine fuel cell motorcycle, 1967. (From Smithonian Institution, neg. EMP059006, from the Science Service Historical Images Collection, courtesy of Union Carbide Corp.)

nitrogen and water will then be the only reaction products according to reaction (6.7).

This implies that catalyst research should focus on finding materials and conditions where anodic hydrazine oxidation is accelerated as much as possible, while cathodic hydrogen evolution would be hindered as much as possible.

Tamura and Kahara (1976) studied in detail the composition of the products evolved in hydrazine–air fuel cells working at a temperature of 25°C when different catalysts were used for the hydrazine electrode. Platinum, palladium, cobalt, and silver supported by a nickel substrate were studied. The highest faradaic efficiency for hydrazine utilization was found for silver, the lowest for platinum. The efficiency was found to increase with increasing anodic current density (for the reasons stated previously). The efficiency decreased with increasing hydrazine concentration in the solution. Obviously, the nonelectrochemical mechanism of catalytic hydrazine decomposition then becomes more important. At temperatures above 60°C, traces of ammonia are detected in the gases evolved. This indicates that a slow catalytic decomposition of hydrazine might occur according to

$$3N_2H_4 \rightarrow N_2 + 4NH_3 \tag{6.10}$$

6.3 ANION-EXCHANGE (HYDROXYL ION–CONDUCTING) MEMBRANES

The development and industrial production of *cation*-exchange (proton-conducting) membranes of the Nafion type induced a "quantum jump" in fuel cell development. A number of important problems could be solved by using such membranes: complete sealing of individual cells in the battery and of the battery as a whole, the avoidance of direct contact between the electrodes, and the prevention of gas bubbles reaching the "wrong" electrode. With these membranes, very compact designs of fuel cell batteries became possible. The cation-exchange membranes have the defect that the rates of a number of electrochemical reactions will decrease considerably when the electrode is in contact with them rather than with an alkaline electrolyte. This is true in particular for cathodic oxygen reduction, anodic methanol oxidation, and the anodic oxidation of many other organic substances.

It is natural then to ask why one could not make *anion*-exchange (hydroxyl ion–conducting) membranes having properties similar to those of Nafion but because they are alkaline in nature, leading to a marked acceleration of the electrode reactions.

Despite great efforts, this has not been achieved until now. Existing types of anion-exchange membranes (used for electrodialysis) are far inferior to Nafion both in their conductivity and in chemical and thermal stability. Therefore, so far, one cannot meaningfully discuss the potential of these membranes as substitutes for the circulating or matrix electrolyte in alkaline hydrogen–oxygen fuel cells.

The situation is different when looking at fuel cells using methanol (or other, similar types of liquid organic fuel). Methanol fuel cells built with proton-conducting membranes of Nafion type (Chapter 4) are currently in an earlier stage of development than that of alkaline hydrogen–oxygen fuel cells. They have the important defect of insufficiently high performance due to a slow anodic oxidation of methanol. Anything that will accelerate this reaction will be justified. In the next section, results of a number of studies are reported that have focused on methanol fuel cells that have anion-exchange (hydroxyl ion–conducting) membranes as an electrolyte. In the next section we also report some basic data about the anion-exchange membranes used.

6.4 METHANOL FUEL CELLS WITH ANION-EXCHANGE MEMBRANES

The reactions occurring in methanol–oxygen fuel cells with alkaline electrolyte can be written as follows:

$$\text{Anode:} \quad CH_3OH + 6OH^- \rightarrow CO_2 + 5H_2O + 6e^- \qquad E^0 = 0.81\,V \quad (6.11)$$

$$\text{Cathode:} \quad \tfrac{3}{2}O_2 + 3H_2O + 6e^- \rightarrow 6OH^- \qquad E^0 = 0.40\,V \quad (6.12)$$

$$\text{Overall:} \quad CH_3OH + \tfrac{3}{2}O_2 \rightarrow CO_2 + 2H_2O \qquad \mathscr{E}^0 = 1.21\,V \quad (6.13)$$

An important difference existing in this reaction between alkaline and acidic media (e.g., the DMFCs discussed in Chapter 4) is water being produced by the cathodic reaction in acidic media but consumed as a reactant in alkaline media. Therefore, when using an alkaline anion-exchange membrane, the flooding of the cathode, representing a strong obstacle to fuel cell operation in cells that have an acidic cation-exchange membrane, is practically absent.

As early as 1951, Kordesch and Marko had put forward the idea of building an alkaline methanol cell. In those years this idea was not taken up. Now, at the beginning of the twenty-first century, experimental models of methanol fuel cells incorporating an anion-exchange membrane with alkaline properties have been reported. Yu and Scott (2004) used membranes Morgane ADP marketed by Solvay S.A. for the electrodialysis of salt solutions. They consist of a cross-linked fluorinated polymer with quaternary ammonium groups. Prior to fuel cell operation, the membranes were thoroughly treated with 1 M NaOH solution in order to exchange the Cl^- counterions for OH^- ions. For fuel cell work, a 2 M methanol solution in 1 M NaOH was used. The temperature was maintained at values not exceeding 60°C (where membranes of this type should not be used). Platinum supported by carbon black (about 2 mg/cm^2) was used as the catalyst for anode and cathode. The open-circuit voltage of the fuel cell had a value of almost 0.8 V, which is markedly more than for similar cells with an acidic cation-exchange membrane. However, the voltage dropped to 0.2 V at a current density as low as 40 mA/cm^2. This strong drop is due to the large ohmic resistance of the membrane used. At the same time the crossover of methanol is much less pronounced than with Nafion-type membranes.

Matsuoka et al. (2005) used a commercial anion-exchange membrane from the Japanese company Tokuyama Co. that contains quaternary ammonium groups in an olefin polymer. They studied the fuel cell behavior of a number of alcohols: methanol and the polyhydric alcohols ethylene glycol, glycerol, xylitol, and others. Laboratory models of alcohol–air fuel cells were tested with Pt–C catalysts in 1 M KOH at 50°C. The open-circuit voltage was always around 0.8 V. At a current density of 30 mA/cm^2, the working voltages were again around 0.2 to 0.3 V. It is interesting to note that the highest energy density obtained in this work with ethylene glycol (9 mW/cm^2) was higher than that obtained with methanol (6 mW/cm^2).

Yang et al. (2006) used a PVA/SSA membrane of their own make obtained by cross-linking of poly(vinyl alcohol) with 10 wt% sulfosuccinic acid. In a methanol–air fuel cell with such a membrane, they used 4 mg/cm^2 Pt–Ru on carbon black as the anode catalyst and an inexpensive catalyst without platinum on the basis of MnO_2 as the cathode catalyst. The electrolyte was 2 M KOH solution + 2 M CH_3OH. At a temperature of 30°C, the current density in the maximum-power point was about 10 mA/cm^2 and the voltage was 0.245 V. At 60°C, the corresponding values were 15.4 mA/cm^2 and 0.268 V.

It can be seen from the data reported that the performance achieved up to now in methanol–oxygen fuel cells (or fuel cells with other alcohols) that have an alkaline ion-exchange membrane is very poor.

6.5 METHANOL FUEL CELL WITH AN INVARIANT ALKALINE ELECTROLYTE

In 1964, Cairns and Bartosik suggested an original way of solving the problem of carbonation of the alkaline electrolyte solution during the oxidation of organic substances to CO_2. This consisted of using concentrated solutions of the alkali metal carbonates Cs_2CO_3 and Rb_2CO_3 instead of the pure alkali solutions of KOH or NaOH. Using such carbonate solutions, it is possible to work at high temperatures (up to 200°C). Bicarbonates are unstable at temperatures above 100°C. Therefore, the composition of these solutions will not change when CO_2 is produced in them. At a current density of 126 mA/cm^2, a specific power of 40 to 45 mW/cm^2 was achieved (on an *IR*-free basis). In a test involving continuous operation for more than 560 hours, no indications of a drop in electrical performance were seen. No information could be found in the literature about further development of this idea.

REFERENCES

Andrew M. R., W. J. Gressler, J. K. Johnson, et al., *J. Appl. Electrochem.*, **2**, 327 (1972).

Cairns E. J., D. C. Bartosik, *J. Electrochem. Soc.*, **111**, 1205 (1964).

Cifrain M., K. Kordesch, *J. Power Sources*, **127**, 234 (2004).

Gillibrand M. I., G. T. Lomax, *Proc. International Symposium Batteries*, Bournemouth, UK, 1962, p. 221.

Gouérec P., L. Poletto, J. Denizot, et al., *J. Power Sources*, **129**, 193 (2004).

Gülzow E., *J. Power Sources*, **61**, 99 (1996).

Gülzow E., M. Schulze, *J. Power Sources*, **127**, 243 (2004).

Jasinski R., *Catalysts for Hydrazine Fuel Cell Anodes*, Res. Div., Allis Chalmers, July 1963; cited from Vielstich W., *Brennstoffelemente*, Verlag Chemie, Weinheim, Germany, 1965, Sec. 4.3.2.2.

Justi E., A. Winsel, *Abhandlungen Mainzer Akademic*, No. 8, Steiner, Wiesbaden, Germany, 1959.

Kordesch K., A. Marko, *Oesterr. Chem. Z.*, **52** (7) 125 (1951).

Kordesch K., J. Gsellmann, M. Cifrain, et al., *J. Power Sources*, **80**, 190 (1999).

Korovin N. V., *Fuel Cells and Electrochemical Power Units* [in Russian], Power Engineering Institute, Moscow, 2005, pp. 125–130.

Lin B. Y. S., D. W. Kirk, S. J. Thorpe, *J. Power Sources*, **161**, 474 (2006).

Matsuoka K., Y. Iriyama, T. Abe, et al., *J. Power Sources*, **150**, 27 (2005).

Raney M., *Ind. Eng. Chem.*, **32**, 1190 (1940); U.S. patent 1,628,190 (1927).

Tamura K., T. Kahara, *J. Electrochem. Soc.*, **123**, 776 (1976).

Tomantschger K., R. Findlay, *J. Power Sources*, **39**, 21 (1992).

Wynveen R. A., T. G. Kirkland, *Proc. 16th Annual Power Sources Conference*, Atlantic City, NJ, 1962, p. 24.

Yang C. C., S.-J. Chiu, W.-C. Chien, *J. Power Sources*, **162**, 21 (2006).

Yu E. H., K. Scott, *J. Power Sources*, **137**, 248 (2004).

CHAPTER 7

MOLTEN CARBONATE FUEL CELLS

7.1 SPECIAL FEATURES OF HIGH-TEMPERATURE FUEL CELLS

Together with the solid-oxide fuel cells (SOFCs) considered in Chapter 8, molten-carbonate fuel cells (MCFCs) are representatives of a class of high-temperature fuel cells that have a working temperature of more than 600°C. High-temperature fuel cells have a number of advantages over other fuel cell types:

- The possibility of efficiently using the reaction heat for generating additional electrical energy
- A high rate of the electrode reactions and relatively little electrode polarization, hence no need to use platinum catalysts
- The possibility of using technical hydrogen with a large concentration of carbon monoxide and other impurities
- The possibility of using carbon monoxide, natural gas, and a number of petroleum products directly through internal conversion of these fuels to hydrogen within the fuel cell itself

Yet changing over to higher temperatures implies a certain diminution of thermodynamic indices of the fuel cells. The Gibbs free energy $-\Delta G$ of hydrogen oxidation by oxygen decreases with increasing temperature. It amounts to 1.23 eV at 25°C but decreases to 1.06 eV at 600°C and to 0.85 eV at 1000°C.

Fuel Cells: Problems and Solutions, By Vladimir S. Bagotsky

(Identical figures in volts are the numerical values for the thermodynamic EMF of hydrogen–oxygen fuel cells at these temperatures.) On the other hand, the reaction enthalpy, $-\Delta H$, which at 25°C has a value of 1.48 eV, will not change significantly with increasing temperature. For this reason the thermodynamic efficiency of the reaction $\eta_{\text{thermod}} = -\Delta G/-\Delta H$ also decreases with increasing temperature. At temperatures of 25, 600, and 1000°C, it has values of 0.83, 0.72, and 0.57, respectively. In fuel cells operated at higher temperatures, numerous problems associated with the limits of chemical and mechanical stability of various materials that are used in them also become important.

7.2 STRUCTURE OF HYDROGEN–OXYGEN MCFCs

The working temperature of MCFCs is around 600 to 650°C. As an electrolyte, mixed melts containing 62 to 70 mol% of Li_2CO_3 and 30 to 38 mol% of K_2CO_3, with compositions close to the eutectic point, are used in MCFCs. Sometimes, Na_2CO_3 and other salts are added to these melts. These liquid melts are immobilized in the pores of a ceramic matrix with fine pores made of sintered MgO or $LiAlO_2$ powders.

Porous metallic gas-diffusion electrodes are used. The anode consists of a nickel alloy with 2 to 10% chromium. The chromium that is added prevents a recrystallization and sintering of the porous nickel while it works as an electrode. This action is based on chromium forming a thin layer of chromium oxide at the nickel grain boundaries, interfering with surface diffusion of the nickel atoms.

The cathode consists of lithiated nickel oxide. Nickel oxide (NiO) is a *p*-type semiconductor that has rather low conductivity. When doped with lithium oxide, its conductivity increases tens of times, owing to a partial change of Ni^{2+} to Ni^{3+} ions. The lithiation is accomplished by treating the porous nickel plates with hot LiOH solution in the presence of air oxygen. The compound produced has a composition given as $Li_x^+Ni_{1-x}^{2+}Ni_x^{3+}O$. This lithiation of nickel oxide was first applied in 1960 by Bacon in his alkaline fuel cell (Section 2.2).

With its fine pores filled with the carbonate melt the matrix is a reliable protection against gases bubbling through and getting into the "wrong" electrode compartment. Therefore, there is no need to provide gas-diffusion electrodes with a special gas barrier layer.

The MCFC components have thicknesses as follows: the anode, 0.8 to 1.5 mm; the cathode, 0.4 to 1.5 mm; and the matrix, 0.5 to 1 mm. In a battery of the filter-press type, the individual cells are separated by bipolar plates made of nickel-plated stainless steel contacting the anode with their nickel side, and the cathode with their steel side. All structural parts are made of nickel or nickel-plated steel. In a working battery, the temperature of the outer part of the matrix electrolyte is lower than that of the inner part, so that in the outer part the electrolyte is solidified. This provides tight sealing around the periphery of the individual battery cells.

In a hydrogen–oxygen MCFC, the following reactions take place:

$$\text{Anode:} \quad H_2 + CO_3^{2-} \rightarrow H_2O + CO_2 + 2e^- \qquad E^0 = 0\ \text{V} \tag{7.1}$$

$$\text{Cathode:} \quad \tfrac{1}{2}O_2 + CO_2 + 2e^- \rightarrow CO_3^{2-} \qquad E^0 = 1.06\ \text{V} \tag{7.2}$$

$$\text{Overall:} \quad H_2 + \tfrac{1}{2}O_2 \rightarrow H_2O \qquad \mathscr{E}^0 = 1.06\ \text{V} \tag{7.3}$$

It is a special feature of the electrode reactions in MCFCs that unlike most other versions of fuel cells, the cathodic reaction consumes not only oxygen (or air) but also carbon dioxide (CO_2). In the anodic reaction, CO_2 is evolved at the anode. It is imperative, therefore, in MCFC design to provide for the possibility of CO_2 evolved at the anode to return to the cathode (Figure 7.1).

The values of electrode potential given in reactions (7.1) and (7.2) refer to the condition where all gases involved in the reactions (H_2, O_2, CO_2, and water vapor) are in their standard states (i.e., have partial pressures of 1 atm). During the reaction, in fact, their amounts and hence also their partial pressures change constantly. This implies that the equilibrium potentials of the electrodes also change. According to the Nernst equation, the potential of the hydrogen anode is

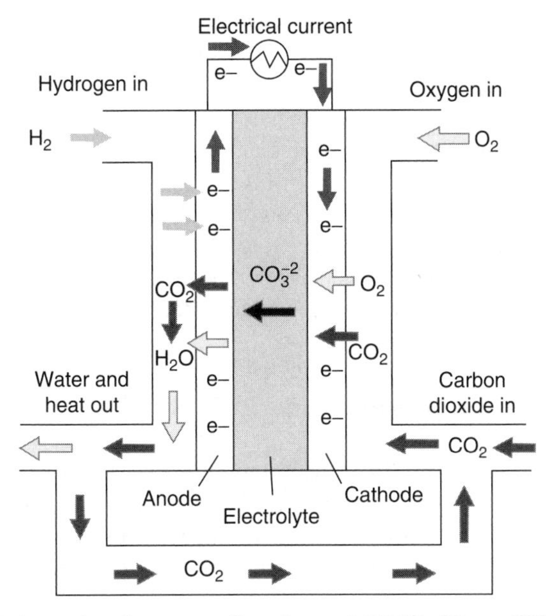

FIGURE 7.1 Schematic of reactant flow in an MCFC. (From Wikipedia, the free online encyclopedia.)

given by

$$E_a = E_a^0 + \frac{RT}{2F} \ln \frac{p_{H_2O}\, p_{CO_2}}{p_{H_2}} \tag{7.4}$$

the potential of the oxygen electrode is given by

$$E_c = E_c^0 + \frac{RT}{2F} \ln pO_2^{1/2} p_{CO_2} \tag{7.5}$$

If the reactant gases were enclosed, the potential of the hydrogen electrode would shift in the positive direction with the progress of H_2 consumption, while the potential of the oxygen electrode would shift in the negative direction with the progress in O_2 and CO_2 consumption. As a result, the overall value of thermodynamic EMF, $\mathscr{E} = E_c - E_a$, would fall below the standard value of 1.06 V. The losses of overall efficiency of the energy conversion process that are associated with these changes have been termed *Nernst losses*. In an operating fuel cell, however, fresh reactant gases are supplied continuously to the gas compartments, so that these losses are mitigated.

Depending on the partial pressures of the reactant gases, the open-circuit voltage of hydrogen–oxygen MCFCs has values of 1.00 to 1.06 V. A practically linear relation between cell voltage U and current density i is a special feature of these fuel cells. This implies a practically constant value of the apparent internal resistance R_{app} in equation (1.11). This leads to a more important voltage decrease with increasing cell discharge current than in other fuel cell types. At a pressure of the reactant gases of 1 bar and a current density of 100 mA/cm^2, the voltage is about 0.85 V; at a current density of 150 mA/cm^2 it is only about 0.6 V. Current densities above 150 mA/cm^2 are practically not used in MCFCs.

The linear relation between voltage and current density is not a result of the internal resistance of the cell being a purely ohmic resistance. Apart from the ohmic voltage drop across the electrolyte kept in the pores of the matrix, a marked contribution to the voltage decrease comes from polarization of the oxygen electrode. At a current density of 100 mA/cm^2, this amounts to 0.5 V. The polarization of the hydrogen electrode is considerably lower.

7.3 MCFCs WITH INTERNAL FUEL REFORMING

From the very outset of MCFC development, research workers were attracted by the fact that not only hydrogen but also carbon monoxide (CO) could be used as a reactant fuel (reducing agent). Carbon monoxide (as *water gas*, a mixture of CO and H_2) is obtained readily by the steam gasification of coal:

$$C + H_2O \rightarrow CO + H_2 \tag{7.6}$$

This opens up possibilities for an indirect electrochemical utilization of huge coal reserves.

The possibility of a direct electrochemical oxidation of carbon monoxide was soon questioned. It was suggested that hydrogen, rather than carbon monoxide, is involved in the electrochemical reaction after being formed from CO by the Boudouard reaction:

$$CO + H_2O \rightarrow CO_2 + H_2 \tag{7.7}$$

which occurs readily under the operating conditions of the carbonate fuel cell.

Sparr et al. (2006) could show that pure CO can actually be oxidized electrochemically according to the reaction

$$CO + CO_3^{2-} \rightarrow 2CO_2 + 2e^- \tag{7.8}$$

although in fact the rate of this reaction is about 20 times lower than the rate of hydrogen oxidation [reaction (7.1)]. The exchange current density (as an index of reaction rates) is about 0.7 mA/cm^2 for the oxidation of water gas (a mixture of 56% H_2 + 8% CO + 28% H_2O); that for the oxidation of pure hydrogen is about 1 mA/cm^2. This implies that the use ("combustion") of CO in MCFCs occurs primarily, but not exclusively, via the intermediate formation of hydrogen. The possibilities of using products of gasification of biomass or residential waste that have a large CO content but include a variety of contaminants have been examined in a paper by Watanabe et al. (2006).

The rates of direct electrochemical oxidation of hydrocarbons (particularly of methane) in MCFCs are negligibly small. Therefore, natural types of fuel can be used in the fuel cells only after prior conversion (reforming) to hydrogen. Since the reforming process is highly endothermic and requires a large heat supply, the idea sprang up that this process should be carried out within the fuel cell itself, where the heat of reaction evolved in the fuel cell could be utilized for the reforming (by analogy with the example of CO conversion to hydrogen reported above).

Two types of internal-reforming fuel cells (IRFCs) are distinguished: the direct internal-reforming fuel cells (DIRFCs), in which the reforming process takes place within the fuel cell at its anode catalyst, and the indirect internal-reforming fuel cells (IIRFCs), in which plates with special reforming catalysts are included within the battery stacks.

The most active company working in this field is FuelCell Energy (FCE) of Danbury, Connecticut, which, starting in the late 1970s, developed and delivered a large series of power plants designated as Direct Fuel Cells. These plants include a combined IRFC system. In the battery, special plates for prior fuel reforming are placed between groups of 8 to 10 DIRFCs. With this system one can achieve a more uniform temperature distribution within the battery; heat is evolved in the fuel cells and is consumed at the reforming plates (Doyon et al., 2003).

A somewhat different version of reforming units combined with a battery of fuel cells was presented by the German company MTU under the designation Hot Modules. All the components of the MCFC system (the horizontally arranged fuel cell batteries, the catalytic burner of the anode tail gases, and a cathode recycle loop), as well as the fuel processing system, are placed into a common, thermally insulated vessel. In this version, special manifolds for the cathodic gases are not needed (Bischoff and Huppman, 2002).

7.4 DEVELOPMENT OF MCFC WORK

The work of Baur et al. (1916) and subsequent work by Baur and co-workers must be counted as the first work in the field of MCFCs. It is important to note that the "solid electrolyte" (a mixture of monazite sand and other components, including alkali metal carbonates) with which O. Davtyan worked in the 1930s, actually represented a carbonate melt immobilized in a solid skeleton of silicates (Davtyan, 1947).

Work on MCFCs was resumed in the 1960s in many places: for example, in the Institute of High-Temperature Electrochemistry in Ekaterinburg, Russia (Stepanov, 1972–1974) and in the Institute of Gas Technology in Chicago (Baker et al., 1980). A large contribution to this field was made by Broers and Ketelaar (1961) in Amsterdam.

Industrial organizations first became interested in MCFCs in about the mid-1960s. At present a large number of companies both in the United States and in other countries (e.g., Germany, Japan, Korea) are working in this field. Large-scale production of power plants operating with such fuel cells has been initiated. Particularly vigorous growth in the production and use of such power plants has been seen since 2002. In the Direct Fuel Cells described in Section 7.3, electrodes with a surface area of about 1 m^2 are used. In total, a battery includes some 350 to 400 individual cells. The average voltage given out by an individual cell is 0.8 V. The overall efficiency of energy conversion in such a power plant (referred to the lower heating value) is about 51%. Starting in 2000, more than 40 power plants in sizes from 250 kW to 1 MW have been delivered and installed in various countries. According to an account ending in April 2005, these plants had produced more than 70 million kilowatthours of electrical energy (Doyon et al., 2003; Farooque and Maru, 2006).

Work to build power plants on the basis of MCFCs has also been done at UTC and a number of companies in different countries: in Germany by MTU CFC Solutions GmbH, Munich (jointly with FCE in the United States), in Italy by Ansaldo S.p.a., in Japan by Ishikawajima Harima Heavy Industries Co., and others.

The largest power plant using MCFCs was a 2-MW installation built in Santa Clara, California, as a joint project of five utilities, the U.S. federal government, and five research organizations. The total cost of the project was \$46 million. The plant consisted of 16 stacks each with 125 kW power, laid out in four modules. Construction started in April 1994 and was finished in June 1995. After 720 hours of operation, problems came up that had to do with the thermal decomposition of insulating materials, leading to the deposition of carbon particles on electrode surfaces and connecting buses. In 1996, operation of this plant was terminated indefinitely. In all, this plant worked for 5290 hours (sometimes at half power) and produced 2500 MWh of electrical energy (Eichenberger, 1998).

7.5 THE LIFETIME OF MCFCs

In view of their ability to work with different types of fuel, molten-carbonate fuel cells are of great interest. The electrical and operating properties of these fuel cells are sufficient for building economically justified stationary power plants with a relatively large power output. The only problem so far is an insufficiently long period of trouble-free operation. The minimum length of time a large (and expensive) power plant should work until replacement is 40,000 hours (4.5 to 5 years).

The first laboratory models of MCFCs built in the 1960s were in the best case operative for only a few months. At present, intense research and engineering efforts have made it possible to build individual units that have worked several hundreds and thousands of hours (Bischoff et al., 2002). Yet the road to a *guaranteed* five-year period of operation is still long. Many causes lead to a gradual decline in the performance of such power plants, or even premature failure. The three most important reasons associated with the fuel cells themselves (rather than with extraneous issues rooted in ancillary equipment or operating errors) are described below.

Gradual Dissolution of Nickel Oxide from the Oxygen Electrode

Nickel oxide (NiO) dissolves in the carbonate melt according to the equation

$$NiO + CO_2 \rightarrow Ni^{2+} + CO_3^{2-} \tag{7.9}$$

Due to this dissolution process, the weight and thickness of the cathode will decrease by about 3% after 1000 operating hours of the fuel cell. The concentration of Ni^{2+} ions in the melt can attain values of 10 to 15 ppm. These ions spread by diffusion all across the electrolyte-filled matrix. The nickel sites are reduced to metallic nickel by diffusing hydrogen:

$$Ni^{2+} + H_2 + CO_3^{2-} \rightarrow Ni + CO_2 + H_2O \tag{7.10}$$

Deposits of metallic nickel have been discovered in different parts of operating fuel cells (Plomp et al., 1992).

Experimental results concerning the influence of various factors on the rate of nickel oxide dissolution and nickel ion transport that were obtained by different workers are contradictory. The *basicity* of the melt has a distinct effect (Minh, 1988). In carbonate melts, the basicity is not determined by the concentration of OH^- ions as in aqueous solutions but by the concentration of O^{2-} ions formed in the equilibrium reaction

$$2CO_3^{2-} \rightleftharpoons O^{2-} + CO_2 \tag{7.11}$$

according to which, the basicity of the melt (or concentration of O^{2-} ions) is a function of CO_2 partial pressure. Under the conditions of MCFC operation, the basicity of the melt is low (the acidity is relatively high), and NiO will dissolve at a relatively high rate. This rate can be reduced by raising the basicity of the melt via lowering of the CO_2 partial pressure. This will lower the rate of the cathode reaction and the overall operating efficiency of the fuel cell. Thus, it will be necessary to select an intermediate value of CO_2 partial pressure that represents a compromise. An acceptable value is 0.15 to 0.20 bar.

Another way to achieve a certain increase in melt basicity is by raising the fraction of Li_2CO_3 in the melt. A similar effect is produced when adding a small amount of alkaline-earth metal ions (Mg, Ca, Sr, Ba) to the melt.

An excellent solution would be to use a material that has a lower solubility in the carbonate melt than that of nickel oxide. Plomp et al. (1992) reported experiments involving oxides of the type of $LiFeO_2$ and $LiCoO_2$. The former had unsatisfactory properties, such as strong polarization with increasing current density. The latter, $LiCoO_2$, looked very promising. The electrical properties were approximately the same as with nickel oxide, while the expected lifetime at a gas pressure of 7 bar was an order of magnitude longer than that of nickel oxide (when calculated by extrapolating the experimental data obtained at lower pressures but not confirmed experimentally).

Anode Creep

When assembling a filter-press battery from individual fuel cells of any type, considerable compression must be applied to minimize the contact resistance between individual battery cells. With a battery in a vertical position, the cells in the lower part of the battery experience an additional load due to the weight of battery parts above. In MCFCs, where the working temperature (in kelvin) is higher than 50% of the melting point of nickel, the prolonged action of mechanical load and temperature leads to plastic deformation of the nickel anodes known as *anode creep*. As a result of this creep, the anode's thickness may decrease by more than 1 to 3%. This compression leads to a lower pore volume in the anode, reflected in the degree of gas filling of the electrode and thus in electrode characteristics. In addition, anode creep may disturb the contact between anode and matrix. Both factors are limiting factors for the lifetime of MCFCs. Nickel is alloyed with chromium and/or aluminum to reduce the plasticity of the nickel anode and make it more rigid (Kim et al., 2001). Another possibility for attaining these goals is by including fine-disperse Al_2O_3 powder when preparing the porous nickel anode (Lee et al., 2006).

Corrosion of Metal Parts

Another factor limiting the lifetime of MCFCs is corrosion of various structural metal parts of the fuel cells. Usually, nickel-plated stainless steel is used to make these parts. The medium in fuel cells of this type—the high-temperature carbonate melt—is highly corrosive. This aggressiveness is particularly high near the anode space, with its reducing properties. Here a carburetion of the steel is possible, which leads to inferior mechanical properties. In addition to this, corrosion leads to a higher contact resistance between bipolar plates and electrodes. Earlier, the bipolar plates were nickel plated on only the anode side. It has now been established that both sides of these plates must be nickel-plated. During prolonged operation of the fuel cells, individual components of the stainless steel may diffuse through the nickel layer. This diffusion detracts from the protective power of the layer (Durante et al., 2005).

Songbo et al. (2002) reported that the intermetallic compound NiAl forms a very strong protective layer under the conditions of MCFC operation, and hence the compound is more highly corrosion resistant than is stainless steel.

REFERENCES

Baker D. S., D. J. Dharia, U.S. patent 4,182,795 (1980).

Baur E., A. Petersen, Q. Füllemann, *Z. Elektrochem.*, **22**, 409 (1916).

Baur E., W. D. Treadwell, G. Trümpler, *Z. Elektrochem.*, **27**, 199 (1921).

Bischoff M., G. Huppmann, *J. Power Sources*, **105**, 216 (2002).

Broers G. H. J., J. A. A. Ketelaar, in: G. C. Young (ed.), *Fuel Cells*, Vol. 1, Reinhold, New York, 1961, p. 78.

Doyon J., M. Farooque, H. Maru, *J. Power Sources*, **118**, 8 (2003).

Durante G., S. Vegni, P. Carobianco, F. Golgovici, *J. Power Sources*, **152**, 204 (2005).

Eichenberger P. H., *J. Power Sources*, **71**, 95 (1998).

Kim Y.-S., K.-Y. Lee, H.-S. Chun, *J. Power Sources*, **99**, 26 (2001).

Lee H., I. Lee, D. Lee, H. Lim, *J. Power Sources*, **162**, 1088 (2006).

Minh N. G., *J. Power Sources*, **24**, 1 (1988).

Plomp J., J. B. J. Veldhuis, E. F. Sitters, S. B. van der Molen, *J. Power Sources*, **39**, 369 (1992).

Songbo X., Z. Yongda, H. Xing, et al., *J. Power Sources*, **103**, 230 (2002).

Sparr M., A. Boden, G. Lindbergh, *J. Electrochem. Soc.*, **153**, A1532 (2006).

Stepanov G. K., *Proc. Institute of High-Temperature Electrochemistry*, No. 18, 1972, p. 129; No. 20, 1973, p. 95; No. 21, 1974, p. 88.

Watanabe T., Y. Izaki, Y. Magikura et al., *J. Power Sources*, **160**, 868 (2006).

Reviews and Monographs

Bischoff M., Molten carbonate fuel cells: a high temperature fuel cell on the edge to commercialization, *J. Power Sources*, **160**, 842 (2006).

Davtyan O. K., *The Problem of Direct Conversion of the Chemical Energy of Fuels to Electrical Energy* [in Russian], Publishing House of the USSR Academy of Sciences, Moscow, 1947.

Farooque M., H. C. Maru, Carbonate fuel cells: milliwatts to megawatts, *J. Power Sources*, **160**, 827 (2006).

Frusteri F., S. Freni, Bio-ethanol, a suitable fuel to produce hydrogen for a molten carbonate fuel cell, *J. Power Sources*, **173**, 200 (2007).

Joon K., Critical issues and future prospects for molten carbonate fuel cells, *J. Power Sources*, **61**, 129 (1996).

Selman J. R., Molten salt fuel cells: technical and economic challenges, *J. Power Sources*, **160**, 852 (2006).

Tomczyk P., MCFC Versus other fuel cells: characteristics, technologies and prospects, *J. Power Sources*, **60**, 858 (2006).

Williams M. C., H. C. Maru, Distributed generation–molten carbonated fuel cells, *J. Power Sources*, **160**, 863 (2006).

CHAPTER 8

SOLID-OXIDE FUEL CELLS

The following acronyms are in common use in the literature for oxide materials used in solid-oxide fuel cells (SOFCs) (x is the doping level).*

BSCF	strontium-doped barium cobaltite ferrite ($Ba_{1-x}Sr_xCo_yFe_{1-y}O_3$)
CGO	gadolinium-doped ceria ($Ce_{1-x}Gd_xO_2$)
CSO	samarium-doped ceria ($Ce_{1-x}Sm_xO_2$)
LDC	lanthanum-doped ceria ($Ce_{1-x}La_xO_2$)
LSC	strontium-doped lanthanum cobaltite ($La_{1-x}Sr_xCoO_3$)
LSCF	strontium-doped lanthanum cobaltite ferrite ($La_xSr_{1-x}Co_yFe_{1-y}O_3$)
LSGM	strontium-doped lanthanum magnesite gallate ($La_xSr_{1-x}Mg_yGa_{1-y}O_3$)
LSM	strontium-doped lanthanum manganite ($La_{1-x}Sr_xMnO_3$)
SDZ	scandium-doped zirconia $(ZrO_2)_{1-x}(Sc_2O_3)_x$
YSZ	yttrium-doped (stabilized) zirconia $(ZrO_2)_{1-x}(Y_2O_3)_x$

* These acronyms represent a compilation from recent literature. The nomenclature is neither systematic (it does not follow the compound names in *Chemical Abstracts*) nor uniform (with different authors, the same letter may stand for different elements, or an element may appear in the "cationic" or "anionic" portion of a name).

Fuel Cells: Problems and Solutions, By Vladimir S. Bagotsky

Fuel cells of this class are built with solid electrolytes that have unipolar O^{2-} ion conduction. Best known among these electrolytes is yttria-stabilized zirconia (YSZ), that is, zirconium dioxide doped with the oxide of trivalent yttrium: $ZrO_2 + 10\% \; Y_2O_3$ or $(ZrO_2)_{0.92}(Y_2O_3)_{0.08}$. This compound, the basis of Nernst's (glower) lamp of 1897, became known as the *Nernst mass* and was regarded as a candidate electrolyte for fuel cells by Baur (1937) and Davtyan (1938) (see Section 2.2). It is commonly found in the oxygen sensors (lambda probes) used to optimize the operation of internal combustion engines.

The Y^{3+} ions introduced into the crystal lattice of ZrO_2 act as dopant ions by giving rise to the formation of oxygen vacancies in the lattice, and electrical conduction comes about by O^{2-} ions jumping from a current position into a neighboring vacancy, thus filling this vacancy while leaving a vacancy behind (one could also visualize this process as vacancy migration, which occurs in a direction opposite to that of the O^{2-} ion migration).

The conductivity of YSZ-type electrolytes becomes acceptable (with values of about 0.15 S/cm) only at temperatures above 900°C. For this reason the working temperature of fuel cells having such an electrolyte is between 900 and 1000°C. Such fuel cells will be called *conventional SOFCs* in what follows.

8.1 SCHEMATIC DESIGN OF CONVENTIONAL SOFCs

Conventional SOFCs exist in several design variants. The basic variants are tubular and planar cells. Monolithic cells joined the first two variants around 1990. The specific design and operating features of these and other variants are described in subsequent sections of this chapter. Those factors that are common for all variants of conventional SOFCs are described in the present section.

The anodes of these cells consist of a cermet (ceramic–metal composite) of nickel and the zirconia electrolyte. This material is made from a mixture of nickel oxide (NiO) and the YSZ electrolyte. The nickel oxide is reduced in situ to metallic nickel, forming highly disperse particles that serve as the catalyst for anodic fuel gas oxidation reactions. These particles are distributed uniformly in the solid electrolyte, and are prevented from agglomerating during fuel cell operation by this electrolyte, thus retaining their catalytic activity. The YSZ material present in the anode also improves the contact between the nickel catalyst and the fuel cell's electrolyte layer.

The cathodes consist of manganites or cobaltites of lanthanum doped with divalent metal ions [e.g., $La_{1-x}Sr_xMnO_3$ (LSM) or $La_{1-x}Sr_xCoO_3$ (LSC), where $0.15 < x < 0.25$]. Apart from their O^{2-} ion conductivity, these cathode materials also have some electronic conductivity that secures a uniform current

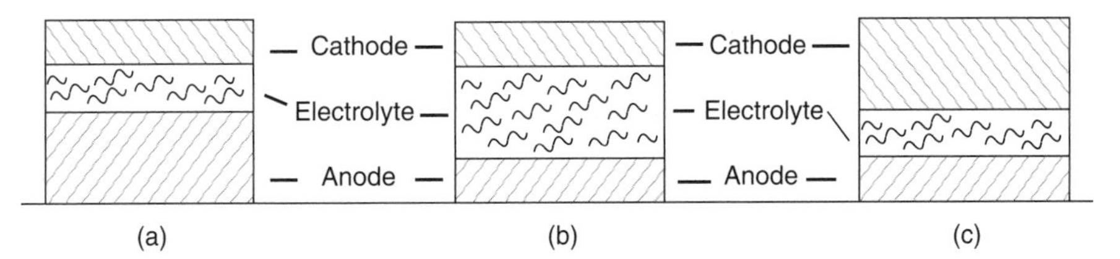

FIGURE 8.1 Different versions of SOFC: (a) anode-supported; (b) electrolye-supported; (c) cathode-supported.

distribution over the entire electrode. LSC has a higher ionic conductivity than LSM but is more expensive and leads to problems, in particular because of a possible chemical interaction with the electrolyte.

These three components of the fuel cell: anode, cathode, and electrolyte, form a membrane–electrolyte assembly (MEA), since by analogy with PEMFCs, one may regard the thin layer of solid electrolyte as a membrane. Any one of the three MEA components can be selected as the entire fuel cell's support, and is then made relatively thick (up to 2 mm) to provide mechanical stability. The other two components are then applied to this support in different ways as thin layers (tenths of a millimeter). Accordingly, one has anode-supported, electrolyte-supported, and cathode-supported cells. Sometimes, though, an independent metal or ceramic substrate is used, to which the three functional layers are then applied (Figure 8.1).

An important component of SOFCs that governs their operating reliability to a large extent (and sometimes their manufacturing cost) is that of the interconnectors needed to combine the individual cells in a battery. These must be purely electronically conducting, chemically sufficiently stable toward the oxidizing and reducing atmospheres within the fuel cells, and free of chemical interactions with the active materials on the electrodes.

Hydrogen and carbon monoxide can be used as the reactive fuels in SOFCs. The reactions occurring in such fuel cells and the thermodynamic parameters associated with the reactions (at 900°C) are

$$\text{Anode:} \quad H_2 + O^{2-} \rightarrow H_2O + 2e^- \qquad E^0 = 0\,\text{V}^* \tag{8.1}$$

$$\text{Cathode:} \quad \tfrac{1}{2}O_2 + 2e^- \rightarrow O^{2-} \qquad E^0 = 0.89\,\text{V}^* \tag{8.2}$$

$$\text{Overall:} \quad H_2 + \tfrac{1}{2}O_2 \rightarrow H_2O \qquad \mathscr{E}^0 = 0.89\,\text{V} \quad \Delta H^0 = -248.8\,\text{kJ/mol} \tag{8.3}$$

when hydrogen is the fuel, and

$$\text{Anode:} \quad CO + O^{2-} \rightarrow CO_2 + 2e^- \qquad E^0 = 0.2\,V^* \tag{8.4}$$

$$\text{Cathode:} \quad \tfrac{1}{2}O_2 + 2e^- \rightarrow O^{2-} \qquad E^0 = 0.89\,V^* \tag{8.2}$$

$$\text{Overall:} \quad CO + \tfrac{1}{2}O_2 \rightarrow CO_2 \qquad \mathscr{E}^0 = 0.87\,V \quad \Delta H^0 = -283.0\,kJ/mol \tag{8.5}$$

when carbon monoxide is the fuel.*

The open-cell voltage (OCV) U^0 in hydrogen–oxygen SOFCs at a temperature of 900°C has a value of about 0.9 V, the exact value depending on reactant composition. The relation between current density and voltage of an operating cell, U_i versus i, is practically linear; that is, the cell's apparent internal resistance R_{app} has a constant value that does not depend on the current density:

$$U_i = U^0 - iR_{app} \tag{8.6}$$

This does not constitute evidence for the voltage drop having merely ohmic origins (e.g., the electrolyte's ohmic resistance). Rather, this function is due in part to special features of polarization of the electrodes. At a current density of 200 mA/cm^2, the cell voltage typically has a value of about 0.7 V; that is, parameter R_{app} has a value on the order of 1 Ω · cm^2 (its exact value will depend on the thickness of the electrolyte layer).

8.2 TUBULAR SOFCs

8.2.1 Tubular Cells of Siemens–Westinghouse

At the start of the new upswing in fuel cell development, Weissbart and Ruka (1962) of the Westinghouse Electric Corporation, Pittsburgh, Pennsylvania, conducted tests with a hydrogen–oxygen fuel cell built around a tube of ion-conducting electrolyte, $(ZrO_2)_{0.85}(CaO)_{0.15}$ (i.e., a material somewhat different from YSZ). Like a test tube, it was closed off on one end. In the bottom part, thin platinum electrodes (less than 25 μm thick) were disposed on the inside and outside. Oxygen was fed into the tube; from the outside it was bathed in a hydrogen stream (Figure 8.2).

* The potentials are referred to the potential of a hydrogen electrode contacting the same electrolyte and kept at the same temperature, and to standard pressure (1 bar) for all components.

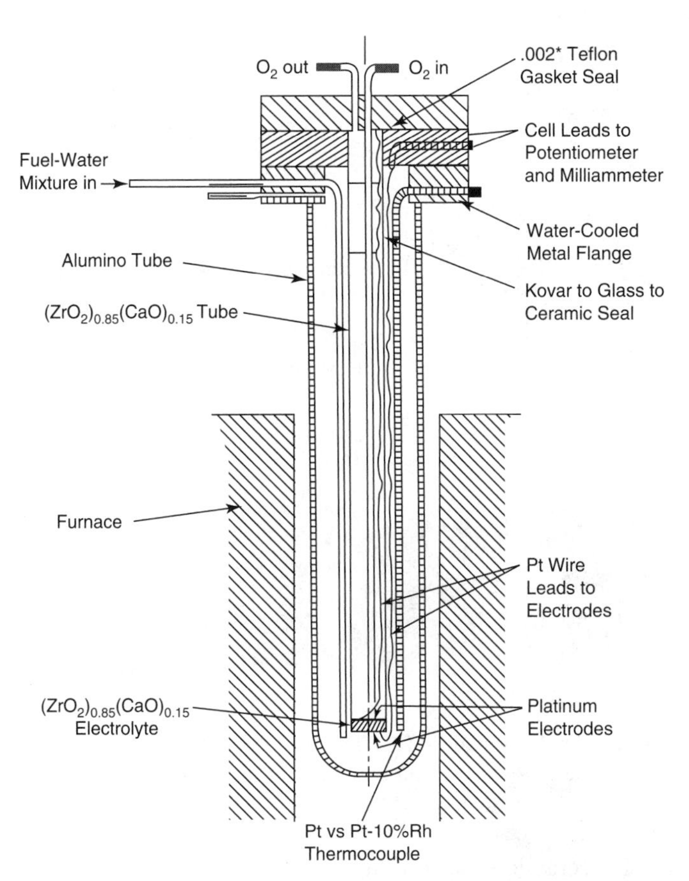

FIGURE 8.2 Schematic of a galvanic cell with a solid electrolyte. (From Weissbart and Ruka, 1962, with permission of The Electrochemical Society.)

At a temperature of 1010°C and gas pressures of about 1 bar, the cell had an OCV of about 1.15 V; at a current density of 110 mA/cm^2, the cell voltage was about 0.55 V. The current–voltage relation was strictly linear. The authors attributed the voltage drop seen with increasing current density to the ohmic resistance of the rather thick electrolyte layer between the electrodes.

Not long after building this first model of a tubular solid-oxide fuel cell of electrolyte-supported design that was associated with large ohmic losses, Westinghouse switched to a new cathode-supported design admitting a much thinner electrolyte layer and thus much lower ohmic losses. Also, the YSZ electrolyte was used for all subsequent work.

In 1998, Westinghouse joined forces with the German company Siemens, which up to then had worked on planar SOFCs, which gave rise to a new enterprise called Siemens–Westinghouse (S-W). This enterprise specialized in the further development and commercialization of tubular SOFCs and soon became the world leader in this field.

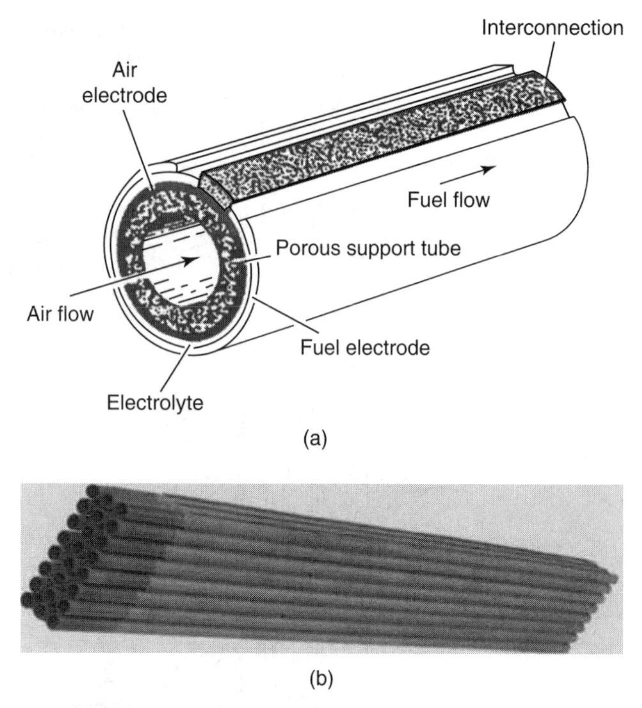

FIGURE 8.3 (a) Single tubular SOFC. (From Yamamoto, 2000, with permission from Elsevier). (b) Bundle of 24 (3 × 8) tubular SOFCs. (Courtesy of Siemens AG, Energy Sector.)

In S-W cells, ceramic tubes produced by extrusion of the cathode material, lanthanum manganite (with some added alkaline-earth metal oxides), are used. They have a porosity of 30%, length between 50 and 150 cm, and a diameter of 22 mm. From the outside, a thin layer of YSZ electrolyte (40 μm) is applied by chemical vapor deposition (CVD) from mixed $ZrCl_4$, YCl_3, and water vapors and oxygen. From inside, a layer of lanthanum manganite with magnesium or strontium dopant is used as cathode material by plasma spraying. An anode material is deposited on top of the electrolyte from a slurry of Ni (or NiO) and YSZ material, and then sintered. A narrow band, 85 μm thick, of lanthanum chromite doped with divalent cations (calcium, magnesium, or strontium), which is an electronically conducting semiconductor material, is plasma-sprayed along the tube's outside to serve as a cell interconnector for series combination of the cells (Figure 8.3a). In a power plant built with such fuel cells, several tubes are combined into a bundle. Such a bundle is shown in Figure 8.3b; it consists of 24 cells (three rows in parallel, each consisting of eight cells in series).

A 100-kW power plant was built by S-W in Westervoort in the Netherlands from tubular cells (Figure 8.4). The fuel cell stacks used in this plant contained four bundles of this type combined in series to form a row, 12 rows then being placed in parallel. Between the rows, units for the conversion of natural gas

FIGURE 8.4 100-kW SOFC CHP plant of Siemens–Westinghouse in The Netherlands. (Courtesy of Siemens AG, Energy Sector.)

were installed. The plant also included units for desulfurization and pre-reforming of the natural gas.

During the years 1998 to 2000, the power plant operated for 16,000 hours, providing local grid power of 105 to 110 kW and an additional 65 kW equivalent as hot water to the district heating scheme. The system had an electrical efficiency of 46% and an overall efficiency of 75%. The operation of this plant gave proof of the high functional reliability of tubular SOFCs under real-world conditions of a large power plant (including several temperature cycles caused by temporary stoppages; the use of natural gas containing sulfur) (George, 2000). In March 2001 the system was moved from the Netherlands to a site in Essen, Germany, where it was operated by the German utility RWE for an additional 3700 hours, for a total of over 20,000 hours. Following the Essen experience, the system was brought to GTT-Turbo Care in Turin, Italy. So far the system has reached an operating time of about 37,000 hours with minimum degradation.

8.2.2 Other Versions with Tubular Electrodes

To achieve higher specific power, Sammes et al. (2005) and Suzuki et al. (2006) proposed using tubular cells of very small diameter (microtubular or submillimeter tubular SOFCs). Tubes of smaller diameter have another important advantage. In fact, the mechanical stresses experienced by all ceramic parts under conditions of drastic temperature change (such as switching the fuel cell on or off) will lead to cracks when these changes occur repeatedly. The stresses

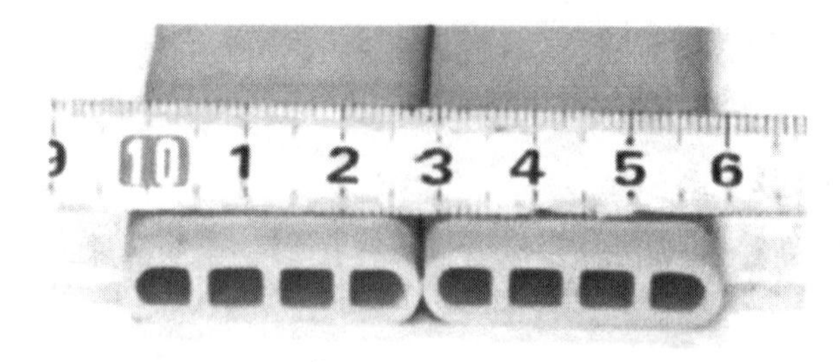

FIGURE 8.5 Anode-supported flat-tube SOFCs (From Kim et al., 2003, with permission from Elsevier.)

will, however, be less significant for the smaller of the linear dimensions (here, the diameter) of the ceramic part.

With the aim of drawing lower currents from the unit surface area of the electrodes, and bringing the cell design closer to the design with the flat electrodes and flat electrolyte generally adopted, Kim et al. (2003) suggested flat-tube SOFCs. A section of such cells is shown in Figure 8.5. Cells of this shape yield a higher specific power of the battery per unit weight and per unit volume.

Flat tubular electrodes are used even now by Siemens for building large SOFC-based power plants (see Chapter 15). There has been some gradual evolutionary change in the shape of these electrodes, as can be seen in Figure 8.6.

FIGURE 8.6 Progress and sequence of advance in the design of Siemens fuel cell electrodes leading up to the 2007 Delta 8 reformation (displayed at the bottom of the stack). (Courtesy of Siemens AG, Energy Sector.)

8.3 PLANAR SOFCs

Flat solid-oxide fuel cells are built analogously to other types of fuel cells, such as PEMFCs. Usually, one of the electrodes (the fuel anode or the oxygen cathode) serves as support for the membrane–electrode assembly (MEA). To this end it is relatively thick (up to 2 mm), and thin layers of the electrolyte and the second electrode are applied to it. In batteries of the filter-press design, the MEA alternates with bipolar plates having systems of channels through which reactant gases are supplied to the electrodes and reaction products are eliminated from them (Figure 8.7). In addition, the bipolar plates act as intercell connectors in the battery by passing the current from a given cell to its neighbor. Alternating with groups consisting of several cells each, special heat exchangers are installed for cooling the operating battery and heating an idle battery during startup.

The development of planar SOFCs started later than that of tubular SOFCs. They have several advantages over the latter. In batteries of planar design, higher values of specific power per unit weight and volume can be realized than in batteries with tubular cells. This is due to the fact that in them, the path of the current from all surface segments of the electrodes to the current collector is shorter. The current path between the individual cells is also shorter. All this leads to an important decrease in the ohmic losses in the battery. Another advantage of the planar version is a much simpler, less expensive manufacturing technology. The technological processes used in making planar solid-oxide fuel cells are more flexible and allow different types of materials to be used for electrodes and electrolyte. For this reason, the basic work on entirely new SOFC versions, particularly those able to work at intermediate and low temperatures, has all been done with cells of the flat type.

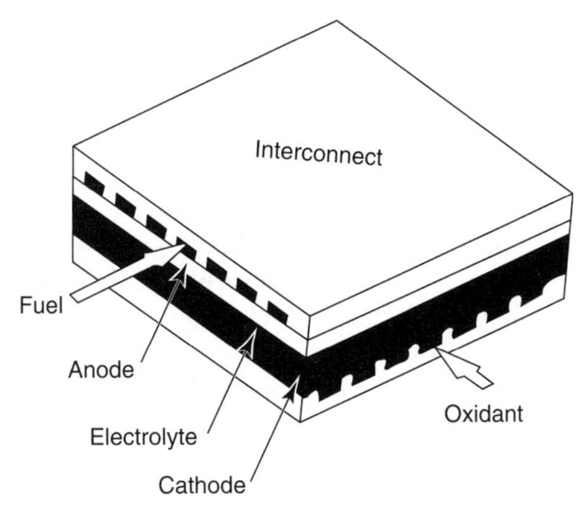

FIGURE 8.7 Flat-plate design of SOFCs. (From Yamamoto, 2000, with permission from Elsevier.)

Considerable difficulties turned up, however, in the development of flat SOFCs and have so far not been definitely resolved, thus preventing broad commercialization of such cells. The difficulties are related primarily to the fact that the selection of materials having sufficient chemical and mechanical strength for operation at temperatures of 900 to 1000°C in the presence of oxygen and/or hydrogen is rather restricted. This holds true both for the materials of electrodes and electrolyte and for various structural materials.

The major problems arising in the development and manufacturing of planar-type SOFC are described below.

8.3.1 Sealing

A large problem in high-temperature fuel cells is careful sealing. The gas compartments of the fuel and oxygen (or air) electrodes should be protected against entry of the "wrong" gas. The joints and welds of all inner channels and gas manifolds should be free of gas leaks. For flat cells, this is a much more difficult problem than for tubular cells, since their entire perimeter must be sealed. Also, flat cells usually have a larger number of gas feeds.

Two types of sealing materials, rigid and compressive, are in use. The elastic materials require constant compression during fuel cell operation. Rigid materials, on the other hand, must meet certain requirements of adhesion (wetting) and of compatible thermal expansion coefficients. Glass or glass ceramics were used initially as the basic sealant. By changes in glass composition and in the conditions of crystallization of glass ceramics, it was possible to adapt the properties of these materials to the operating conditions prevalent in fuel cells, but they have the basic defect of brittleness. In recent years, therefore, development of rigid and elastic sealants on the basis of metals and ceramics was initiated. By using multiphase materials, one can adjust the elastic properties and wetting of surfaces in contact. More details concerning the various sealing materials under study today may be found in a review by Fergus (2005).

8.3.2 Bipolar Plates

The bipolar plates that function as intercell connectors are the most expensive and at the same time the most vulnerable component of planar SOFCs. The basic requirements that must be met by these plates are high chemical stability under fuel cell operating conditions, high electronic conductivity, and complete impermeability to gases. Usually, two types of material are used to make bipolar plates: ceramics and thermally stable high-alloy steels. The ceramics generally used for bipolar plates are oxides based on lanthanum chromite ($LaCrO_3$) doped with MgO, CaO, or SrO. It is a defect of this material that at elevated temperatures, structural changes occur that cause internal stresses in the plates. Liu et al. (2006) suggest using ceramics based on praseodymium oxides ($PrCrO_3$) of the perovskite type to overcome this defect. A common defect of ceramic materials is their brittleness and insufficient mechanical strength (high sensitivity) under mechanical and heat shocks.

Chemically and thermally resistant steels contain considerable quantities of chromium (more than 20%). On the side of the oxygen cathode, metallic chromium is oxidized to Cr_2O_3. Depending on the temperature and oxygen partial pressure, this oxide may oxidize further to volatile compounds CrO_3 and $CrO_2(OH)_2$, which could then settle at the cathode–electrolyte interface and hinder oxygen reduction. Such an "evaporation" of chromium is the basic difficulty in the use of metallic bipolar plates (Hilpert et al., 1996). It was found by Stanislowski et al. (2007) that the rate of this evaporation can be reduced substantially by covering the surface of the chrome steel with thin layers (about 10 μm) of cobalt, nickel, or copper. These metal layers are converted completely to the corresponding oxides, which become firmly bound to the substrate and have good electronic conductivity.

8.3.3 Stresses in Planar SOFCs

Considerable internal stresses often develop when making and using ceramic parts. Strong temperature changes as well as temperature gradients within a given part are the most important reasons for the development of these stresses. These factors develop in SOFCs operated at temperatures going up to 1000°C. When such fuel cells are started or stopped, the temperature may change at a rate of hundreds of degrees per minute. Local temperature gradients arise when colder reactant gases enter a hot fuel cell. Fischer et al. (2005) measured the values of these stresses quantitatively with x-ray powder diffraction. They found that the compressive stresses in anodically supported MEA would be as high as 500 MPa.

The risk that such stresses develop is particularly high in ceramic parts consisting of a number of layers of different materials having a mismatch of their coefficients of thermal expansion (CTEs). The membrane–electrode assemblies (MEAs) in SOFCs are exactly in this category. For this reason, all materials used to make high-temperature fuel cells should have identical or at least very close CTEs. For many of the materials used, this coefficient has values between 9 and 14 K^{-1} (Apfel et al., 2006). A somewhat lower probability of internal stress development in ceramic parts can be achieved when making them from mixtures of coarse and fine powders of given materials.

All the problems and difficulties listed will arise precisely at the high working temperatures of conventional SOFCs. For this reason, many research groups working in different countries have tried over the last decade to develop new variants working at lower temperatures. This work is discussed in Sections 8.7 and 8.8.

8.4 MONOLITHIC SOFCs

A new design variant of the SOFC was developed at Argonne National Laboratory in the United States (Myles and McPheeters, 1990). This variant does not differ from conventional SOFCs, particularly the flat version, in the choice of materials to be used for the electrolyte (YSZ), cathode (LSM), anode (Ni cermet), or bipolar plates (lanthanum chromite). The design suggested is more compact than other versions, and hence was named monolithic. All cell components listed (the electrolyte, the electrodes, the bipolar plates) are corrugated in their "green" state (prior to being calcined) to a ∨∧∨∧-shape, calcined together, and then assembled to a structure reminiscent of a honeycomb (Figure 8.8).

This structure, on the one hand, provides the highest possible ratio of reacting surface area to total volume, and on the other hand, free access of the reactant gases to all segments of the working surface area. This structure is relatively thin-walled, yet very strong. The ohmic voltage losses in the individual parts of the cell are minimized as a result of these thin walls. For the same reason, the probability of the development of internal stresses during temperature changes is minimized (with the proviso of having selected materials with similar CTEs). A 60-kW battery consisting of such cells takes up a volume of about 60 L (2.12 ft^3). The specific power of 1 kW/L attained in this manner is the highest figure among all fuel cells.

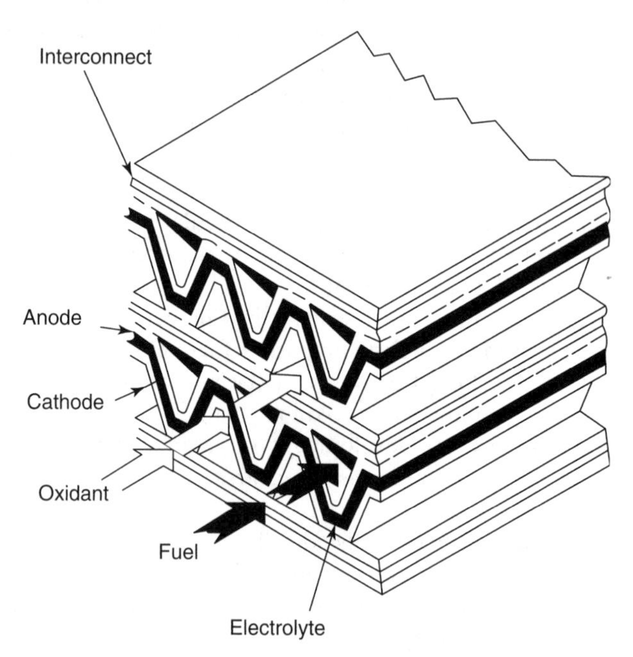

FIGURE 8.8 Monolithic design of SOFCs. (From Yamamoto, 2000, with permission from Elsevier.)

8.5 VARIETIES OF SOFCs

8.5.1 Single-Chambered SOFCs

In the design variant with a single chamber, the input streams of fuel and oxidant gases are not separate but enter the common reaction chamber together. A potential difference between the electrodes is able to develop since highly selective catalysts are used in them. The catalyst in the negative electrode reacts only with the reducing agent (reformed hydrocarbon, syngas), and oxidizes it electrochemically to the final product, while the catalyst in the positive electrode reacts only with oxygen (usually, air oxygen), reducing it to water. This design variant is much simpler than the design with separate reactant supply usually adopted. It is essential that the need for reliable heat-resistant sealing encountered with the usual designs does not exist here.

Yet this variant has the important defect of a lower fuel utilization efficiency, since parasitic side reactions not related to the generation of electrical energy may occur because of a direct chemical interaction between the reactants or because of a reaction of a reactant at the wrong electrode. In many applications, particularly in small power plants for mobile and portable equipment, a simple design and simple gas management are more important than a high fuel utilization efficiency. To a certain extent, the ratio of the rates of the current-generating and parasitic reactions can be adjusted by changing the amount of oxygen in the gas mixture entering the fuel cell compartment.

A single-chamber solid-oxide fuel cell (SC-SOFC) working with methane was described by Shao et al. (2006). In this cell the electrolyte was samaria-doped ceria (CSO), the anode was a Ni + CSO cermet, and the cathode material was a mixture of a perovskite-type compound $Ba_{0.5}Sm_{0.5}Co_{0.8}O_{3-\vartheta}$ with a CSO electrolyte. The fuel cell cubicle was placed into a furnace heated to 650°C, but a higher temperature was set up when the cell was operating. When a 87 : 75 gas mixture was fed to the cell, its OCV was 0.71 V and the limiting current was about 3.1 A/cm^2. In the point of maximum power, the specific power was as high as 760 mW/cm^2. This high value is evidence for the high activity of the BSCF cathode for oxygen reduction or for its extremely low activity for methane oxidation.

8.5.2 Direct-Flame SOFCs

In direct-flame SOFCs, part of the fuel gas is combusted in an ordinary burner with an open flame. When the combustible fuel-oxygen mixture is fuel-rich, then a partial fuel reforming will occur in the flame yielding hydrogen and carbon monoxide CO. The SOFC anode is set up in the immediate vicinity of the flame, within a few millimeters. There these reforming products are oxidized electrochemically. The surrounding air has immediate access to the cathode surface, located on the opposite side of the cell. The fuel cell itself is heated to its working temperature by the flame.

This version of a fuel cell has many of the advantages offered by the single-chamber cell described in the preceding section: simple design (gas compartments are practically absent) and no sealing problems. Another important advantage of this variant is its rapid startup, since heating the cell to a working temperature with the flame takes a matter of seconds. In contrast to other models, startup does not require an external heater. One more advantage is the possibility of using practically any type of gaseous, liquid, or solid fuel that contains carbon and/or hydrogen atoms. In contrast to the single-chamber variant described above, direct-flame SOFCs do not need selective catalysts on their electrodes.

Defects of this fuel cell variant are the relatively low fuel utilization efficiency and the high heat losses, just as in the case described above. Also, considerable mechanical stresses may arise in the fuel cell itself because of its considerable temperature gradients and because of the rapid changes in temperature.

Fuel cell performance was studied by Kronemayer et al. (2007) as a function of a number of factors: the fuel/air ratio in the gas mixture fed to the burner, the distance of the anode from the tip of the flame, and the temperature of the fuel cell itself. Methane, propane, and butane were used as fuels in these experiments. Values of the specific energy density of up to 120 mW/cm^2 were attained under optimum conditions.

8.5.3 Ammonia SOFCs

In recent years, ammonia NH_3 was suggested as a fuel for solid-oxide fuel cells by a number of workers (Maffei et al., 2005; Fournier et al., 2006). Ammonia is a large-scale product of the chemical industry. At temperatures above 500°C, ammonia is readily decomposed to nitrogen and hydrogen at nickel catalysts. Therefore, when ammonia is introduced into such a cell, it is completely converted to nitrogen and hydrogen at the nickel-containing anode, the hydrogen then undergoing electrochemical oxidation. This direct ammonia fuel cell is actually a direct internal ammonia-reforming fuel cell.

Relative to hydrogen, ammonia has many advantages as a primary fuel for fuel cells. It is much simpler to handle, store, and ship. It poses practically no explosion hazard. At room temperature it can be kept and moved as a liquid under a pressure of 8.6 bar. Unlike methanol, it does not need added steam for steam reforming. Unlike natural liquid fuels (petroleum products), its conversion to hydrogen is not attended by the formation of harmful contaminants (sulfur compounds), and charring of the catalyst will not occur in the fuel cells. The other product of ammonia decomposition, nitrogen, is completely harmless for the environment and for operation of the fuel cell. The specific energy content of ammonia is rather high, with a value of 1.45 kWh/L, yet ammonia is cheaper than many other types of fuel. The cost of ammonia producing 1 kWh of electrical energy is about $1.20, while for methanol and for pure (electrolytic) hydrogen, the cost figures are $3.80 and $25.40, respectively (Fournier et al., 2006).

8.6 UTILIZATION OF NATURAL FUELS IN SOFCs

8.6.1 Ways to Utilize Natural Types of Fuel

The major types of fuel for solid-oxide fuel cells (as the reactants being oxidized) are hydrogen and carbon monoxide. An important difference exists between these fuel cells and other types of fuel cells, in that various natural fuels or products of relatively simple processing of such fuels may also be utilized directly.

As we know, the original aim of all work on fuel cells has actually been precisely the direct transformation of the chemical energy of natural fuels to electrical energy. In seeking solutions to this problem of a direct utilization of natural fuels in fuel cells, workers have encountered numerous difficulties that in many cases could practically not be overcome. These difficulties were associated with the very low rates of electrochemical oxidation of these fuels and with the presence of various contaminants that hinder and sometimes block these reactions completely.

For this reason, the most realistic way of utilizing these natural fuels in a fuel cell includes their prior chemical (catalytic) conversion to other substances, primarily hydrogen, that are more readily oxidized electrochemically. In addition to conversion, the final processing product must also be freed carefully from all contaminants that could hinder the electrochemical reaction. These processes of conversion and purification are described in greater detail in Chapter 11.

In view of the chief aim—of attaining a highly efficient utilization of the fuel's chemical energy—it will be obvious that installations combining conversion units, purification units, and fuel cells will not be advantageous, apart from their very large space requirements. Also, fuel conversion processes are usually highly endothermal, so that appreciable thermal energies must be supplied, usually by burning part of the fuel. On the other hand, in fuel cell operation, a large amount of heat is evolved, which must be rejected to the environment by means of system of heat exchangers.

The conversion processes proceed at elevated temperatures close to those of SOFC operation. Therefore, an important aspect of the work on SOFCs has been the attempt to build unified plants combining conversion processes and fuel cell operation. In this way, the heat from the fuel cells could be transferred directly to converters, with a much lower loss of thermal energy. This combination has been called *internal reforming*.

As pointed out in Chapter 7, there are two ways of conducting internal reforming in high-temperature fuel cells: direct internal reforming, when the reforming occurs directly at the fuel cell electrodes, and indirect internal reforming, when the reforming is conducted in a separate unit located within the fuel cell stack. In SOFCs direct reforming is possible, since their nickel anodes are good reforming catalysts at the cell's operating temperature. However, two problems arise in the operation of such DIRSOFCs: (1) coking of the catalyst may occur, and (2) the catalyst may become poisoned by sulfur contaminants of the natural fuel. These two problems are considered in the following two sections.

8.6.2 The Problem of Carbon Formation

The formation of carbon deposits (coking) is a serious problem in SOFC operation when natural hydrocarbon fuels are used. Carbon deposits on the surface of nickel-containing anodic catalysts will drastically lower their activity. The formation of such deposits was studied by T. Kim et al. (2006). Two mechanisms exist. According to the first mechanism, the organic molecule that is the carbon source becomes adsorbed on the catalyst surface. While interacting with the metal, carbon atoms penetrate the metal, then evaporate from there to form fibrous deposits on the surface. This mechanism produces a dry (pitting-type) corrosion or "dusting" of the metals in contact with the hydrocarbons at high temperatures. According to the second mechanism, the organic molecules undergo pyrolysis in the gas phase, producing free radicals that polymerize to resinous substances. These block the pores of the anode and in this way interfere with reactant access to the reaction sites.

Coking of the catalyst is particularly pronounced when dry hydrocarbons are supplied to the fuel cells. A simultaneous supply of steam mixed into the hydrocarbons (e.g., in an H_2O/C ratio of 1.2) is the best way of fighting coking. Coking can also be prevented by using copper instead of nickel in the anodes. A copper surface is practically insensitive to all phenomena producing the coking, but copper is also not sufficiently active as a catalyst for the electrochemical oxidation of hydrogen and carbon monoxide. It has been suggested as a way of raising this activity, to add cerium oxide (CeO_2) to the anode mass as an active component (Gorte et al., 2002; Fuerte et al., 2007).

8.6.3 The Problem of Sulfur Compounds in Natural Fuels

All natural types of fuel contain a certain amount of sulfur compounds [sulfides (H_2S)]. The amount is relatively small in natural gas (less than 1 ppm), but in syngas and biogases as well as in diesel fuel there may be as many as 50 to 300 ppm. Sulfur compounds will poison catalysts consisting of nickel and many other materials. In SOFCs of the conventional type operated at temperatures of about 900°C, the poisoning may be reversible, but at lower working temperatures the catalysts are poisoned irreversibly, and lose their activity completely. In many cases, therefore, the gases entering a fuel cell must undergo desulfurization in special units.

Electrode materials exist (e.g., sulfide-based materials) which are insensitive to any sulfur compounds that may be present, but their catalytic activity for hydrogen oxidation is relatively low. Anodes made from copper–cerium oxide composites (Section 8.6.2) are quite active, yet will tolerate sulfur compounds to some extent (He et al., 2005). Quite recently, another anode material having such qualities has been developed; it is a binary perovskite ($Sr_2MgMoO_{6-\delta}$) (Huang et al., 2006). Aspects of the poisoning effects produced by sulfur compounds during SOFC operation have been considered in greater detail in a review by Gong et al. (2007).

8.7 INTERIM-TEMPERATURE SOFCs

8.7.1 The Problem of Lowering the Operating Temperature of SOFCs

The high working temperatures of solid-oxide fuel cells between 900 and 1000°C lead to numerous problems in the development, manufacture, and practical use of such fuel cells.

Materials Selection

A very limited number of materials exist that can be used to fabricate the various functional and structural components of SOFCs and that are sufficiently stable thermally and chemically to withstand their operating environment. As an example, take the interconnectors between cells for which mechanically weak lanthanum chromite ceramics or thermally stable high-alloy steels must be used. These materials are very expensive, so the interconnectors make a considerable contribution to the fuel cell's total cost. Numerous difficulties also arise in the selection of sealing materials.

Startup and Shutdown of Power Plants

Since numerous ceramic components that are sensitive to thermal shocks are used in SOFCs, brute force must not be used when heating these cells from ambient temperature to the high working temperatures during startup. The same holds for cooling down during shutdown of the power plant. This implies that a rather long time is required for these processes. This is not very important in large stationary power plants built for long-term uninterrupted operation, but constitutes an important factor in smaller units such as those intended for transport applications, where possible uses are limited by this factor.

Effects on the Lifetime of SOFCs

Various incidental processes leading to gradual degradation of fuel cells are accelerated as the temperature increases (see Section 8.9). This is true not only for corrosion processes but also for the diffusion of individual components from the electrodes and electrolyte into phases with which they are in contact, which lowers the conductivity of these phases. In addition, the likelihood of thermal shock and thermal gradients increases at high temperatures (e.g., when colder reactants hit hot segments of the electrodes), leading to mechanical stresses and possibly to cracking of electrodes and electrolyte.

Many of the problems listed could be eliminated if it were possible to lower the operating temperature of SOFCs without any significant loss of performance and without losing the valuable possibility of direct internal reforming of the hydrocarbon fuel. Therefore, starting in the 1990s, many research groups initiated concerted efforts to develop SOFCs operating at lower temperatures. The first aim was that of lowering the operating temperature from 900°C to at least temperatures in the range 600 to 700°C. Cells having such an operating temperature were called *interim-temperature SOFCs* (IT-SOFCs). Attempts to lower the operating temperature further, to temperatures below 600°C, were initiated a little later, approximately at the start of the twenty-first century. Cells with such operating temperatures were called *low-temperature SOFCs* (LT-SOFCs). The temperature boundary between these two categories is conditional and may differ somewhat for different authors.

The task of lowering the working temperature of SOFCs is made difficult by two facts: (1) the conductivity of the conventional, YSZ-type electrolyte drops off sharply with decreasing temperature; and (2) the rates of the electrochemical reactions occurring at the electrodes used in conventional SOFCs decrease with decreasing temperature, and the polarization of these electrodes (particularly that of the cathode) increases accordingly.

To some extent, the first factor could be overcome by using thinner electrolytes. The ohmic resistance of a layer of YSZ electrolyte less than 5 μm thick is not a limiting factor for designing fuel cells that have a large specific power. However, it is a rather difficult task to make defect-free electrolyte membranes so thin, yet sufficiently stable and reliable. For this reason, the search for SOFCs operable at lower temperatures has been primarily via the development of new types of material for the electrolyte and for electrodes that could work at these temperatures.

8.7.2 New Types of Solid Electrolytes

Materials that have relatively high ionic (oxide ion) conductivity in the temperature range considered have been found in numerous studies in the field of solid-state physics. Two such materials, doped ceria (CeO_2) and doped lanthanum gallate ($LaGaO_3$), have been particularly attractive for work on SOFC development. Figure 8.9 shows plots of log σT against $1/T$ for the electrical conductivity σ as a function of temperature for four electrolyte materials: yttrium-doped zirconium (YSZ), scandium-doped zirconia (SDC), gadolinium-doped ceria (CGO), and strontium-doped lanthanum magnesite gallate (LSGM). (Such coordinates are commonly used for such plots.) It can be seen from the figure that over the temperature range 1000 to 400°C, the last two electrolytes have a markedly higher conductivity than YSZ. It can also be seen from the figure that scandium-doped zirconia has a markedly higher conductivity than YSZ, but it is very rarely used, because of its high price.

Doped Ceria

For higher conductivity, cerium dioxide (CeO_2) can be doped either with gadolinium [$Ce_{1-x}Gd_xO_2$ (CGO)] or with samarium [$Ce_{1-x}Sm_xO_2$ (CSO)].

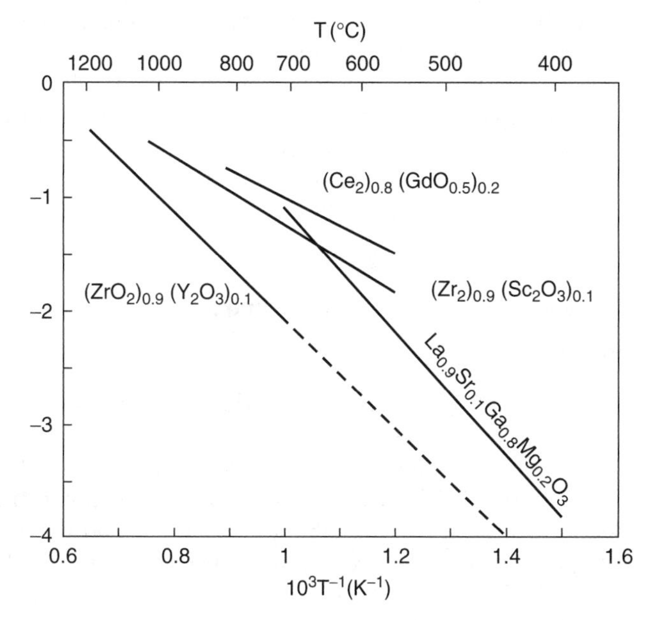

FIGURE 8.9 Temperature dependence of electrical conductivity for some solid oxide electrolytes: $(ZrO_2)_{0.9}(Y_2O_3)_{0.1}$ (YSZ), $(ZrO_2)_{0.9}(Sc_2O_3)_{0.1}$ (SDZ), $Ce_{0.8}Gd_{0.2}O_2$ (CGO), and $La_{0.9}Sr_{0.1}Ga_{0.8}Mg_{0.2}O_3$ (LSGM).

Doped cerias are quite stable chemically. In SOFCs, they lack the effect of interactions between the electrolyte and the cathode materials that would lead to the formation of poorly conducting compounds. However, doped cerias have an important defect, in that at low oxygen partial pressures (such as those existing close to the anode), they develop a marked electronic conduction. This is entirely inadmissible for the electrolyte, since it leads to internal self-discharge currents and even a complete internal short circuit. The electronic conduction comes about when Ce^{4+} ions in the lattice are partially reduced to Ce^{3+} ions and create a possibility of electrons hopping between ions of different valency.

Doped Lanthanum Gallate

Lanthanum gallate ($LaGaO_3$) is doped with divalent metal ions [e.g., strontium and/or magnesium: $La_{1-x}Sr_xMg_yGa_{1-y}O_3$ (LSGM); usually, x and y have values between 0 and 1]. Unlike ceria-based electrolytes, this electrolyte can also be used at low oxygen partial pressures without the menace of developing electronic conduction. When in contact with cathodes of the LSM type, some diffusion of manganese and/or cobalt from the cathode into the electrolyte is possible, but this has little effect on fuel cell performance. One defect of this electrolyte material is its complicated preparation.

In a review by Hui et al. (2007), various approaches to enhancing the ionic conductivity of zirconia- and ceria-based solid electrolytes in the light of composition, microstructure, and processing are described.

In a review by Fergus (2006), several new types of electrolytes for IT-SOFCs are compared. The focus is on conductivity, but other issues, such as compatibility with electrode materials, are also discussed.

8.7.3 New Types of Materials for the Electrodes

Cathode Materials

The conductivity of LSM, generally used as the cathode material in SOFCs decreases notably with decreasing temperature, the properties of the electrode and of the cell as a whole falling off accordingly. Doshi et al. (1999) reported the following data. If at 1000°C the contribution of cathode polarization to the cell's total apparent resistance R_{app} is about $1\,\Omega \cdot cm^2$, at 500°C this contribution has risen to $2000\,\Omega \cdot cm^2$. For such reasons, new types of cathode material were suggested for IT-SOFCs.

Shao and Haile (2004) suggested a material based on strontium-doped barium cobaltite ferrite [$BaFeO_3$: $Ba_{1-x}Sr_xCo_yFe_{1-y}O_3$ (BSCF) ($0.1 \leq x \leq 0.5$, $0.2 \leq y \leq 0.8$]. Electrodes made of this material exhibited very good properties. However, according to data reported by Wei et al. (2005), difficulties arise because of the complex temperature dependence of the thermal expansion coefficient. According to results obtained by Lim et al. (2007), this coefficient matches that of the electrolyte when $x = 0.2$.

In the work of Gong et al. (2006), an analogous material based on lanthanum cobaltite ferrite [$La_{0.6}Sr_{0.4}Co_{0.8}Fe_{0.2}O_3$ (LSCF)] was used.

Anode Materials

The anodes of nickel catalyst and of cermet with a YSZ electrolyte that are used in conventional SOFCs also lose their ability to work at lower temperatures, because of a loss of conductivity by the ceramic. This suggests that for a ceramic in the anode, a material having higher conductivity at intermediate temperatures should be used. It was, in fact, shown by Setoguchi et al. (1992) that an anode made with a nickel/samarium-doped ceria cermet has a much lower polarization than that of the conventional variant. It is very important to note that the electronic conduction appearing in this material in the anode compartment at low oxygen partial pressures (see Section 8.7.2) is helpful for the anode's function. Maric et al. (1998) used such an anode in IT-SOFCs working at a temperature of 800°C.

In recent reviews by Goodenough and Huang (2007) and Sun and Stimming (2007), advances in alternative anode materials for SOFCs are described.

8.7.4 Prototypes of IT-SOFCs

Gong et al. (2006) described the prototype of an intermediate-temperature solid-oxide fuel cell. In this cell a 1-mm-thick electrolyte of the LSGM ($La_{0.9}Sr_{0.1}Gd_{0.8}Mg_{0.2}O_3$) type was used. The anode was Ni–LDC ($Ce_{0.6}$-$La_{0.4}O_2$) cermet with a thickness of 30 to 40 μm and a porosity of 25 to 35%. Between the anode and the electrolyte, a thin layer (1.5 μm) of LDC was arranged to prevent direct interaction between the nickel anode and the electrolyte, which would lead to the formation of a poorly conducting lanthanum nickelate phase. The cathode was a layer of composite material (50 vol% each) of LSCF ($La_{0.6}Sr_{0.4}Co_{0.8}Fe_{0.2}O_3$) and LSGM of thickness 30 to 40 μm and with a porosity of 25 to 30%.

The cell worked with air and moistened (about 3%) hydrogen. The cell's OCV at 800°C was 1.118 V, which is rather close to the theoretical EMF value. At 800°C and a current density of 400 mA/cm^2, the cell voltage was 0.45 V. The maximum energy density was 190 mW/cm^2. At a lower temperature of 600°C and a current density of 60 mA/cm^2, the cell voltage was 0.5 V and the maximum energy density was 30 mW/cm^2. [These numbers are approximate, as they are taken from a figure reproduced in a paper by Gong et al. (2006).]

A cell with an LSGM-type electrolyte 200 μm thick and an LSCF-type cathode was described by Huijsmans et al. (1998). At a temperature of 700°C and a current density of 550 mA/cm^2, the cell voltage was 0.7 V.

8.8 LOW-TEMPERATURE SOFCs

Efforts to lower the operating temperature of SOFCs to values below 600°C have continued to the present. Two directions are followed in this work: reducing the thickness of electrolytes having higher conductivities at lower temperatures that were mentioned in Section 8.7.1, and developing new composite electrolytes having even higher conductivites.

8.8.1 Thin-Film LT-SOFCs

Tsai et al. (1997) reported building solid-oxide fuel cells with an electrolyte of yttrium-doped ceria that was 4 to 8 μm thick. A thin layer (1 to 1.5 μm) of a second electrolyte, of the YSZ type, was deposited between the electrolyte and the anode to eliminate the influence of electronic conductivity of the electrolyte. A cell having this bilayer electrolyte had an OCV of 98% of the theoretical EMF, which shows that there is no effect of the electronic component of conduction (without the second YSZ electrolyte, the OCV was only 50% of the theoretical value). At 600°C the electrolyte had a resistivity of 0.25 Ω · cm^2. At a current density of 400 mA/cm^2, the cell voltage was 0.4 V. The maximum energy density was 210 mW/cm^2. At 550°C, the resistivity of the electrolyte was 0.45 Ω · cm^2 and the cell voltage was 0.3 V at a current density of 300 mA/cm^2.

Doshi et al. (1999) used an electrolyte of ceria-doped gallate (CeGO) 30 μm thick. In the cell, a two-phase cathode 250 μm thick was used that consisted of a cermet of 45 to 55% silver and electrolyte material in the form of yttria-doped bismuth oxide (YDB). An Ni–CeGO cermet served as the anode. At a working temperature of 500°C and a cell voltage of 0.6 V, the current density was 100 mA/cm^2; at a cell voltage of 0.4 V, the current density was 300 mA/cm^2. It is interesting to note that in this cell, the contributions of the ohmic resistance of the electrolyte (0.7 $\Omega \cdot cm^2$) and of cathode polarization (0.6 $\Omega \cdot cm^2$) to the total apparent internal resistance of the cell, R_{app}, of 1.3 $\Omega \cdot cm^2$, were almost identical. The contribution of anode polarization was very small.

Another principle of building a cell with a thin electrolyte layer was used by Ito et al. (2005). An ultrathin layer of electrolyte (0.7 μm) of the type of $BaCe_{0.8}Y_{0.2}O_3$ was deposited onto a 40-μm palladium membrane that is hydrogen-permeable. It served as the anode, hydrogen was supplied from the side opposite to the electrolyte. A cathode layer was deposited on top of the electrolyte layer. At a cell voltage of 0.6 V, current densities of about 1.5 A/cm^2 were obtained at 400°C, and of 2.5 A/cm^2 at 600°C.

8.8.2 LT-SOFCs with Composite Electrolytes

In collaboration between groups of scientists at the Royal Institute of Technology, Stockholm, Sweden and the University of Science and Technology, Hefei, China, systematic investigations of composite electrolytes based on doped ceria and different nanosized, inorganic additives were carried out. It was shown in papers by Zhu et al. (1999, 2003) and Zhu (2001, 2003) that certain properties of the ceria electrolyte undergo substantial change in such composites. First, electronic conduction that is harmful to fuel cell operation is suppressed. Second, the conductivity increases dramatically at, both high and lower temperatures. Additives studied included both such obvious dielectrics as aluminum oxide (Al_2O_3) and lanthanum oxide (La_2O_3) and certain salts, such as NaCl and K_2CO_3, giving pronounced indications of protonic conduction. In many cases, stunning effects were seen. For instance, conductivities between 10^{-2} and 1 $S \cdot cm^{-1}$ (i.e., 10 to 100 times higher than those of the pure ceria electrolyte and 10^4 times higher than those of YSZ-type electrolytes) were obtained at temperatures between 400 and 600°C.

Test samples of LT-SOFCs having a composite electrolyte containing a mixture of lithium and sodium carbonates at a cell voltage of 0.5 V produced current densities of more than 2 A/cm^2 at 600°C, of about 1.25 A/cm^2 at 540°C of about 0.75 A/cm^2 at 500°C, and of about 0.5 A/cm^2 at 480°C) (from data of Zhu (2003, Fig. 7).

A possible explanation for the sharp increase in conductivity of the ceria electrolytes that occurs in contact with the proton-conducting alkali–metal carbonates or with nonconducting oxides was offered in a paper by Zhu et al. (2003). Adsorption of an excess number of oxygen ions (O^{2-}) is possible at the contact surface areas between the nanoparticles of the two phases. In this way a new, additional interfacial conduction pathway is opened up for the ionic current.

Similar views concerning the mechanism of conductivity enhancement in ceria electrolytes were expressed in a paper by Huang et al. (2005), where it was pointed out the interaction between the flows of negative oxygen ions and positive protons directed in opposite directions.

A similar two-phase electrolyte which in addition to the ceria electrolyte contained powdered nickel–aluminum alloy to which NaOH had been added was suggested by Hu et al. (2006). This electrolyte had the distinguishing feature of being solid at room temperature but had a liquid phase that appeared in it at working temperatures of the fuel cell above 318°C. In this case again, mixed conduction by O^{2-} and H^+ appears in the solid phase, and by Na^+ and OH^- in the liquid phase.

The work on composite electrolytes mentioned in this section is of great interest and may bring SOFCs closer to commercialization.

8.9 FACTORS INFLUENCING THE LIFETIME OF SOFCs

In Section 8.2.1 we pointed out that a power plant with a battery of tubular SOFCs had been operated more or less successfully for about 16,000 hours. Apart from the communication cited, very few data can be found in the literature as to the results of long-term testing of solid-oxide fuel cells. Since this type of high-temperature fuel cell is intended primarily for large stationary power plants and thus has large investment needs, information as to potential lifetimes and reasons for gradual performance degradation or possible cases of sudden failure are extremely important.

In Section 8.3, fundamental problems that arise in the development of planar solid-oxide fuel cells have been listed: sealing, corrosion of the bipolar plates, and the development of mechanical stresses in the numerous ceramic components found in such fuel cells. Solutions depend not only on the selection of suitable materials for the individual component parts of the fuel cells but also on respecting certain principles of design and operation.

Thermal shocks have a highly negative effect on lifetime. These are sudden temperature changes in individual segments of a fuel cell. The development of considerable temperature gradients within given segments is also detrimental.

In fuel cell operation, considerable heat is set free. In SOFCs with internal fuel conversion, part of the heat evolved is absorbed by the endothermal reforming reaction. Yet for the elimination of excess heat, artificial cooling is usually required. In power plants with SOFCs, this is often achieved by blowing large amounts of air (large relative to the stoichiometric needs of the reaction). Owing to the temperature gradients that develop, the point of entry of the cold air into the hot fuel cell is very vulnerable. Various schemes to preheat the incoming air with heat from the exhaust gases have been developed (Costamagna, 1997). The same problem arises when feeding fuel and the components (water or steam) needed for fuel reforming into the fuel cell. According to reaction (1.17) heat evolution is stronger the lower the operating voltage of the fuel cell. From this point of view, it will be preferable to work not with a cell voltage of 0.7 V, which is often used, but rather, with cell voltages not lower than 0.75 to 0.80 V and lower currents. This will not only lower the influence of thermal shocks but will also lead to large savings in the energy needed to operate the cooling system (pumps, ventilators, etc.).

Another important aspect is that of thermal management during startup and shutdown of a power plant. This problem has been discussed in detail by Apfel et al. (2006).

Various processes taking place in the electrodes and in the electrolyte of a fuel cell also influence the lifetime of SOFCs, in addition to the factors already mentioned. As in other types of fuel cells, the active working surface area of highly disperse metal catalysts has a tendency to shrink gradually with time. This decrease leads to a gradual increase in polarization of the corresponding electrode.

The phenomenon is seen in particular in the highly disperse nickel used in the anodes of most SOFC variants. Tomita et al. (2005) could show that one of the reasons for the degradation of Ni–YSZ anodes during long-term operation are changes in their volume occurring during the periodic shutdowns and startups of the cells always attended by sharp temperature changes. Jørgensen et al. (2000) showed that the degradation of an LSM cathode is due not so much to its being maintained for a certain length of time at elevated temperature but to its actively working during this time (under current load).

A serious problem arising during SOFC operation is the interaction between the materials of electrodes and electrolyte by diffusion of individual components from a given phase to a neighboring phase in contact with it. This interaction often gives rise to the formation of new phases or compounds having a low conductivity. Many of the factors influencing SOFC lifetime that are associated with aging and degradation of these cells have been discussed in a review by Tu and Stimming (2004).

REFERENCES

Apfel H., M. Rzepka, H. Tu, U. Stimming, *J. Power Sources*, **154**, 370 (2006).

Costamagna P., *J. Power Sources*, **69**, 1 (1997).

Doshi R., V. L. Richards, J. D. Carter, et al., *J. Electrochem. Soc.*, **146**, 1273 (1999).

Fischer W., J. Malzbender, G. Blass, R. W. Steinbrech, *J. Power Sources*, **150**, 73 (2005).

Fournier G. G. M., I. W. Cumming, K. Hellgardt, *J. Power Sources*, **162**, 198 (2006).

Fuerte A., R. X. Valenzuela, L. Daza, *J. Power Sources*, **169**, 47 (2007).

George R. A., *J. Power Sources*, **86**, 134 (2000).

Gong W., S. Gopalan, U. B. Pal, *J. Power Sources*, **160**, 305 (2006).

Gorte R. J., H. Kim, J. M. Vohs, *J. Power Sources*, **106**, 10 (2002).

He H., R. J. Gorte, J. M. Vohs, *Electrochem Solid-State Lett.*, **8**, A279 (2005).

Hilpert K., D. Dos, M. Miller, et al., *J. Electrochem. Soc.*, **143**, 3643 (1996).

Hu J., S. Tosto, Z. Guo, Y. Wang, *J. Power Sources*, **154**, 106 (2006).

Huang J., Z. Mao, L. Yang, R. Peng, *Electrochem. Solid-State Lett.*, **8**, A440 (2005).

Huang Y.-H., R. I. Dass, J. C. Denyszyn, J. B. Goodenough, *J. Electrochem. Soc.*, **153**, A1266 (2006).

Huijsmans J. P. P., F. P. F. van Berkel, G. M. Christie, *J. Power Sources*, **71**, 107 (1998).

Ito N., M. Iijima, K. Kimura, S. Iguchi, *J. Power Sources*, **152**, 200 (2005).

Jørgensen M. J., P. Holtappels, C. C. Appel, *J. Appl. Electrochem.*, **30**, 411 (2000).

Kim J.-H., R.-H. Song, K.-S. Song, et al., *J. Power Sources*, **122**, 138 (2003).

Kim T., G. Liu, M. Boaro, et al., *J. Power Sources*, **155**, 231 (2006).

Kronemayer H., D. Barzan, M. Horiuchi, et al., *J. Power Sources*, **166**, 120 (2007).

Lim Y. H., J. Lee, J. S. Yoon, et al., *J. Power Sources*, **171**, 79 (2007).

Liu Q. L., K. A. Khor, S. H. Chan, *J. Power Sources*, **161**, 123 (2006).

Maffei N., L. Pelletier, J. P. Charland, A. McFarlan, *J. Power Sources*, **140**, 264 (2005).

Maric R., S. Ohara, T. Fukui, *Electrochem. Solid-State Lett.*, **1**, 201 (1998).

Myles K. M., C. C. McPheeters, *J. Power Sources*, **29**, 311 (1990).

Sammes N. M., Y. Du, R. Bove, *J. Power Sources*, **145**, 428 (2005).

Setoguchi T., K. Okamoto, K. Eguchi, H. Arai, *J. Electrochem. Soc.*, **139**, 2875 (1992).

Shao Z. P., S. M. Haile, *Nature* (*London*), **431**, 170 (2004).

Shao Z. P., J. Mederos, W. C. Chueh, S. M. Haile, *J. Power Sources*, **162**, 589 (2006).

Stanislowski M., J. Froitzheim, L. Niewolak, et al., *J. Power Sources*, **164**, 578 (2007).

Suzuki T., T. Yamaguchi, Y. Fujishiro, M. Awano, *J. Power Sources*, **160**, 73 (2006).

Tomita A., T. Hibino, M. Sano, *Electrochem. Solid-State Lett.*, **8**, A333 (2005).

Tsai T., E. Perry, S. Barnett, *J. Electrochem. Soc*, **144**, L130 (1997).

Wei B., Z. Lü, S. Li, et al., *Electrochem. Solid-State Lett.*, **8**, A428 (2005).

Weissbart J., R. Ruka, *J. Electrochem. Soc.*, **109**, 723 (1962).

Yakabe H., T. Ogiwara, M. Hishinuma, I. Yasuda, *J. Power Sources*, **102**, 144 (2001).

Zhu B., *J. Power Sources*, **93**, 82 (2001); **114**, 1 (2003).

Zhu B., Q. Meng, B.-E. Mellander, *J. Power Sources*, **79**, 30 (1999).

Zhu B., X. T. Yang, J. Xu, et al., *J. Power Sources*, **118**, 47 (2003).

Reviews and Monographs

Fergus J. F., Sealants for solid oxide fuel cells, *J. Power Sources*, **147**, 46 (2005).

Fergus J. F., Electrolytes for solid oxide fuel cells, *J. Power Sources*, **162**, 30 (2006).

Gong M., X. Liu, J. Trembly, C. Johnson, Sulfur-tolerant materials for solid oxide fuel cell application, *J. Power Sources*, **168**, 289 (2007).

Goodenough J. B., Y.-H. Huang, Alternative anode materials for solid oxide fuel cells, *J. Power Sources*, **173**, 1 (2007).

Hui R., J. Roller, S. Yick, et al., A brief review of the ionic conductivity enhancement for selected oxide electrolytes, *J. Power Sources*, **172**, 493 (2007).

Hui R., Z. Wang, O. Kesler, et al., Thermal plasma spraying for SOFCs: Applications, potential advantages, and challenges, *J. Power Sources*, **170**, 308 (2007).

Liu Z., M.-F. Han, W.-T. Miao, Preparation and characterization of graded cathode $La_{0.6}Sr_{0.4}Co_{0.2}Fe_{0.8}O_{3-\delta}$, *J. Power Sources*, **173**, 837 (2007).).

Riley B., Solid oxide fuel cells: the next stage, *J. Power Sources*, **29**, 223 (1990).

Singhal S. C., K. Kendall, *High Temperature Solid Oxide Fuel Cells*, Elsevier, Amsterdam, 2003.

Sun Ch., U. Stimming, Recent anode advances in solid oxide fuel cells, *J. Power Sources*, **171**, 247 (2007).

Tu H., U. Stimming, Advances, aging mechanisms and lifetime in solid-oxide fuel cells, *J. Power Sources*, **127**, 284 (2004).

Yamamoto O., Solid oxide fuel cells: fundamental aspects and prospects, *Electrochim. Acta*, **45**, 2423 (2000).

CHAPTER 9

OTHER TYPES OF FUEL CELLS

9.1 REDOX FLOW CELLS

By definition, redox flow cells are a type of fuel cell set up in loops of circulating electrolyte solutions containing reversible redox systems. By design they are intended for a temporary accumulation of electrical energy that they can give off later to a user. In this way they fill the function of an ordinary electrical storage battery. Unlike fuel cells, they produce the "fuel" themselves, by accepting power (charging up their redox systems). Like fuel cells, they produce power as long as their redox systems are pumped through and give off charge to the electrodes. As in ordinary fuel cells, the fuel (the redox solutions) is stored outside, and the fuel reserve (their capacity, as it were) can be as large as feasible in the overall context. The redox reactions occur at suitable inert electrodes.

9.1.1 Iron–Chromium Redox Flow Cells

Redox flow cells working with redox systems consisting of iron and chromium ions are the simplest representatives of redox-type fuel cells and constitute a convenient example for describing their working principles. A cell consists of two halves separated by an ion-exchange membrane. The positive electrode (or positive half-cell) resides in one half and the negative electrode (or negative half-cell) resides in the other half. An electrolyte solution containing divalent

Fuel Cells: Problems and Solutions, By Vladimir S. Bagotsky

and trivalent iron ions (i.e., the ions Fe^{2+} and Fe^{3+}) flows through the positive half-cell. When the cell delivers charge, the cathodic reaction occurring at the positive electrode is

$$Fe^{3+} + e^- \rightarrow Fe^{2+} \qquad E^0 = 0.77\,V \tag{9.1}$$

When the cell is recharged, this reaction occurs in the opposite (anodic) direction.

An electrolyte solution containing divalent and trivalent chromium ions (i.e., the ions Cr^{3+} and Cr^{2+}) flows through the negative half-cell. When the cell delivers charge, the anodic reaction occurring at the negative electrode is

$$Cr^{2+} \rightarrow Cr^{3+} + e^- \qquad E^0 = -0.41V \tag{9.2}$$

When the cell is recharged, this reaction occurs in the opposite (cathodic) direction. The two reactions occur simultaneously when the external circuit is closed. The overall current-generating reaction that occurs when the cell delivers charge is given by

$$Fe^{3+} + Cr^{2+} \rightarrow Fe^{2+} + Cr^{3+} \qquad \mathcal{E}^0 = 1.18\,V \tag{9.3}$$

When the cell is recharged, it occurs in the opposite direction. The cathode and anode where reactions (9.1) and (9.2) take place are inert but "catalytically active" (i.e., selected so that the reactions will be fast).

The anion-exchange (polymer) membrane separating the two half-cells contains positive functional groups (most often the cations, NR_4^+, of quaternary alkylammonium compounds). These positive groups electrostatically repel the positive iron and chromium ions, preventing their crossover. It is therefore not possible for these ions to get to the "wrong" electrode, and their direct chemical interaction is excluded as well. No hindrance exists for the anions of the iron and chromium salts to pass through the membrane. Therefore, each elementary act of reaction (9.3) is then attended by transfer of an electron through the external circuit from the negative to the positive electrode, and also by transfer of one negative charge in the form of anions through the membrane (through the internal circuit) from the positive to the negative half-cell or electrode to make reactions (9.1) and (9.2) possible and close the electrical circuit.

The equilibrium potentials set up by the iron and chromium redox systems at the inert electrodes depend on the concentration ratios of the divalent $[M^{2+}]$ and trivalent $[M^{3+}]$ ions (M standing for Fe and Cr, respectively), according to a Nernst equation:

$$E_M = E_M^0 + \frac{RT}{nF} \ln \frac{[M^{3+}]}{[M^{2+}]} \tag{9.4}$$

When divalent and trivalent ions are present in identical concentrations, $E_M = E_M^0$. The EMF of the cell, $\mathcal{E} = E_{Fe} - E_{Cr}$, depends on the four ion concentrations through two equations (9.4).

When the cell delivers charge, the Fe^{3+} ion concentration decreases while the Cr^{3+} ion concentration increases. Accordingly, the potential of the positive electrode becomes more negative, that of the negative electrode becomes more positive, and the cell voltage $U_{discharge}$ decreases gradually. When the cell is recharged, the Fe^{3+} ion concentration increases, the Cr^{3+} concentration decreases, the potentials of the electrodes move in the opposite directions, and the cell voltage U_{charge} increases gradually.

In iron–chromium redox flow cells, carbon materials are commonly used as the (inert) electrodes (carbon fiber cloth, felt, etc.). Johnson and Reid (1985) suggested depositing traces of lead and gold ($\mu g/cm^2$) on the carbon material of the electrode in the chromium redox system (the "chromium electrode") in order to accelerate the electrode reaction. The solution for the positive half-cell usually contains a certain concentration of hydrochloric acid (HCl) in addition to $FeCl_3$ and $FeCl_2$.

Many investigations into iron–chromium redox flow cells were carried out by a research group at Alicante University, Spain. Lopez-Atalaya et al. (1992) optimized the operating conditions (electrolyte composition, temperature, etc.) of a battery of this type. At a temperature of 44°C with 2.3 M HCl + 1.25 M $FeCl_2$ + 1.25 M $CrCl_3$ solution, they attained a maximum specific power of 73 mW/cm^2. Codina et al. (1994) reported building a 20-cell battery delivering 0.1 kW of power. Murthy and Srivastava (1989) and later Wen et al. (2006) tried to see whether the electrical performance of such redox flow cells could be improved by adding various complexing agents to the solution. NASA considered using such batteries for spacecraft.

Redox flow cells have certain advantages over ordinary electrical storage batteries such as the lead-acid batteries: (1) the electrode reactions are relatively simple; (2) the energy losses due to polarization of the electrodes are low; (3) aging phenomena at the electrodes do not exist, hence the lifetime is long; (4) the electrical capacity and energy content of the system can be increased simply by using larger redox reactant solution volumes (and related storage tanks); and (5) no gas is evolved during overdischarge or overcharge. A certain disadvantage of redox flow cells is the need to spend part of the energy, both when delivering power and when recharging, to operate the pumps to move the solutions containing the redox systems.

9.1.2 All-Vanadium Redox Flow Cells

In vanadium redox flow cells, a redox system of penta- and tetravalent vanadium ions is used in the positive half-cell, and a redox system of di- and trivalent vanadium ions is used in the negative half-cell. When the cell delivers charge, the following reactions take place:

$$(+): \quad VO_2^+ + 2H^+ + e^- \rightarrow VO^{2+} + H_2O \qquad E^0 = 1.00V \tag{9.5}$$

$$(-): \quad V^{2+} \rightarrow V^{3+} + e^- \qquad E^0 = -0.26V \tag{9.6}$$

$$\text{Overall}: \quad VO_2^+ + V^{2+} + 2H^+ \rightarrow VO^{2+} + V^{3+} + H_2O \qquad \mathscr{E}^0 = 1.26\,V \tag{9.7}$$

During recharging, the reactions occur in the opposite direction. As before, the values cited for the thermodynamic parameters of the reactions refer to the situation where the concentrations of the vanadium ions in the higher and lower oxidation states are equal.

A large program of investigations into such vanadium redox flow cells was conducted between 1985 and 2003 by workers at the University of New South Wales in Kensington, Australia. Battery prototypes with 1 kW power were developed (Skyllas-Kazacos et al., 1991). In the two half-cells, solutions 1.5 to 2 M in vanadium sulfate and 2.6 M in sulfuric acid were used. Carbon felt was the electrode material. A battery consisting of 10 cells in series could be cycled between voltages of 8 V (the terminal discharge voltage) and 17 V (the terminal charging voltage). At currents for discharge and charging of 20 A, the faradaic efficiency was 92.6%, the voltage efficiency 95%, and the overall energy efficiency 88%. At a discharge current of 120 A and a charging current of 45 A, the voltage efficiency dropped to 73%, the energy efficiency to 72%. For electrochemical energy storage, such performance figures must be acknowledged as being quite high.

Joerissen et al. (2004) made a detailed technical and economic analysis of the potential of vanadium redox flow batteries in various low-power energy systems in which the primary source of electrical energy would be solar batteries and wind power generators, both highly variable. The authors noted as a drawback in long-term operation of the battery at elevated temperatures that part of the pentavalent vanadium may precipitate as insoluble oxide (V_2O_5).

More detailed data about the two systems described above and a number of other redox flow cells may be found in a detailed review by Ponce de León (2006).

9.2 BIOLOGICAL FUEL CELLS

Biological fuel cells, or semi-fuel cells (see Section 9.3), are cells in which at least one of the following two conditions is met:

1. At one of the electrodes at least, the electrochemical reactant is a substance found in biological fluids (e.g., in blood) or in other biological materials (e.g., in biomasses).
2. At one of the electrodes at least, the catalyst for the electrochemical reaction are microorganisms (microbial fuel cells) or enzymes (enzymatic fuel cells).

At present, biological fuel cells are a subject of great interest for the following two reasons:

1. The range of medical devices being implanted into the human body, such as pacemakers, defibrillators, insulin micropumps, and analyzers, has

been increasing in recent years. A time will soon come when artificial hearts and kidneys will be implanted. All these devices need electrical energy. Implanted batteries have a limited capacity and must be replaced periodically, which requires renewed surgical intervention. Long, uninterrupted operation of such devices could become possible if implantable fuel cells could be built that use products of the metabolism as the reducing agent, and oxygen of the hemoglobin as the oxidizing agent.

2. While energy resources are becoming really scarce in many countries, huge energy reserves in the form of biowaste (i.e., biomass) accumulate everywhere. Existing ways of utilizing them, such as, by enzymatic transformation to ethanol and subsequent use of the ethanol in heat engines, are complicated and involve many individual steps. Direct processing of these biomasses in biological fuel cells would lead to significantly more efficient utilization of their energy content.

Definite solutions to attaining these goals are still far away. So far, research has focused on individual, relatively narrow aspects of the problems.

9.2.1 Enzymatic Fuel Cells

Enzymes are natural very active and efficient catalysts for the chemical reactions that occur in the bodies of humans and animals. The distinguishing feature of enzymes is their high selectivity. A given enzyme is active in the reaction of only one particular reactant, the *substrate*. The reactions occurring in the body are homogeneous, proceeding in aqueous (physiological) solutions. A number of attempts have been made in recent decades to use enzymes for the catalytic acceleration of electrochemical reactions (bioelectrocatalysis). To this end, enzymes are immobilized on the surface of supporting electrodes (usually, a carbon support). Special intermediate redox systems (mediators) serving as shuttles between the enzyme and the substrate are used to secure electron transfer from the enzyme to the substrate (or vice versa). As a rule, the same ambient conditions that are found in biological systems should be maintained in electrochemical systems, such as a pH value of about 7 and a moderate temperature.

Topcagic and Minteer (2006) reported building a fuel cell that worked using simple reactants, ethanol and oxygen, but where the platinum catalysts that had commonly been used were replaced by enzymes.

Alcohol dehydrogenase and aldehyde dehydrogenase were used as the enzymes for anodic ethanol oxidation, together with coenzyme NAD^+ (nicotinamide-adenine dinucleotide) forming the $NAD^+/NADH$ redox system. Aided by these enzymes, this reaction occurs in two steps. First, the former enzyme oxidizes the ethanol to acetaldehyde, then the latter enzyme oxidizes the aldehyde to acetic acid.

Bilirubin and bilirubin oxidase were used for cathodic oxygen reduction, and $Ru(bpy)_3^{3+}/Ru(bpy)_3^{2+}$ acted as the mediator redox system. In the electrodes, these enzymes were immobilized using a Nafion solution treated with quaternary ammonium salts and put on a support of carbonized cloth serving as the current collector. The treated Nafion helped to maintain enzyme activity for a long time.

As the enzymes are highly selective, one can use these electrodes in two types of fuel cells: (1) cells with cathode and anode compartments separated by an ion-exchange membrane and individual reactant supplies; or (2) cells without a separator where the reactants are added as a mixture.

The membrane version had an OCV of 0.68 V. The highest specific power achieved was 83 mW/cm^2. In the undivided cell, a buffer solution of pH 7.15 containing 1.0 mM ethanol and 10 mM NAD$^+$ was used. The solution was kept in contact with air, and after a number of days had the equilibrium concentration of dissolved oxygen. The starting OCV value was 0.51 V. The highest energy density attained was 0.39 mW/cm^2. The cell worked for more than 30 days, at which point the specific power had decreased by 20%. The reason was insufficient chemical stability of the redox system used. The enzyme systems themselves retained their activity for more than 90 days. Other workers (Gao et al., 2007) have reported much lower values of the maximum specific power attained (i.e. values of 30 to 100 μW/cm^2).

Results from investigations of electrochemical reactions occurring at individual electrodes with enzyme catalysts have been reported in many older papers. Yaropolov et al. (1976) studied cathodic oxygen reduction at a pyrolytic graphite electrode with peroxidase as an enzyme. As the mediator of electron transfer, a quinone–hydroquinone redox system was selected. The electrode's polarization decreased by 30 mV when the enzyme and mediator were present in the solution. Betso et al. (1971) observed an increase in the rate of oxygen reduction at a mercury electrode in the presence of cytochrome *C*. Wingard et al. (1971) saw an increase in the current of glucose oxidation at a platinum electrode when glucose oxidase was present as an enzyme. The influence of a number of structure factors (including factors related to the orientation of the enzyme molecules on the disperse carbon support) has been studied by Tarasevich et al. (2005).

One of the properties of enzyme systems that is of great interest for fuel cells in general is their high selectivity. If the universal but absolutely nonselective platinum catalysts could be replaced by other, highly selective catalysts, fuel cells with a mixed reactant supply that have considerably higher technical and economic efficiency could be built. This problem is discussed in Chapter 18.

9.2.2 Bacterial Fuel Cells

Bacterial biofuel cells are much less selective than enzyme fuel cells and can be used for reactions with a great variety of reactants. A large group of microorganisms can be used for the oxidation of any carbohydrate compound in biomasses. Most often, a mixture of different types of microorganisms is used rather than a particular microorganism. In some papers on bacterial fuel cells, it was reported that samples of microorganisms were scooped up from ponds close to the laboratory (Zhang et al., 2006).

In their mechanism and nature of their action, however, bacterial and enzyme fuel cells have much in common. In bacterial fuel cells, intermediate redox systems are often used to facilitate electron transfer to (or from) the substrate, but as the effect of microorganisms is much less specific than that of enzymes, a much wider selection of redox systems can be used, down to the simplest Fe^{3+}/Fe^{2+} system. The working conditions of these two types of biofuel cells are also similar: a solution pH around 7 and moderate temperatures close to room temperature.

Tests involving a bacterial fuel cell with anode and cathode compartments separated by a cation-exchange membrane were reported by Zhang et al. (2006). The compartments were filled with phosphate buffer solution at pH 7. In addition, the anode compartment was made 20 mM in acetate or 10 mM in glucose. The cathode compartment was 50 mM in the salt, $K_3Fe(CN)_6$. Carbonized cloth carefully freed of metallic contaminants was used as the electrode material. Air was passed through the cathode compartment during the tests. A culture of the pond microorganisms was grown in standard nutrient solutions and added to the compartments prior to the tests.

The OCV in the cell with glucose was about 0.55 V. At a load of 1000 Ω, the voltage was about 0.34 V (the maximum power was about 5 $\mu W/cm^2$). Figures of the same order of magnitude were obtained in the cell with acetate. It is interesting to note that by electron microscopy, a morphological difference between the bacterial beds was noted on the anodes after the tests with glucose and acetate.

The effects of common baker's yeast were studied by Walker and Walker (2006). Here again, the cell was divided into two halves by an ion-exchange membrane. The anodic part of the cell contained phosphate buffer solution with the yeast *Saccharomyces cerevisiae*, 0.1 M $K_3Fe(CN)_6$, and 0.2 M glucose. The cathodic part contained a mediator solution of the Fe^{3+}/Fe^{2+} type. When charge is drawn, the Fe^{3+} ions are reduced to Fe^{2+} ions; these are, in turn, regenerated by the oxygen brought into this part. At a temperature of 45°C, the maximum energy density was 130 $\mu W/cm^2$. At 10°C this value dropped to less than 20 $\mu W/cm^2$.

Another version of a bacterial fuel cell that worked with sugar industry effluents was described by Prasada et al. (2006). A culture of *Clostridium sporogenes* was used in the anodic part of the cell, and a culture of *Thiobacillus ferrooxidans* was used in the cathodic part. The OCV was 0.83 V and the maximum energy density was about 4 $\mu W/cm^2$.

9.3 SEMI-FUEL CELLS

Semi-fuel cells is an unfortunate name sometimes given to electrochemical power sources (galvanic cells for single use or rechargeable storage batteries) where one of the electrodes is reacting (is consumed) while the other, functioning like a fuel cell electrode, will work as long as it is supplied with a reactant.

9.3.1 Metal–Air Cells

The best known example of a semi-fuel cell is the disposable galvanic zinc–air cell widely used in hearing aids and other small electronic devices. Metallic zinc is the negative electrode in these cells (usually in the form of a highly disperse powder). When current is drawn, the zinc dissolves anodically in a concentrated alkaline solution according to the equation

$$(-): \quad Zn + 2OH^- \rightarrow ZnO + H_2O + 2e^- \qquad E^0 = -1.254\,V \tag{9.8}$$

Electrodes built according to the principles of the oxygen (air) electrodes of alkaline fuel cells are used as the positive electrode in these cells:

$$(+): \quad \tfrac{1}{2}O_2 + H_2O + 2e^- \rightarrow 2OH^- \qquad E^0 = 0.40\,V \tag{9.9}$$

$$\text{Overall:} \quad Zn + \tfrac{1}{2}O_2 \rightarrow ZnO \qquad \mathscr{E}^0 = 1.654\,V \tag{9.10}$$

Since the currents drawn from the small zinc–air cells in hearing aids are very low, it is not necessary to use expensive catalysts in the air electrodes of these cells. The catalytic activity of activated carbon is quite sufficient to sustain these loads. These cells are designed for a moderate service life (several weeks), so the risk of carbonation of the alkaline electrolyte solution by traces of CO_2 from the air is not large. The hole providing access of air to the electrode is closed off with adhesive tape or paper that is removed prior to using the cell, to provide protection against carbonation during prolonged storage.

When the supply of metallic zinc is used up, the zinc–air cell stops working and cannot be reanimated. Repeated attempts have been made to build rechargeable zinc–air cells (zinc–air storage cells). The reactions reported above are basically revertible, that is, could also occur in the opposite (charging) direction. For a number of reasons associated with both the zinc and air electrodes such rechargeable zinc–air cells are very unreliable and have so far not found any practical application.

Another metal–air power source is the iron–air cell. The reactions occurring in these cells are analogous to those occurring in zinc–air cells, the only difference being that the iron oxides Fe_2O_3 and/or Fe_3O_4 are the final product of anodic iron oxidation. This means that this cell is based on the well-known iron corrosion reaction and may be assumed to be the cheapest type of electrochemical power source. Earlier, iron–air cells with rather massive iron and carbon–air electrodes had been built. The electrical capacity (and size) of these cells was relatively large (tens or hundreds of amperehours). These cells were very simple to maintain. In the former Soviet Union, such cells were used as a power supply to unmanned installations such as railroad signaling equipment.

9.3.2 Nickel–Hydrogen Storage Cells

Nickel–hydrogen storage cells were developed in the Soviet Union in 1964. There and in a number of other countries, a small production output of batteries including such storage cells was started to satisfy the needs of space technology. The positive nickel oxide electrodes used in the very common alkaline (nickel–cadmium) storage batteries were used in these batteries. Hydrogen electrodes from alkaline fuel cells, which in these years had already found some applications were used as negative electrodes. When current is, drawn (during discharge), the following reactions occur in these storage batteries:

$$(+): \quad NiOOH + H_2O + e^- \rightarrow Ni(OH)_2 + OH^- \qquad E^0 = 0.49\,V \tag{9.11}$$

$$(-): \quad \tfrac{1}{2}H_2 + OH^- \rightarrow H_2O + e^- \qquad E^0 = -0.828\,V \tag{9.12}$$

$$\text{Overall}: \quad NiOOH + \tfrac{1}{2}H_2 \rightarrow Ni(OH)_2 \qquad \mathscr{E}^0 = 1.22\,V \tag{9.13}$$

During charging, reactions occur in the opposite direction. Therefore, hydrogen gas is evolved at the negative electrode during charging, is recovered into tanks, and is kept there (usually under higher pressure) until needed for discharge.

Such storage batteries became feasible when it was realized that gaseous hydrogen is almost completely inert (unreactive) toward trivalent nickel oxide. It was therefore possible to place a block of charged positive electrodes into the hydrogen tank, greatly simplifying battery design. A very convenient aspect in the use of these batteries was the discovery that from the hydrogen pressure in the tank, the state of charging of the battery could be estimated. The number of discharge–recharge cycles sustained is exceptionally high. Batteries of this type are used in the ISS and in NASA's Mars Global Surveyor. For general applications, these batteries were later replaced by nickel hydride storage batteries, which are much easier to handle.

In the literature on electrochemical power sources, "semi-fuel cells" are generally regarded as a variety of ordinary batteries (galvanic cells or storage batteries), rather than as a variety of fuel cells (which have the distinguishing feature of a continuous supply of all reactants). In this section, brief information was given on these "half-fuel cells" to show the connections between their development and the development of "real" fuel cells.

9.4 DIRECT CARBON FUEL CELLS

9.4.1 Purpose and First Efforts

When toward the end of the nineteenth century Wilhelm Ostwald formulated his idea of using an electrochemical mechanism for direct conversion of the chemical energy of natural fuels to electrical energy, coal was the chief fuel in human hands. Even today the widespread use of petroleum products and the development of nuclear power notwithstanding, coal remains a very important component of world energy supply. Its share of all known natural fuel reserves worldwide is about 60%. In China today about 80% of the electrical energy is produced by coal-fired power stations, these being responsible for 70% of CO_2 emissions and 90% of SO_2 emissions in that country (Cao et al., 2007).

Electrochemical utilization of coal's energy would provide huge gains not only and not so much because of higher conversion efficiencies but also for other reasons. The thermal power stations emit CO_2 mixed with air and other gases as well as with uncombusted coal particles into the atmosphere. In a fuel cell, coal would be oxidized anodically in separate compartments closed off from the air. From the gases evolved in these compartments, by-product hydrocarbons could readily be separated and then utilized, while particulates could be filtered off. The remainder, almost pure CO_2, could be sent to underground storage. This would make a huge contribution to solving the problem of global warming caused by CO_2 emissions. The power plants with direct carbon fuel cells (DCFCs) could in principle be set up directly in coal mines to eliminate the numerous economic and ecological problems associated with long-distance coal transport.

There are two possibilities for electrochemical utilization of the chemical energy of coal: (1) via prior coal gasification and use of the hydrogen and/or carbon monoxide produced in this process in various fuel cells, and (2) by direct electrochemical oxidation within the fuel cell. Various methods of coal gasification are discussed in Chapter 11. Some attempts to realize the second approach, which is basically a much simpler one-step process, are discussed in the present section. [Note that *coal* and *carbon* are somewhat interchangeable terms when discussing them as fuels, coal being a natural carbon material, carbon being a derived material (not necessarily from coal).]

Antoine César Becquerel in France in 1855 and Pavel Yablochkov in Russia in 1877 built electrochemical devices using coal anodes in a molten KNO_3 electrolyte (see Howard, 1945, and Liebhafsky and Cairns, 1968). In 1896, William Jacques obtained a U.S. patent for his invention of a "coal battery" with a coal anode and an iron cathode immersed in molten alkali NaOH (this battery was mentioned in Chapter 2). Despite the great doubts raised as to the nature of the processes taking place in it, the electrical performance of his 100-cell battery operating at 400 to 500°C was rather impressive: total power of 1.5 kW and current densities up to 100 mA/cm^2. Much later, the prototype of a carbon fuel cell with a solid electrolyte working at a temperature of 1000°C was built by Baur and Preis in Switzerland, 1938).

9.4.2 Reactions and Thermodynamic Parameters

The reactions ideally occurring in DCFCs with an acidic electrolyte solution and their thermodynamic parameters at 25°C are as follows:

$$\text{Anode:}\quad C + 2H_2O \rightarrow CO_2 + 4H^+ + 4e^- \quad E^0 = 0.21\,\text{V} \tag{9.14}$$

$$\text{Cathode:}\quad O_2 + 4H^+ + 4e^- \rightarrow 2H_2O \quad E^0 = 1.23\,\text{V} \tag{9.15}$$

$$\text{Overall:}\quad C + O_2 \rightarrow CO_2 \quad \mathscr{E}^0 = 1.02\,\text{V} \tag{9.16}$$

$$-\Delta G^0 = 394.3\,\text{kJ/mol} = 1.02\,\text{eV} \qquad -\Delta H^0 = 393.5\,\text{kJ/mol} = 1.02\,\text{eV}$$

It will be seen below that these reactions actually take place not in aqueous solutions but at high temperatures and in other electrolytes (e.g., in molten carbonates or alkalies). Thus, in molten carbonates at 600°C, the equations can be formulated as follows:

$$\text{Anode:}\quad C + 2CO_3^{2-} \rightarrow 3CO_2 + 4e^- \quad E^0 = -0.02\text{V}^* \tag{9.17}$$

$$\text{Cathode:}\quad O_2 + 2CO_2 + 4e^- \rightarrow 2CO_3^{2-} \quad E^0 = 1.10\text{V}^* \tag{9.18}$$

Equation (9.16) for the overall reaction remains unchanged:

$$-\Delta G^0 = 395.4\,\text{kJ/mol} = 1.02\,\text{eV} \qquad -\Delta H^0 = 394.0\,\text{kJ/mol} = 1.02\,\text{eV}$$

It can be seen that the thermodynamic parameters of this reaction are practically independent of temperature. It is a remarkable feature of the reaction that its Gibbs free energy and its enthalpy have practically identical values, which implies that entropic losses are absent. This means that a 100% conversion of the chemical energy of carbon to electrical energy is theoretically possible.

At temperatures above 750°C, the Boudouard equilibrium can become established:

$$C + CO_2 \rightleftharpoons 2CO \tag{9.19}$$

which during carbon oxidation would lead to the formation of carbon monoxide (CO), entailing some energy loss (only two electrons are liberated instead of four when a carbon atom is merely oxidized to CO).

* These values of the electrode potentials are referred to the equilibrium potential of the hydrogen electrode under the same conditions (carbonate melt at 600°C).

9.4.3 Work on Carbon Fuel Cells Since 1960

Despite the numerous problems encountered in earlier research, attempts to build versions of carbon fuel cells were continued when the fuel cell "boom" began in the 1960s.

Two characteristics are of importence when carbon (or coal) is used in fuel cells: the nature of the carbon material to be used and the physical state of this material. The notion of "carbon" is highly indefinite; it covers a large number of carbon materials, both natural and human-made. It includes various types of graphite, coke, and carbon black, differing very strongly in their structure and in the content of additional components, both volatile (hydrogen, oxygen, nitrogen, sulfur, organic compounds, etc.) and nonvolatile (mineral salts). Carbon materials of natural origin could be used as such, or after having been subjected to pretreatments to eliminate undesired components.

The physical state of a carbon material in a fuel cell can be of two kinds: (1) relatively massive electrodes in the shape of plates or rods which may be cut out of graphite or pressed from powdered carbon materials (usually with different binders) and may serve as the current collector and as a consumable electrode; or (2) highly disperse carbon powders, present as a slurry in a liquid electrolyte such as carbonate melt, which are in constant contact with a metal electrode serving as the current collector when they take part in the electrochemical reactions and electrons must be transferred.

In the work on DCFCs, different directions were followed with respect to the electrolytes: aqueous solutions (at temperatures below 100°C), high-temperature melts of carbonates or sodium hydroxide, and high-temperature solid electrolytes.

Aqueous Solution

In 1979 Coughlin and Farooque advanced the idea of electrochemical gasification of coal by electrolysis of a suspension of carbon materials in a solution of sulfuric acid with platinum electrodes. Here, the cathodic reaction of hydrogen evolution,

$$2H^+ + 2e^- \rightarrow H_2 \qquad E^0 = 0\,V \tag{9.20}$$

occurs together with the anodic reaction (9.14), so that the overall reaction;

$$C + 2H_2O \rightarrow CO_2 + 2H_2 \qquad \mathscr{E}^0 = 0.21V \tag{9.21}$$

leads to electrolytic hydrogen evolution at a voltage of 0.21 V (instead of the 1.23 V needed in ordinary water electrolyzers). The device suggested is not a fuel cell but makes use of anodic carbon oxidation in aqueous solution. Details about this reaction were not communicated by the authors. It was shown later by Okada et al. (1981) and Dhooge et al. (1982) that at a temperature of 80°C, such a reaction is actually possible. It proceeds since carbon samples contain iron as an impurity dissolving in the acid, thus producing the Fe^{3+}/Fe^{2+} redox system in the solution. This served as a mediator for coal oxidation. When iron was removed, the rate of carbon dropped to values too low for the practical generation of electrical energy in a fuel cell.

In 2006, Patil et al. continued research into this reaction, using carbon suspensions in a sulfuric acid solution to which 100 mM each of Fe^{3+} and Fe^{2+} ions had been added on purpose. It could be shown that with a voltage of 0.6 to 1.0 V applied across a cell with platinum or platinum-alloy electrodes, even at 40°C the reaction occurs with an acceptable current density of 30 mA/cm^2. This is a rather promising result, since in ordinary electrolyzers a voltage of at least 1.7 V is needed for hydrogen evolution. The authors pointed out, though, that a number of aspects requiring further study are not clear with respect to the reaction.

Low-temperature oxidation (below 200°C) of carbon has received attention from research workers not only because of its potential utility in DCFCs but also because of the harm it does in many types of fuel cells and other electrochemical devices where platinum catalysts on carbon supports are used. It was in this connection that Choo et al. (2007) studied the anodic oxidation mechanism of graphite in sulfuric acid.

Melts

Studies of electrochemical carbon oxidation in carbonate melts at 700°C were performed by Weaver et al. (1981) at the Stanford Research Institute (SRI) in Menlo Park, California. They used rods of different carbon materials as the electrodes. The electrode potentials were measured relative to a gold reference electrode in an atmosphere of $CO_2 + O_2$ at the same temperature. The electrodes proved to be more active the lower the degree of crystallinity of the original carbon powder used to press the rods. The electrodes had open-circuit potentials around 1.1 V. At a current density of 100 mA/cm^2, the potential of the most active sample was 0.8 V (and as high as 0.9 V when the temperature was raised to 900°C).

At the Lawrence Livermoore National Laboratory, a fuel cell using a semisolid suspension of carbon powder in a molten carbonate was developed. Porous nickel was used as the cathode material. At a temperature of 800°C and a voltage of 0.8 V, current densities of 50 to 125 mA/cm^2 were obtained. In one test, a current density of 27 mA/cm^2 was drawn for 30 hours (Cherepy et al., 2005).

Hackett et al. (2007) reported experiments with DCFCs where rods prepared from different carbon materials were used as the anodes. The cathodes were made from iron–titanium alloy. Melts of NaOH at temperatures in the range 600 to 700°C were used as the electrolyte. The most stable operation with high performance figures was obtained with graphite anodes. The OCV value was 0.788 V. At the optimum temperature of 675°C, the voltage was 0.45 V at a current density of 140 mA/cm^2. Anodes made of other carbon materials had higher OCVs (up to 1.044 V), but exhibited inferior and less stable performance when current was drawn. The equation for the anodic reaction in NaOH melt can be written as

$$C + 4OH^- \rightarrow CO_2 + 2H_2O + 4e^- \tag{9.22}$$

An undesirable side reaction is carbonation of the alkali by the CO_2 evolved, which involves the equilibrium

$$CO_2 + 2OH^- \rightleftharpoons CO_3^{2-} + H_2O \tag{9.23}$$

When steam is added to the oxygen flow, the equilibrium (9.22) can be shifted somewhat to the left.

Solid-Oxide Electrolytes

A difficulty of principle arises when using a solid electrolyte in DCFCs. In fact, in cells with a liquid electrolyte (solution or melt), the entire surface area of the carbon material is in contact with the electrolyte (is wetted by the electrolyte). In cells with a solid electrolyte, to the contrary, the contact between the solid carbon material and the solid electrolyte is a mere point contact, and the working surface area is much smaller.

Two ways to overcome this difficulty have been suggested. The first (Pointon et al., 2006) makes combined use of a solid-oxide electrolyte and liquid (molten) carbonate electrolyte. The oxygen electrode was separated by the solid electrolyte from the carbonate melt that contained the suspended carbon particles. The second (Gür and Huggins, 1992) used a cell consisting of two halves separated by a solid electrolyte in the shape of a tube. The inside and outside surface of the tube were coated with a layer of platinum. The inside of the tube was in contact with ambient air. The outside was in contact with a closed compartment holding air and the carbon samples. The solid electrolyte was maintained at a temperature of 932°C, while the compartment holding the carbon was maintained (in one experiment) at a level of 955°C. Because of the reaction between oxygen and carbon, a reduced oxygen equilibrium pressure was set up in the compartment. This led to a considerable oxygen pressure gradient between the two sides of the solid electrolyte, and hence to a potential difference of about 1.05 V. When a current of 10 mA/cm^2 was drawn, this difference (the discharge voltage) dropped to a value of about 0.4 V.

REFERENCES

Section 9.1

Codina G., J. R. Perez, M. Lopez-Atalaya, et al., *J. Power Sources*, **48**, 293 (1994).
Joerissen L., J. Garche, Ch. Fabian, G. Yomazic, *J. Power Sources*, **127**, 98 (2004).
Johnson D. A., M. A. Reid, *J. Electrochem. Soc.*, **132**, 1058 (1985).
Lopez-Atalaya M., G. Codina, J. R. Perez, et al., *J. Power Sources*, **39**, 147 (1992).

Murthy A. S. N., T. Srivastava, *J. Power Sources*, **27**, 119 (1989).

Skyllas-Kazacos M., D. Kasherman, et al., *J. Power Sources*, **35**, 399 (1991).

Wen Y. H., H. M. Zhang, et al., *Electrochim. Acta*, **51**, 3775 (2006).

Section 9.2

Betso S. R., M. H. Klapper, L. B. Andreasson, in: *Biological Aspects of Electrochemistry*, Birkhäuser, Basel and Stuttgart, Germany, 1971, p. 162.

Gao F., Y. Yan, L. Sub, et al., *Electrochem. Commun.*, **9**, 989 (2007).

Prasad, D., T. K. Sivaram, S. Berchmans, V. Yegnaraman, *J. Power Sources*, **160**, 991 (2006).

Tarasevich M. R., Yu. G. Chirkov, V. A. Bogdanovskaya, A. V. Kapustin, *Electrochim. Acta*, **51**, 418 (2005).

Topcagic S., S. D. Minteer, *Electrochim. Acta*, **51**, 2168 (2006).

Walker A. L., C. W. Walker, *J. Power Sources*, **160**, 123 (2006).

Wingard L. W., C. C. Liu, N. L. Nagda, *Biotechnol. Bioeng.*, **13**, 629 (1971).

Yaropolov A. I., S. D. Varfolomeev, I. V. Berezin, et al., FEBS Lett., **71**, 306 (1976).

Zhang E., W. Xu, G. Diao, C. Shuang, *J. Power Sources*, **161**, 820 (2006).

Section 9.4

Baur E., H. Preis, *Z. Elektrochem.*, **43**, 727 (1937); **44**, 695 (1938).

Cao D., Y. Sun, G. Wang, *J. Power Sources*, **167**, 250 (2007).

Cherepy N. J., R. Krueger, K. J. Fiet, et al., *J. Electrochem. Soc.*, **152**, A80 (2005).

Choo H.-S., T. Kinumoto, S.-K. Jeong, et al., *J. Electrochem. Soc.*, **154**, B1017 (2007).

Coughlin R.W., M. Farooque, *Nature*, **279**, 301 (1979).

Dhooge P. M., D. L. Stillwell, S. M. Park, *J. Electrochem. Soc.*, **129**, 1719 (1982); **130**, 1029 (1983).

Gür N. M., R. A. Huggins, *J. Electrochem. Soc.*, **139**, L95 (1992).

Hackett G. A., J. W. Zondlo, R. Svensson, *J. Power Sources*, **168**, 111 (2007).

Howard H. C., *Direct Generation of Electricity from Coal and Gas (Fuel Cells)*, Wiley, New York, 1945.

Jacques W. W., U.S. patent 555,511 (1896); *Z. Elektrochem.*, **4**, 286 (1910).

Liebhafsky H. A., E. J. Cairns, *Fuel Cells and Batteries*, Wiley, New York, 1968.

Okada G., V. Guruswamy, J. O'M. Bockris, *J. Electrochem. Soc.*, **128**, 2097 (1981).

Patil P., Y. D. Abreu, G. G. Botte, *J. Power Sources*, **158**, 368 (2006).

Pointon K., J. Irvine, J. Bradley, S. Jain, *J. Power Sources*, **162**, 750 (2006).

Weaver R. D., S. C. Leach, L. Nanis, *Proc. 16th Intersociety Energy Conversion Engineering Conference*, ASME, New York, 1981, p. 717.

Reviews and Monographs

Bagotsky V. S., A. M. Skundin, *Chemical Power Sources*, Academic Press, London, 1980.

Cao D., Y. Sun, G. Wang, Direct carbon fuel cells: Fundamentals and recent developments, *J. Power Sources*, **167**, 250 (2007).

Ponce de León C., A. Frías-Ferrer, J. González-García, D. A. Szánto, F. C. Walsh, Redox flow cells for energy conversion, *J. Power Sources*, **160**, 716 (2006).

Vincent C. A., B. Scrosati, *Modern Batteries: An Introduction to Electrochemical Power Sources*, Edward Arnold, London, 1997.

CHAPTER 10

FUEL CELLS AND ELECTROLYSIS PROCESSES

Electrolyzers, used widely in industry to synthesize various types of products, in a sense are fuel cells in reverse. The two devices are built analogously, including two electrodes in contact with an ion-conducting electrolyte, but they operate according to opposite principles. When electric current is forced through an electrolyzer from an outside source, chemical reactions generating new products occur at the electrodes. When reactants are fed into a fuel cell from an outside source, chemical reactions generating a current occur at the electrodes. Often, the same electrochemical reactions occur in the two devices, but they go in opposite directions. Analogous designs can be used for the two devices: for example, a battery or electrolyzer consisting of jar-type cells, or a battery or electrolyzer in filter-press arrangement. Numerous other analogies could be listed, but many specific differences exist as well.

In the present chapter, by considering a few examples, we explore how developments in these two fields of applied electrochemistry have been interrelated and influenced each other. Some examples of devices having mixed functions are also examined: functions typical of fuel cells and functions typical of electrolyzers.

10.1 WATER ELECTROLYSIS

Water was first electrolyzed in 1801, as soon as the Volta pile had become known. Grove's discovery of a fuel cell (described in 1839) was the result of

Fuel Cells: Problems and Solutions, By Vladimir S. Bagotsky

research into water electrolysis. It took another 100 years until large electrolyzers with alkaline electrolyte for the production of hydrogen and oxygen were operative, which have seen broad development in the second half of the twentieth century. They are equipped with electrodes consisting of iron-group metals. It is quite likely that Bacon, who spent nearly three decades developing his model of a relatively powerful hydrogen–oxygen fuel cell with nickel electrodes that was demonstrated in 1959, made use of experience gathered in the field of water electrolyzers. However, the major phases of development in these two fields followed independent pathways. The situation was quite different in the development of later systems of water electrolyzers, considered next.

10.1.1 Electrolyzers with Proton-Conducting Membranes

The structure of membrane electrolyzers differs little from that of the membrane fuel cells described in Chapter 3; their work differs only in the direction of current flow through the device and in the directions of all electrochemical reactions. Certain structural differences between fuel cells and electrolyzers arise from the intended function and properties of the gas-diffusion layers (GDLs). In a fuel cell, the GDLs must provide uniform access of the reactant gases to the catalytically active layer of the MEA. A GDL is built with highly hydrophobic materials to keep out any water that could interfere with the gas supply. In an electrolyzer, water is the reactant, and the corresponding layer of the MEA must be hydrophilic to secure a water supply. On the other hand, gases are the reaction products and must be able to get away from the MEA. For these reasons, the GDL in an electrolyzer needs to have a carefully balanced ratio of hydrophilic to hydrophobic pores.

Work on water electrolyzers with proton-conducting membranes rather than an alkaline electrolyte solution between the electrodes began at practically the same time as work on the first hydrogen–oxygen membrane fuel cells. In the early 1960s, General Electric began development of the first membrane water electrolyzers designated to produce oxygen for submarine crews.

Then, as PEMFC experienced important improvements, membrane water electrolyzers were also improved. Iridium dioxide (IrO_2) was suggested as the catalyst for anodic oxygen evolution (see, e.g., Rasten et al., 2003). It was suggested (Song et al., 2007) that this catalyst be applied directly to the membrane [the technique of catalyst-coated membranes (CCMs)]. Test models built in this way could be operated at 1 A/cm^2 when working with voltages of 1.6 to 1.8 V. Stucki et al. (1998) reported long-term tests of two 100-kW industrial membrane electrolyzers working at 80°C with a current density of 1 A/cm^2, the average voltage being 1.75 V per cell. The tests were terminated prematurely because of hydrogen leaking through the membrane into the oxygen stream. Local stresses in the membranes were cited as a possible reason.

10.1.2 High-Temperature Electrolyzers

At high temperatures, steam, rather than water, undergoes electrolysis. High-temperature solid-electrolyte electrolyzers started to be developed much later than the corresponding hydrogen–oxygen fuel cells. Accorsi and Bergmann (1980) investigated steam electrolysis with a cell consisting of an ytterbia-stabilized zirconia electrolyte, a tin-doped india anode having the composition $(In_2O_3)_{0.96}(SnO_2)_{0.04}$, and a cathode consisting of sputter-deposited nickel cermet. As expected from the Nernst equation, the cell's OCV (the cell voltage measured when interrupting electrolysis) was a function of the ratio of partial pressures of hydrogen and steam at the cathode. At equal partial pressures, it had a value of 0.903 V. Such electrolysis cells have been operated for more than 1000 hours at current densities of up to 1000 mA/cm^2 while applying voltages below 1.5 V.

Tests of a high-temperature hydrogen–oxygen fuel cell with a solid electrolyte based on strontium-doped cerium trioxide ($SrCeO_3$) were reported in 1982 by Iwahara et al. This electrolyte differs from other solid oxide electrolytes by developing noticeable proton conduction in a hydrogen atmosphere at high temperature. Such a fuel cell design will also be useful for high-temperature steam electrolysis. In it, the reactions would be:

$$\text{Anode:} \quad 2H_2O \rightarrow O_2 + 4H^+_{(\text{electrolyte})} + 4e^- \tag{10.1}$$

$$\text{Cathode:} \quad 4H^+_{(\text{electrolyte})} + 4e^- \rightarrow 2H_2 \tag{10.2}$$

$$\text{Overall:} \quad 2H_2O \rightarrow O_2 + 2H_2 \tag{10.3}$$

10.1.3 Hydrogen–Oxygen Systems for Storing Electrical Energy

The idea of using fuel cells, not only as a primary power source but also for temporary energy storage, has been put forward repeatedly. To this end, combined units would be built that include electrolyzers producing hydrogen and oxygen, tanks storing these gases (or merely the hydrogen), and hydrogen–oxygen (or hydrogen–air) fuel cells producing power on demand. Large installations of this type could be used for load leveling in wide-area grids. Smaller installations could be used in combination with power generators of variable output, such as wind farms or solar collectors. When storing sufficiently large amounts of energy, such installations will be much more compact and markedly cheaper than ordinary storage batteries. When storing small amounts of energy, ordinary battery storage systems will be more convenient.

10.1.4 Reversible (Unitized) PEMFC Systems

Combining the functions of an electrolyzer and a fuel cell such as a PEMFC into regenerative fuel cells, also known as unitized regenerative fuel cells (URFCs), has been attempted repeatedly. Fundamentally, any hydrogen–oxygen fuel cell will work as an electrolyzer when a current is passed through it in the opposite direction from an outside source, and water is supplied. At the hydrogen electrode (which was the anode and now becomes the cathode), hydrogen will then be evolved (rather than being oxidized), and at the oxygen electrode (which was the cathode and now becomes the anode), oxygen will be evolved (rather than being reduced). The reactions are (1.7) and (1.8) in the opposite direction (i.e., with an arrow to the left).

In the development of URFCs that could work as both an electrolyzer and a fuel cell, depending on the demand, two difficulties became apparent. One of them was associated with the gas-diffusion layer. The way to overcome it is along the same lines as described in Section 10.1.1 for membrane water electrolyzers (i.e., by a compromise in the hydrophilic/hydrophobic pore structure).

The second difficulty was associated with the catalysts. Platinum, a good catalyst for cathodic oxygen reduction in fuel cells, will also promote anodic oxygen evolution in electrolyzers. However, while this anodic reaction takes place, the platinum surface becomes oxidized, a layer of platinum oxide two or three atom layers thick forming on it. This oxidized platinum is no longer active for cathodic oxygen reduction. The oxidation is irreversible, so an electrode with a platinum catalyst will no longer function as a fuel cell electrode.

One could attempt to overcome this difficulty by using a three-electrode cell having one hydrogen electrode but two oxygen electrodes, one for oxygen evolution and the other for oxygen reduction, but in a multicell battery, it would be difficult to switch over between the two sets of electrodes in all the individual cells in a synchronized fashion.

Repeated attempts have been made to develop a stable catalyst for bifunctional oxygen electrodes that could sustain a continued alternation between the electrolyzer and fuel cell mode. In 1994, Swette et al. showed that a catalyst consisting of a 1:1 mixture of highly disperse particles of platinum and iridium oxide IrO_2 had such bifunctional properties. The iridium oxide is a poor conductor, but when in contact with platinum it develops marked electronic conduction. This mixed catalyst is rather active for both reactions, the evolution and the reduction of oxygen. How it works has not yet been established. Possibly, anodic oxygen evolution takes place primarily at the iridium oxide particles, and cathodic oxygen reduction takes place at the platinum particles, now less strongly oxidized.

A model of a URFC with thin catalytic layers on the electrodes was built by Zhigang et al. (1999). The positive electrode was coated with 0.02 mg/cm^2 each of platinum and iridium oxide, and the hydrogen electrode was coated with 0.4 mg/cm^2 of pure platinum. Figure 10.1 shows the current–voltage curves

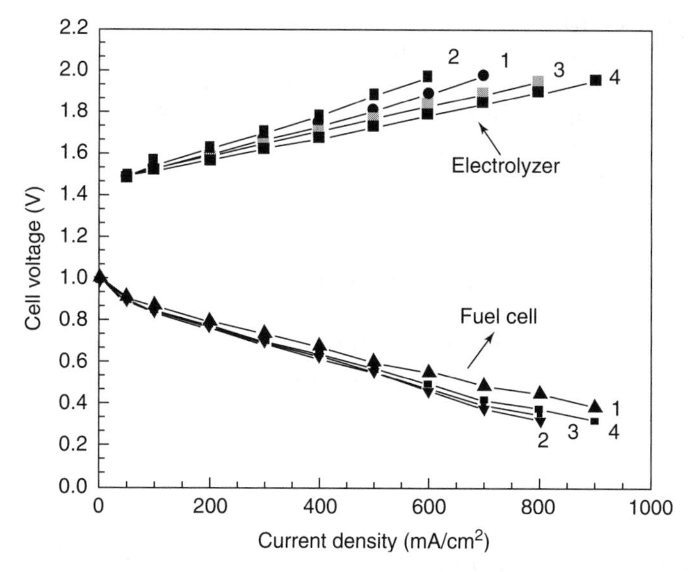

FIGURE 10.1 Performance of reversible PEMFCs after the four cycles indicated. Fuel cell mode: $P_{O_2} = 0.3\,\text{MPa}$, $P_{H_2} = 0.3\,\text{MPa}$; electrolysis mode: ambient pressure, Nafion 115, 80°C. (From Zhigang et al., 1999, with permission from Elsevier.)

recorded over four cycles of alternating operation of this device as an electrolyzer and as a fuel cell. It can be seen from the figure that at a current density of 400 mA/cm^2, a voltage of about 1.71 V was needed in the electrolysis mode, but a voltage of only about 0.7 V was produced in the fuel cell mode. This means that the voltage efficiency of this device was about 40%.

Liu et al. (2004) used multilayer electrodes. Closer to the membrane, a thin layer (about 5 μm) with Pt–IrO_2 catalyst for oxygen evolution was arranged. This layer is completely hydrophilic, so that good contact of water with the catalyst is secured. Adjacent to this inner catalytic layer is an outer catalytic layer containing platinum and Nafion that is supported by a hydrophobic gas-diffusion layer as in an ordinary fuel cell. In this way, oxygen evolution at the inner layer has little impact on the activity of the outer catalytic layer. The authors reported 25 cycles of successful alternative operations of this device.

These devices are still in their early stages of development. Further studies are needed, including observations of the long-term behavior of such devices under conditions of frequent changes between the electrolysis and the fuel cell mode (see the review of Pettersson et al., 2006).

Workers at the Lawrence Livermore National Laboratory (Mitlitski et al., 1999) see good chances for these developments: When lighter gas tanks are used, storage installations having a specific energy density of up to 400 Wh/kg could be built, which is three to four times more than the same figure for the best storage batteries. It is a great advantage of such installations over conventional storage batteries that they can be recharged, not only for a longer time by electric power input but also almost instantaneously, merely by replacing (or refilling) the gas tanks. Workers at NASA (Bents et al., 2006) also see good prospects for such devices in combination with solar collectors for future long-term space expeditions.

10.2 CHLOR-ALKALI ELECTROLYSIS

The electrolysis of sodium chloride solutions (brine) producing chlorine at the anode and sodium hydroxide (NaOH, caustic soda) in the catholyte via the overall reaction

$$2Cl^- + 2H_2O \rightarrow Cl_2 + H_2 + 2OH^- \qquad \mathcal{E}^0 = 2.17\,V \tag{10.4}$$

is a very important industrial process. The hydrogen generated as the third reaction product in chlor-alkali electrolysis is generally utilized, merely as a fuel for heat generation. Current annual world production of chlorine by electrolysis is over 30 megatons and that of alkali is 35 megatons, with annual increases of 2 to 3%. This industry consumes about 100 billion kilowatthours of electrical energy per year. Cell voltages of most modern electrolyzers are 3.3 to 3.8 V at a current density of 100 mA/cm^2 (1 kA/m^2). With current yields of 95% for the anodic reaction, this corresponds to an energy consumption of 2600 to 3000 kWh per ton of chlorine.

In the past, graphite had been used as the anode in chlorine cells. During electrolysis, it was gradually consumed by oxidation so that the gap between the electrodes increased, producing an important increase in ohmic resistance, cell voltage, and power consumption. With the general introduction of dimensionally stable anodes, DSA working with layers of mixed titanium–ruthenium oxides, this problem was overcome completely.

It is important for the chlorine industry to lower the cell voltages even further. Every 0.1-V decrease in cell voltage would produce a savings of 0.3% of the total electrical power, or an annual 300 million kilowatthours worldwide. Local use of the hydrogen produced in chlorine cells for power generation could considerably reduce the grid power consumed. Since this hydrogen is quite pure, it would be feasible to use it more efficiently in fuel cells rather than in heat engines coupled to conventional power generators. Another, alternative possibility for achieving power savings would be to replace cathodic hydrogen generation by the cathodic reduction of air oxygen (see Section 10.2.1).

Chlorine is a very active oxidizing agent and readily undergoes electrochemical reactions. The question arises, then, whether chlorine could not be used as an oxidizing agent in fuel cells. Some attempts undertaken in this direction are discussed in Section 10.2.2.

10.2.1 The Problem of Oxygen Electrodes for Chlorine Cells

In existing chlorine cells, the cathodic process is hydrogen evolution

$$2H_2O + 2e^- \rightarrow H_2 + 2OH^- \qquad E^0 = -0.828\,V \tag{10.5}$$

As this reaction proceeds, the layer of electrolyte solution (brine) next to the cathode becomes strongly alkaline, so that for the reaction formula and for the thermodynamic potential, the versions applicable to alkaline solutions must be used.

When hydrogen evolution is replaced by oxygen reduction, the cathodic process is

$$\tfrac{1}{2}O_2 + H_2O + 2e^- \rightarrow 2OH^- \qquad E^0 = 0.4\,V \tag{10.6}$$

Therefore, the equilibrium potential of the cathode will then be 1.23 V more positive than formerly, and the cell's thermodynamic EMF will be 0.94 V instead of 2.17 V. Under the assumption that other losses (ohmic losses and polarization) will not be much higher than now, then theoretically, cell voltages would be expected to decrease by almost 1 V.

It must be pointed out that this problem is real not only in the chlorine industry but in many other electrochemical industries where the product wanted is generated anodically while hydrogen is a cathodic by-product.

A practical realization of these ideas is beset with many difficulties. The electrodes used in the chlorine industry, for instance, should secure many years of uninterrupted, highly reliable cell operation where replacements are practically impossible. Nonporous electrodes are used for cathodic hydrogen evolution, with the attending problem of hydrogen bubbles filling the gap between the electrodes, which apart from excellent stirring action lead to a higher internal ohmic resistance in the cells. If cathodic oxygen reduction were to be used instead, porous gas-diffusion electrodes would be required. A lot of experience has been accumulated in many years of work on different fuel cell versions using oxygen-diffusion electrodes. Numerous problems have been revealed in this work: gradual catalyst degradation, gradual losses or changes in the hydrophobicity of the materials involved, gradual changes in pore structure, and so on. It must be taken into account as well that the electrodes that would be used in chlorine cells are much larger than those used in fuel cells. In cells having considerable height, changes in hydrostatic pressure of the liquid electrolyte in the vertical direction will lead to a nonuniform electrode function.

All these problems require further analysis. Solving them would be of great value for the future of both industrial electrolysis and fuel cells.

10.2.2 Chlorine Fuel Cells

In a hydrogen–chlorine fuel cell, the reactions are as follows:

$$\text{Anode:} \quad Cl_2 + 2e^- \rightarrow 2Cl^- \quad E^0 = 1.36\,V \tag{10.7}$$

$$\text{Cathode:} \quad H_2 \rightarrow 2H^+ + 2e^- \quad E^0 = 0\,V \tag{10.8}$$

$$\text{Overall:} \quad Cl_2 + H_2 \rightarrow 2HCl \quad \mathcal{E}^0 = 1.36\,V \tag{10.9}$$

The chief advantage of these hydrogen–chlorine fuel cells over hydrogen–oxygen fuel cells is the higher electrochemical activity of chlorine. The thermodynamic equilibrium potential of chlorine has a more positive value than that of oxygen (by 0.13 V), but even more important, it lacks the major defects of the oxygen electrode (where the potential of zero current is 0.4 V more negative than the thermodynamic value, and strong polarization occurs at nonzero currents). Despite these advantages, hydrogen–chlorine fuel cells have not yet found widespread use as primary power sources, since at least in small installations, the handling of chlorine is inconvenient and corrosion problems arise. Probably the only situation in which their use can be justified is for power generation to satisfy additional power needs in chemical industries producing excess chlorine.

Another application exists, however, where using such fuel cells is not only possible but even highly appropriate: in the storage of electrical energy. Results of a technical and economic analysis of hydrogen–chlorine fuel cells used for load leveling in the electric utilities were published by Gileadi et al. (1977–1978). Their basis was the scientific and technical level that had been attained at that time, which for instance did not include the use of DSA in the chlorine industry. They showed that because of the high reversibility of reactions (10.7)–(10.9), overall values of the electrical efficiency of more than 70% could be attained, since at current densities of 300 to 400 mA/cm^2, the electrolyzer voltage is not over 1.50 V, the fuel cell voltage not under 1.20 V, and thus the voltage efficiency about 80%. In this respect, this system of energy storage has obvious advantages over storage schemes involving the hydrogen–oxygen system or redox systems. They also pointed out that the same electrodes and catalysts might be used in the electrolysis mode and in the fuel cell mode, so that the two processes might be realized in a single unit. [It should be pointed out that their paper was published prior to the appearance of the first models of reversible (unitized) PEMFC systems, described in Section 10.1.4.]

Chin et al. (1979) conducted a detailed analysis of mass and heat-transfer processes occurring in such a regenerative hydrogen–chlorine system for power storage. Establishing the thermal balance is complicated by the fact that the overall reaction (10.9) involves a considerable change in entropy of the system.

The latent heat of the reaction is about 37 kJ/mol. This heat is evolved when the reaction occurs from left to right (discharge), but is absorbed (i.e., the installation cools down) when the reaction occurs from right to left (recharge). It was shown again that with such an installation, electrical energy could be stored with an overall efficiency of over 70%.

10.3 ELECTROCHEMICAL SYNTHESIS REACTIONS

10.3.1 Reactions with Carbon Dioxide

Work on methanol fuel cells has advanced considerably over the last 10 years (Chapter 4). In such cells the electrochemical oxidation of methanol goes directly to CO_2:

$$CH_3OH + H_2O \rightarrow CO_2 + 6H^+ + 6e^- \quad E^0 = 0.02\,V \tag{10.10}$$

The reverse process, cathodic reduction of CO_2 to methanol [reaction (10.10), going from right to left] or to other organic compounds, has also been the subject of numerous studies in recent decades. These studies had basically two practical aims, as described below.

Utilization or Elimination of CO_2 in Life-Support Systems

In closed living spaces such as submarines and spacecraft, there is always a problem of accumulating CO_2 gas exhaled by the crew. For various reasons it is usually not possible to drain this gas into the environment. In the life-support systems of such objects, electrolyzers are sometimes used where CO_2 is reduced anodically to liquid products (such as formic acid), while at the cathode, oxygen needed for breathing is evolved. The final target, so far still very remote, is that of reducing CO_2 to carbohydrates that could be used to feed the crew during extended space voyages.

Synthesis of Useful Intermediates for Organic Synthesis and of Synthetic Fuels

Useful intermediates to be produced by CO_2 reduction could include aldehydes, ketones, and oxalates, among others. Methanol itself could be a very useful synthetic fuel.

Apart from these practical aims, an electrochemical reduction of CO_2 is of great cognitive interest, since it is analogous to the photochemical reduction of CO_2 to carbohydrates accomplished by chlorophyll in plants. The two fields of work—development of methanol fuel cells and investigations into cathodic CO_2 reduction—have practically no mutual connections and have most often been associated with different groups of scientists.

10.3.2 Fuel Cells and Chemical Cogeneration

The basic purpose of fuel cells is power generation from the chemical energy set free in the reactions of oxidizing and reducing agents. Apart from current, a chemical product is generated as well and in a number of cases will have practical value. In hydrogen–oxygen fuel cells, the product is water, which from the fuel cell batteries in the *Apollo* spacecraft and the space shuttles was withdrawn as drinking water for the crew. Similarly, any hydrochloric acid generated during operation of the hydrogen–chlorine fuel cells discussed above could be used for local needs unless it is recycled in the storage mode.

Brillas et al. (2002) described a hydrogen–oxygen fuel cell with circulating alkaline electrolyte where the oxygen gas-diffusion electrode was made of graphitized cloth coated with a PTFE–carbon mixture and free of additional catalysts. At such an electrode, the oxygen is not reduced to water [reaction (1.8)] but to hydrogen peroxide:

$$\text{Cathode:}\quad O_2 + H_2O + 2e^- \rightarrow HO_2^- + OH^- \quad E^0 = -0.17\,\text{V} \tag{10.11}$$

$$\text{Anode:}\quad H_2 + 2OH^- \rightarrow 2H_2O + 2e^- \quad E^0 = -0.8\,\text{V} \tag{10.12}$$

$$\text{Overall:}\quad O_2 + H_2 + OH^- \rightarrow HO_2^- + H_2O \quad \mathscr{E}^0 = 0.66\,\text{V} \tag{10.13}$$

At a temperature of 25°C and with an external load of 10 Ω, the cell produced a current density of 100 mA/cm^2 and generated hydrogen peroxide with a productivity of about 2 mmol · cm^{-2} · h^{-1}.

A fuel cell representing great interest is the sulfur dioxide–oxygen fuel cell, where the following reactions take place:

$$\text{Anode:}\quad SO_2 + 2H_2O \rightarrow H_2SO_4 + 2H^+ + 2e^- \quad E^0 = 0.17\,\text{V} \tag{10.14}$$

$$\text{Cathode:}\quad \tfrac{1}{2}O_2 + 2H^+ + 2e^- \rightarrow H_2O \quad E^0 = 1.23\,\text{V} \tag{10.15}$$

$$\text{Overall:}\quad SO_2 + \tfrac{1}{2}O_2 + H_2O \rightarrow H_2SO_4 \quad \mathscr{E}^0 = 1.06\,\text{V} \tag{10.16}$$

This fuel cell produces the very valuable product sulfuric acid from SO_2, a gas ejected in huge amounts into the atmosphere by coal-fired power plants and by many metallurgical and chemical industries, so a way of capturing and using this SO_2 would be of great environmental benefit. The process can be accomplished in the electrode chambers of cells divided by a cation-exchange membrane and filled with circulating 3 M sulfuric acid solution. In tests, current densities of up to 2 A/cm^2 were realized at 60°C. With platinum catalysts it was possible to attain up to a 90% conversion of SO_2, even when its concentration in the gas stream to the anode was low (see the review of Alcaide et al., 2006).

To cite an early example, Langer and Yurchak (1969) reported tests with a fuel cell where at the cathode, activated with a platinum catalyst, benzene (C_6H_6) was reduced to cyclohexane while hydrogen was evolved at the cathode. The thermodynamic EMF of this cell was about 0.14 V, and at a current density of 6 mA/cm^2, the working voltage was 0.07 V.

The hydrogenation of a number of unsaturated alcohols and acids in fuel cells with a hydrogen anode was studied by Yuan et al. (2005). Here the thermodynamic EMF of the cells had values between 0.4 and 0.6 V, and the (zero-current) OCV was about 2.0 V. Current densities of up to 40 mA/cm^2 were attained at room temperature. If synthesis or elimination of a particular product is the main purpose, it would be realistic to put an external power source in series with the fuel cell to speed up the process.

Further examples where fuel cells are used to produce power and useful chemical processes simultaneously have been reported in the review by Alcaide et al. (2006).

REFERENCES

Accorsi R., E. Bergmann, *J. Electrochem. Soc.*, **127**, 804 (1980).

Bents D. J., J. Vincent, J. Scullin, B. J. Chang, et al., *Fuel Cell Bull.*, **2006**, No. 1, p. 12 (2006).

Brillas E., F. Alcaide, P.-L. Cabot, *Electrochim. Acta*, **48**, 331 (2002).

Chin D.-T., R. S. Yeo, J. McBreen, S. Srinivasan, *J. Electrochem. Soc.*, **126**, 713 (1979).

Gileadi E., S. Srinivasan, F. J. Salzano, C. Braun, A. Beaufrere, S. Gottesfeld, *J. Power Sources*, **2**, 191 (1977–1978).

Iwahara H., H. Uchida, N. Maeda, *J. Power Sources*, **7**, 293 (1982).

Langer S. H., S. Yurchak, *J. Electrochem Soc.*, **116**, 1228 (1969).

Liu H., B. Yi, M. Hou, et al., *Electrochem. Solid-State Lett.*, **7**, A56 (2004).

Mitlitski F., B. Myers, A. H. Weisberg, et al., *Fuel Cell Bull.*, **2**, 6 (1999).

Rasten E., G. Hagen, R. Tunold, *Electrochim. Acta*, **48**, 3945 (2003).

Song S., H. Zhang, B. Liu, et al., *Electrochem. Solid-State Lett.*, **10**, B122 (2007).

Stucki S., G. G. Scherer, S. Schlagowski, E. Fischer, *J. Appl. Electrochem.*, **28**, 1041 (1998).

Swette L. L., A. B. LaConti, S. A. McCatty, *J. Power Sources*, **47**, 343 (1994).

Yuan X., Z. Ma, H. Bueb, et al., *Electrochim. Acta*, **50**, 5172 (2005).

Zhigang Sh., Y. Baolian, H. Ming, *J. Power Sources*, **79**, 82 (1999).

Reviews

Alcaide F., P.-L. Cabot, E. Brillas, Fuel cells for chemicals and energy cogeneration, *J. Power Sources*, **153**, 47 (2006).

Pettersson J., B. Ramsey, D. Harrison, A review of the latest developments in electrodes for unitised polymer electrolyte fuel cells, *J. Power Sources*, **157**, 28 (2006).

PART III

INHERENT SCIENTIFIC AND ENGINEERING PROBLEMS

CHAPTER 11

FUEL MANAGEMENT

Hydrogen is the true working reactant, that is, the reducing agent being oxidized in the current-producing reaction at the anode of most types of fuel cells. It is only in DMFCs and DLFCs that methanol and some other liquid reducing agents constitute the working reactant; in high-temperature MCFCs and SOFCs, carbon monoxide may serve as a working reactant in addition to hydrogen.

Theoretically, 0.03 kg or 357 L of hydrogen (at a pressure of 1 bar and a temperature of 25°C) is needed to produce 1 kWh of electrical energy. Practically, though, because of ohmic and polarization losses, when the working voltage of the fuel cell is 0.8 V, for example, 0.046 kg or 548 L of hydrogen is needed.

As a working reactant in fuel cells, hydrogen has unique properties. It is highly active electrochemically, and when it is oxidized anodically, the polarization of electrodes is very low (usually below 10 mV). Therefore, energy losses associated with its use as a reducing agent (or fuel) at the anode are very low as well. These losses are much lower than those associated with the use of other reducing agents at the anode and all oxidizing agents at the cathode.

Still, two problems must be remembered when considering the use of hydrogen in fuel cells:

Fuel Cells: Problems and Solutions, By Vladimir S. Bagotsky

1. Hydrogen is not a natural fuel, which means that its use is *inconsistent* with the *original* aim of all work on fuel cells: to achieve the direct conversion of the chemical energy of natural fuels to electrical energy.
2. Hydrogen is rather complicated in its handling, storage, and transport from places of production to places of use, which means that its use is *difficult* when considering the *alternative* aim of work on fuel cells: to provide efficient and clean autonomous power sources, including mobile and portable power sources.

In view of the points made above, the following targets were set for fuel cell development:

1. Finding ways to convert natural fuels to hydrogen and to carbon monoxide
2. Finding chemical ways to produce hydrogen under conditions convenient for fuel cell–powered autonomous power plants
3. Finding ways to purify hydrogen and carbon monoxide obtained as in targets 1 and 2, so as to be fit to be used in the various types of fuel cells
4. Finding ways for convenient handling, storage, and transport of hydrogen

11.1 REFORMING OF NATURAL FUELS

The most realistic way to use natural types of fuel in fuel cells is their prior chemical (catalytic) reforming to gas mixtures rich in hydrogen. Natural fuels that are of basic interest for fuel cells include the various hydrocarbons: natural gas (methane), higher hydrocarbons (propane, butane), petroleum products (gasoline, diesel fuel), and several kinds of fossil carbons, such as coal. In addition, bioethanol produced today on a large scale by enzymatic processing of biomass must be counted among natural energy resources.

The gas mixtures produced by reforming contain products such as hydrogen and carbon monoxide that can be used to generate electrical energy in various fuel cells, but also products that are useless or even harmful for fuel cell purposes: nitrogen, carbon dioxide, sulfur compounds, and others.

Fuels are characterized by their thermal energy of combustion (oxidation), $Q_{ox} \equiv -\Delta H$. An important criterion of the reforming process is its efficiency η^0: the ratio of the thermal energy of combustion (oxidation) of the products generated, Q_{ox}^{prod}, diminished by all the energy Q^{loss} spent for the reforming process (including particularly the high-temperature steam needed for reforming), and divided by the heat of combustion of the starting materials, Q_{ox}^{fuel}; that is, $\eta^0 = (Q_{ox}^{prod} - Q^{loss})/Q_{ox}^{fuel}$. In low- and intermediate-temperature fuel cells, only the hydrogen is a useful product in the gas mixture produced by reforming. Then an efficiency η^H with respect to hydrogen is calculated where instead of Q_{ox}^{prod}, only Q_{ox}^{H}, the heat of combustion of hydrogen, is counted.

Most natural fuels contain sulfur compounds: sulfides, mercaptans, and others. These may be harmful to the activity of the catalysts, so that prior to their reforming, the fuels usually undergo a desulfurization process.

11.1.1 Natural Gas (Methane)

Natural gas consists primarily of methane (CH_4). Methane trapped as hydrate in marine sediments and as frozen hydrate in cold regions of the Earth should be counted in addition to the natural gas found in independent gas fields and in coal and oil fields. Methane is a major source for the mass production of cheap technical hydrogen. Several production methods exist, and we describe these next.

Steam Reforming (SR)

The steam reforming (SR) of methane yields hydrogen and carbon monoxide. It is an endothermic reaction: a reaction requiring an input of thermal energy (the heat of this reaction, Q_{react}, is negative):

$$CH_4 + H_2O \rightarrow 3H_2 + CO \qquad Q_{react} = -206.2\,\mathrm{kJ/mol} \tag{11.1}$$

This reaction may be accompanied by the water-gas shift reaction (WGSR):

$$CO + H_2O \rightarrow H_2 + CO_2 \qquad Q_{react} = -41.1\,\mathrm{kJ/mol} \tag{11.2}$$

boosting the share of hydrogen in the gas mixture generated.

Steam reforming is the most efficient reaction for producing hydrogen from methane. Its overall energy conversion efficiency is as high as 60%. This reaction occurs at temperatures of 500 to 800°C over catalysts such as Co (5 to 9 wt%)/Al_2O_3 or Ni/Al_2O_3. It may be accompanied by coking of the catalyst during the Boudouard reaction producing carbon:

$$2CO \rightarrow C + CO_2 \qquad Q_{react} = 171.5\,\mathrm{kJ/mol} \tag{11.3}$$

The partial pressure of the steam introduced into the reactor together with the methane should be kept rather high, to prevent coking and to preserve CO. For the same purposes, various promoters, such as rare-earth metals, are added to the catalysts.

Reforming by Partial Oxidation (POX)

For partial oxidation, a limited amount of oxygen is introduced into the reactor together with the methane. Then carbon monoxide and hydrogen are produced according to

$$CH_4 + \tfrac{1}{2}O_2 \rightarrow CO + 2H_2 \qquad Q_{react} = 35.6\,kJ/mol \tag{11.4}$$

In addition, a certain amount of methane will undergo complete oxidation to CO_2 and H_2O. Unlike SR, POX is an exothermic reaction; that is, heat is evolved as the reaction proceeds. Less complicated and bulky equipment is needed for POX than for SR, but the gas mixture produced contains not as many combustible gases, so that the overall energy conversion efficiency of POX is lower. In the opinion of Peters et al. (2002), this process would be applicable to power plants of low output, such as those used to power portable devices. In large power plants where a high energy-conversion efficiency is the prime consideration, the SR of natural gas will be preferred.

Autothermal Reforming

This process is the combination of POX and SR. A measured amount of oxygen is added to the steam-gas mixture introduced into the SR reactor:

$$\begin{aligned} &CH_4 + (x/2)O_2 + (1-x)H_2O \rightarrow CO + (3-x)H_2 \\ &Q_{react} = 241.8x - 206.2\,kJ/mol \end{aligned} \tag{11.5}$$

It is the advantage of this combination that endothermic SR and exothermic POX occur within the same reactor, so that heat need not be transferred from one reactor to another. In the case of $x = 1$, one has $\Delta H = 35.6\,kJ/mol$, which implies that an insignificant amount of heat needs to be supplied or withdrawn for the reaction to proceed. ATR was proposed by workers at the Argonne National Laboratory (Ahmed and Kumpelt, 2001). A highly efficient reactor was developed for ATR, the catalyst being 5 wt% of Pt or Ru on a support of Gd-doped CeO_2 (CGO).

Thermal Decomposition

Thermal decomposition or thermal cracking of methane is another endothermic reaction by which hydrogen can be produced from methane:

$$CH_4 \rightarrow C + 2H_2 \qquad Q_{react} = -74.6\,kJ/mol \tag{11.6}$$

which occurs over activated-carbon catalysts at a temperature of about 50°C. It is the great advantage of this reaction that pure hydrogen free of CO and CO_2 can be produced. Fast deactivation of the catalyst is a considerable defect (Pinilla et al., 2007).

Dry Reforming

Endothermic dry reforming with carbonic acid:

$$CH_4 + CO_2 \rightarrow 2CO + 2H_2 \qquad Q_{\text{react}} = -274\,\text{kJ/mol} \tag{11.7}$$

is a further possibility when a source of carbon dioxide is available, as in biogases produced by enzymatic decomposition of various wastes and containing CH_4 and CO_2.

11.1.2 Carbon Resources

The processes of gasification of the various carbon resources (fossil coal, shale, peat, coke, etc.) are well developed and widely used in the economy. Here, too, one distinguishes the endothermic processing of carbon with steam, producing water gas or syngas (a mixture of CO and H_2):

$$C + H_2O \rightarrow H_2 + CO \qquad Q_{\text{react}} = -130.2\,\text{kJ/mol} \tag{11.8}$$

and the exothermic partial oxidation of carbon with oxygen:

$$C + \tfrac{1}{2}O_2 \rightarrow CO \qquad Q_{\text{react}} = -110.5\,\text{kJ/mol} \tag{11.9}$$

For the gasification of carbon materials, most often a combined steam–air gasification of carbon at a temperature of about 600°C, sometimes with elevated pressure of the steam–air mixture, is used:

$$\begin{aligned} &C + (x/2)O_2 + (1-x)H_2O \rightarrow CO + (1-x)H_2 \\ &\quad Q_{\text{react}} = 19.7 - x \cdot 130.2\,\text{kJ/mol} \end{aligned} \tag{11.10}$$

11.1.3 Bioethanol

Bioethanol that is obtained by enzymatic decomposition of biomasses such as corn or wheat waste comes as an aqueous solution that contains about 12 vol% of ethanol (H_2O/EtOH ≈ 2:1). Pure ethanol can be obtained from this solution by distillation.

The pure bioethanol can be converted to hydrogen by steam reforming. Efficient catalysts of this process are cobalt on supports such as Al_2O_3 and ZrO_2 or nickel on supports such as Y_2O_3 and La_2O_3. The process is conducted at temperatures of 500 to 700°C. The first step is dehydrogenation of the ethanol to acetaldehyde, which under the given conditions is unstable and decomposes to CO and CH_4. Therefore, the process continues as in methanol reforming.

The conditions used in bioethanol reforming are similar to those existing in the anode compartment of MCFCs. It will therefore be possible to use bioethanol directly in direct internal reforming MCFCs (DIR-MCFCs) (see the review by Frusteri and Freni, 2007).

It must be remembered that any reforming of natural fuels will not yield pure hydrogen but a gas mixture that in addition to hydrogen contains considerable amounts of carbon monoxide and many other compounds, such as carbon dioxide, nitrogen, and sulfur compounds. Such mixtures are not fit for direct use in fuel cells, they must first be subjected to a purification by which all those contaminants are removed that could be harmful to fuel cells of a given type. Aspects of the purification of hydrogen from CO and other undesirable contaminants are discussed in Section 11.3.

The selection of catalysts for the various reforming reactions used for natural fuels has been discussed in a review by Cheekatamarla and Finnerty (2006).

11.2 PRODUCTION OF HYDROGEN FOR AUTONOMOUS POWER PLANTS

Fuel cell–based power plants can be supplied with hydrogen in two ways: (1) from special hydrogen storage means [tanks with compressed hydrogen gas, cryogenic vessels with liquefied hydrogen, and hydrogen absorbers (see Section 11.4); the hydrogen thus stored has most often been produced by electrolysis of aqueous solutions (see Section 11.2.1)]; and (2) from a chemical reaction involving products containing hydrogen. For the second, those products are of particular interest for which a distribution infrastructure covering all places interested in local power generation exists. These include natural gas (methane), liquefied petroleum gas (LPG; a mixture of propane and butane), gasoline, and diesel fuel. Products readily converted to hydrogen by chemical means are of interest as well: methanol and certain hydride compounds, including ammonia.

11.2.1 Water Electrolysis

The production of hydrogen and oxygen by water electrolysis has long been known and practiced. It is usually conducted at about 90°C in aqueous alkaline solutions (4 to 10 M KOH or NaOH) using iron alloy electrodes. Sometimes the gases produced are maintained under elevated pressures of up to 40 bar, for instance. Current densities are 3 to 4 kA/m^2. An individual electrolysis cell works at a voltage of 1.7 to 2.0 V. The electrical energy consumed per cubic meter of hydrogen gas (at atmospheric pressure) is 4.0 to 4.8 kWh. The hydrogen produced is very pure (99.5 to 99.7%) and merely contains traces of oxygen that have diffused from the oxygen to the hydrogen chamber. The hydrogen produced by electrolysis can be used in any type of fuel cell without further processing. Onda et al. (2004) reported that electrolysis would be feasible at gas pressures of up to 700 bar and a temperature of 250°C. The energy consumption would then be somewhat smaller than in ordinary electrolysis, followed by compression of the gases to the stated pressure.

11.2.2 Hydrocarbon Reforming for Autonomous Power Plants

The reforming of higher hydrocarbons to be used in autonomous power plants is not basically different from that of methane described above and to be used on a large industrial scale. The same reactions—SR, POX, and ATR—are used. Depending on the hydrocarbon actually used, particular catalysts and processing conditions (temperature, pressure) will be selected.

Special needs and features in plant operation will, of course, require adjustments in reactor design. The reforming reactors used for autonomous power plants will have to be more compact. They will often be integrated into the fuel cell plant itself, forming indirect reforming fuel cells (IR-FCs). Through coordination and optimization of the heat flows in the reactor and fuel cells, a marked weight and volume reduction of the entire plant can be attained as well as a reduction of the energy losses.

Different types of compact reforming reactors have been described in the literature. Megede (2002) described a reactor ME75-5 specifically built for a Necar-3 electric car developed by Daimler-Chrysler. This reactor was able to function in an intermediate temperature range from 250 to 300°C for methanol reforming, and in a high-temperature range of up to 800°C for LPG reforming.

11.2.3 Methanol Reforming

Methanol as a starting material for producing hydrogen has the great advantage that its reforming is possible at somewhat lower temperatures (250 to 300°C) than that of other organic materials. For its steam reforming, catalysts such as $Cu/ZnO/Al_2O_3$ (with the metallic copper in a highly dispersed state), CeO_2-containing intercalated Si^{4+}, Y^{3+}, Mg^{2+}, and other ions, and Pt- or Pd-containing catalysts are used. The scheme of a methanol reforming plant used for hydrogen supply to a PEMFC that was reported by Kamarudin et al. (2006) is shown as an example in Figure 11.1.

11.2.4 Hydrogen from Inorganic Products

A number of ways exist for local hydrogen generation when relatively small amounts are needed. A well-known device is the Kipp generator, which used to be found in chemical laboratories, where the hydrogen is produced by chemical reaction of zinc with hydrochloric acid. For fuel cells, Soler et al. (2007) suggested an analogous method of generating hydrogen by reacting aluminum

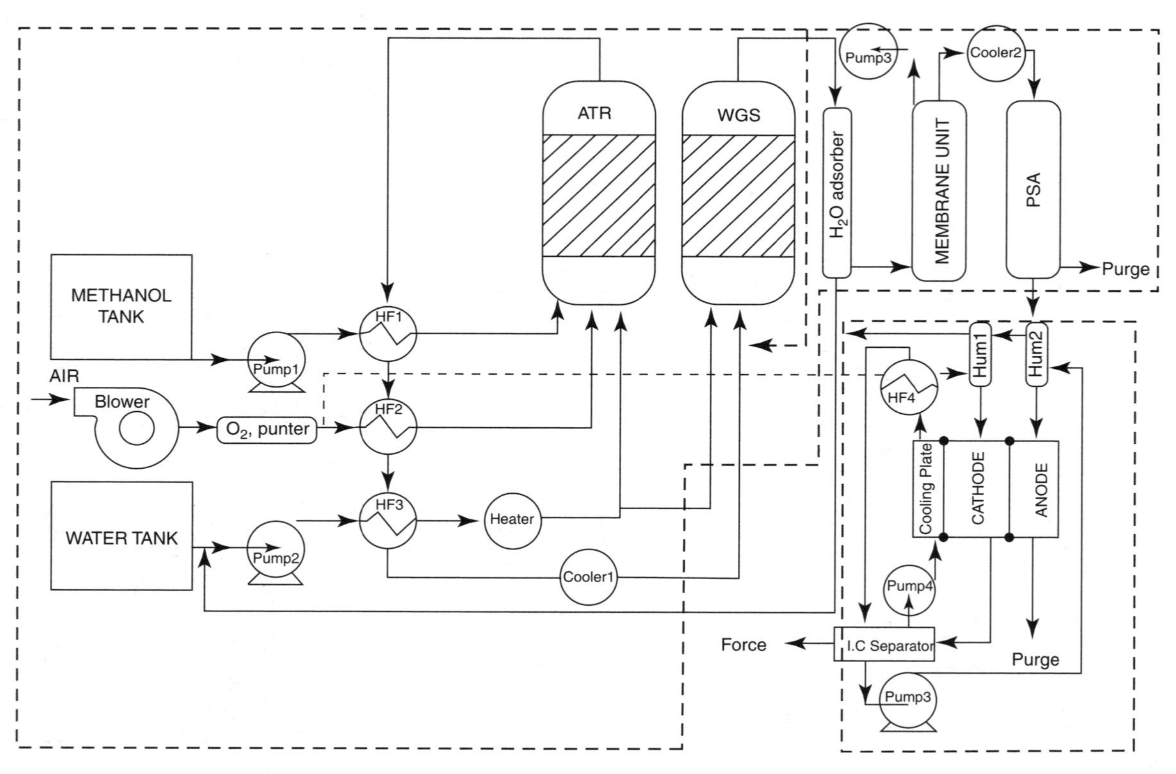

FIGURE 11.1 Schematic of a PEMFC system with methanol reforming. (From Karamudin et al., 2006, with permission from Elsevier.)

or aluminum alloys with an alkaline solution. The aluminates left in the solution could be regenerated to alkali. These authors thought that this was the cheapest way of generating hydrogen. For the same purpose, Chandra and Xu (2006) had suggested the hydrolysis of ammonia borane (NH_3BH_3), which is a substance containing 19.5 wt% of hydrogen. Aqueous solutions are stable when oxygen is absent. The material decomposes and liberates hydrogen when platinum catalysts are present.

Amendola et al. (2000) had suggested the hydrolysis of an aqueous solution of sodium borohydride ($NaBH_4$), a material containing 10.9 wt% of hydrogen, for local hydrogen generation:

$$NaBH_4 + 2H_2O \rightarrow 4H_2 + NaBO_2 \tag{11.11}$$

An alkaline solution of this hydride is stable; hydrolysis proceeds in the presence of ruthenium catalyst. It is a great virtue of this process (and also of the process of ammonia borane decomposition mentioned above) that the reaction stops instantly when the catalyst is lifted from the solution. Therefore, hydrogen generation can be controlled on demand: that is, in harmony with fuel cell needs.

Other papers describing the use of sodium borohydride have appeared in the literature. Thus, Kojima et al. (2004) suggested that the reaction be run in a closed pressure vessel, to collect hydrogen under higher pressure.

11.3 PURIFICATION OF TECHNICAL HYDROGEN

11.3.1 Desulfurization

Natural gas, liquefied petroleum gas (LPG), and carbon materials serving as the raw materials for cheap technical hydrogen usually contain marked amounts of sulfur compounds. In addition, sulfur-containing odorants (tetrahydrothiophene or ethylmercaptan) often are purposely added to the gas distribution network so that gas leaks may be detected in a timely manner, because natural gas is usually odorless.

Even in small amounts, sulfur compounds are extremely harmful to the activity of most catalysts used in fuel cells. The admissible amounts of these impurities are in the ppb (parts per billion) range. In larger amounts, sulfur contaminants will also be harmful to hydrocarbon fuel reforming.

On an industrial scale, sulfur compounds are eliminated from natural gas by their hydrogenation to H_2S and subsequent absorption by zinc oxide (ZnO). For small fuel cell power plants, this method is too unwieldy and is not used. De Wild et al. (2006) suggested using commercial absorbents of the NGDM-1 type (doped, although details have not been reported) for desulfurization in small power plants. This is a rather simple but very effective method, leaving less than 20 ppb of residual sulfur after the treatment.

11.3.2 Purification from CO

Apart from carbon dioxide, nitrogen, and unreacted hydrocarbons, the technical hydrogen produced by the reforming of hydrocarbon fuels also contains marked quantities (up to 1% = 10,000 ppm) of carbon monoxide. Before being used in low- and intermediate-temperature fuel cells, the CO level in hydrogen should be brought down to values of about 10 ppm. This can be achieved in different ways, as noted below.

The Water-Gas Shift Reaction

This reaction (11.2) not only lowers the CO level but produces additional hydrogen. It is a defect of this reaction that the equilibrium tends toward the side of CO formation when the CO_2 levels are considerable. A variant of this method where this endothermic reaction is run with electrical energy but without thermal energy from outside has been developed by Huang et al. (2006). They used an electrolyzer with platinized electrodes operated with pulsed current (Figure 11.2). The gas, which contains CO (not in very large quantities), is fed into the anode chamber. During idle periods, the platinum surfaces saturate with adsorbed CO. During active periods the adsorbed CO is oxidized electrochemically to CO_2, while hydrogen oxidation that should occur in the anode chamber is blocked by residual adsorbed CO. At the cathode, hydrogen is evolved. Two such electrolyzers working in counterphase could be used simultaneously. Their overall reaction is equivalent to WGSR. The electrochemical water-gas shift reaction (EWGSR) can be run at any

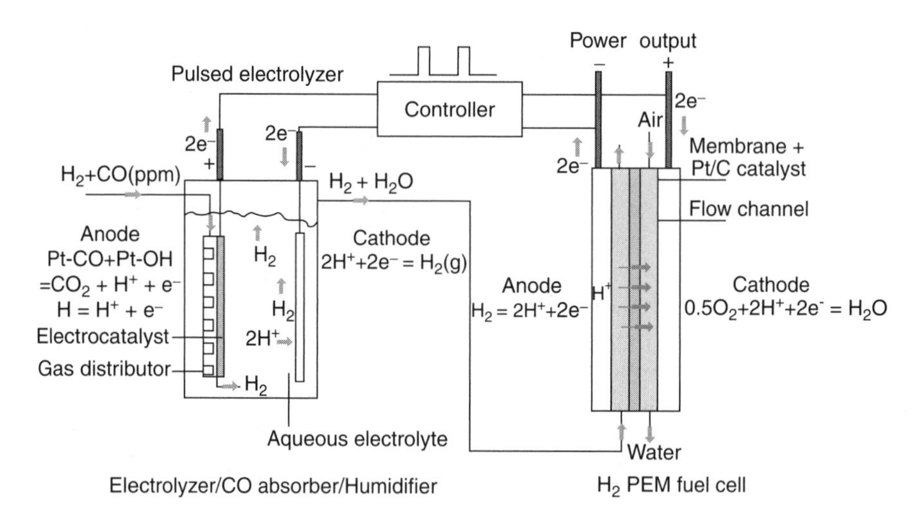

FIGURE 11.2 Flow diagram depicting the removal of low-level CO from the PEMFC anode H_2 feedstream by pulse electrolysis. (From Huang et al., 2006, with permission from Elsevier.)

temperature: for example, close to the conditions of PEMFC operation (60 to 80°C, 1 bar). This approach would also work for the removal of some other kinds of impurities in hydrogen.

The Methanation Reaction

The exothermic reaction producing methane:

$$CO + 3H_2 \rightarrow CH_4 + H_2O \qquad Q_{react} = 206.2\,kJ/mol \qquad (11.12)$$

will for thermodynamic reasons, occur as written (to the right) only at temperatures below 530°C. At higher temperatures it will proceed in the other (endothermic) direction, as for instance during the steam conversion of methane (11.1). It is a disadvantage that hydrogen is not generated but, to the contrary, is consumed in large amounts when the methanation reaction is used to eliminate CO.

The two reactions discussed provide a marked, although not a complete elimination of CO from hydrogen. A more thorough purification can be attained by running the two reactions simultaneously (Batista et al., 2005).

Palladium Diaphragm Pumps

A complete i.e., practically 100% purification of hydrogen from CO and any other contaminants can be achieved with thin membranes (30 to 70 μm) of palladium or a palladium alloy with 10% silver. Palladium absorbs hydrogen, and when a pressure gradient is present across the membrane, the absorbed hydrogen will diffuse through the membrane, leaving behind all other substances to which the membrane is practically completely impermeable. This method has the defect of being slow as a result of its diffusion mechanism. Some acceleration is possible at higher temperatures and pressure gradients.

This membrane process of hydrogen purification can be simplified considerably and can be made more productive by using the electrochemical pump principle. An electrochemical cell holding two auxiliary electrodes and an acidic electrolyte solution (e.g., hydrochloric acid) is divided into two halves by a palladium membrane (Figure 11.3). An electric current is sent through the cell, making the two sides of the palladium membrane electrodes of opposite polarities (a bipolar palladium electrode). Contaminated hydrogen is supplied to the positive auxiliary electrode A1, where it is oxidized anodically to hydrogen ions. The electric field drives these ions toward the metal membrane C1, where they are reduced cathodically to hydrogen atoms. These penetrate through the membrane by diffusion, to be reoxidized to hydrogen ions on the other side (A2) of the membrane. These ions travel to the negative auxiliary electrode (C2), where they undergo cathodic reduction to pure hydrogen, ready to be withdrawn from the cell. When an electric current is passed, an electrochemical potential gradient of the hydrogen ions is set up which is

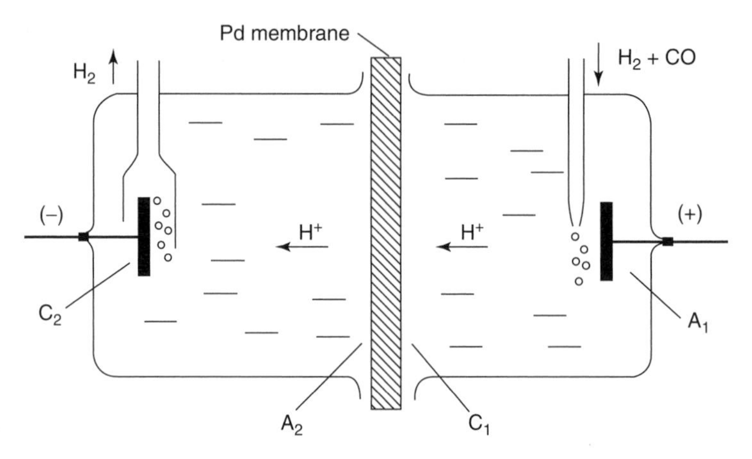

FIGURE 11.3 Schematic of an electrochemical pump with a palladium membrane for hydrogen purification.

such that diffusion through the membrane is accelerated (the same acceleration is produced when a pressure differential of hydrogen gas exists across the membrane). Here hydrogen diffusion through the membrane is accelerated not by a gas pressure gradient, but by an electrochemical potential gradient of the hydrogen ions.

In recent years, purely electrochemical versions of hydrogen pumps working without any palladium membrane were reported (Lee et al., 2004; Onda et al., 2007). They employ ordinary fuel cells (e.g., of the PEMFC type) operated symmetrically, with two hydrogen electrodes, so when current is made to flow through the fuel cell and contaminated hydrogen is used as fuel at the anode, clean hydrogen is produced at the cathode. This is a cheap method, even though the degree of purification is much lower than that achieved with a palladium membrane, since various contaminants may diffuse through the fuel cell's MEA.

11.4 HYDROGEN TRANSPORT AND STORAGE

Problems with the storage and transport of hydrogen arise primarily when using electrolytic hydrogen. Technical hydrogen is produced by the reforming or gasification of products (natural gas, LPG, coal, and carbon products) for which distribution and storage infrastructures have long been established and will reach even very remote areas. The technical hydrogen consumed in stationary power plants as a rule is generated right next to these plants or even in a unit that is part of the plant. Even in vehicles (electric cars, locomotives, ships), technical hydrogen is supposed to be generated right on board. The production of electrolytic hydrogen, to the contrary, is concentrated in electrolysis plants next to sources of cheap power but sometimes quite far from the power plants that need it.

Installations for temporary hydrogen storage and for hydrogen transport must satisfy a number of requirements. They have to be compact [i.e., have a small volume and weight (mass)]. They should readily give off the hydrogen and readily refuel with it. These characteristics are described by quantitative parameters: the weight-specific (gravimetric) capacity C_{grav} (in kWh/kg or kg H_2/kg or wt% H_2), the volume-specific capacity C_{vol} (in kWh/L or kg H_2/L), the rate of hydrogen supply, u_{suppl}, and the rate of refueling, u_{refuel} (in kg H_2/min). The U.S. Department of Energy (DOE) has set the following target figures to be met by vehicle power plants by the year 2010 (2015): $C_{grav} = 2.0\ (3.0)$ kWh/kg or 6 (9) wt%; $C_{vol} = 1.5\ (2.7)$ kWh/L; $u_{refuel} = 1.5\ (2.0)$ kg H_2/min.

11.4.1 Compressed Hydrogen

The best known and most widely used method of hydrogen storage and shipping is with steel cylinders of capacity 50 to 100 L, where the gas is compressed (at up to 700 bar). The cylinders or tanks are filled with hydrogen readily and rapidly and give it off as needed, just as readily and rapidly. The major drawback is the large weight (more than 40 kg) of such a cylinder and the low specific capacity that follows: $C_{grav} = 0.3$ to 1.0 kWh/kg, $C_{vol} = 0.3$ to 0.6 kWh/L.

Such tanks with compressed hydrogen are said to have been used to supply relatively small stationary fuel cell power plants. Considering the unfavorable weight and, more so, volume figures, it will be difficult to use them in electric cars, unless a compromise is struck with the driving range.

In recent years, versions of lighter tanks have been developed, such as tanks made of fiber-reinforced plastic. The weight figures of such tanks are twice as good as those of steel tanks, but the volume figures are practically the same.

11.4.2 Liquefied Hydrogen

Improved methods for storing hydrogen liquefied at 20 K (−253°C) have developed along with rocket technology and the use of relatively large amounts of hydrogen as a fuel in certain types of rockets as well as in the power plants of U.S. space shuttles. To reduce losses, the cryogenic tanks holding the liquid hydrogen are usually kept in cryogenic tanks with liquid nitrogen. In addition, the entire storage plant is provided with multilayer screens to ward off heat radiation.

Hydrogen liquefaction is a highly energy-consuming process requiring up to 20 kWh/kg, that is, about 65% of the energy content of the liquid hydrogen produced. The numerous successful shuttle flights have shown that cryogenic hydrogen storage is entirely suitable for such applications.

The use of cryogenic storage of hydrogen for fuel cells in electric cars and other vehicles has been discussed repeatedly. Three points cause concern here: (1) During refueling of a car, very large amounts of hydrogen are lost until the "hot" automobile tank has cooled down to the low temperature required; (2) despite all thermal insulation, some evaporation and leakage of hydrogen is unavoidable and necessitates regular refueling of parked cars; and (3) all possible consequences of car accidents have not yet been clarified.

11.4.3 Metal Hydrides

In the mid-1960s, scientists working in the laboratories of the Dutch Philips company demonstrated that certain alloys containing rare-earth metals, particularly the alloy $LaNi_5$, were able to absorb and desorb hydrogen reversibly. During absorption, the intercalation of hydrogen atoms yields *intercalation compounds* such as $LaNi_5H_x$. This discovery immediately drew the attention of many workers, instigating a large amount of research in this field. It was shown that the hydrogen intercalated in the alloy and present as host compounds remains electrochemically active, having potentials close to that of gaseous hydrogen. Subsequently, new batteries were developed on this basis: nickel-hydride batteries, which have reached multimillion production figures in many countries around the world.

In recent years, the hydrogen intercalation compounds of various metals and alloys have been studied with the aim of building reversible systems for temporary hydrogen storage. The important special feature of these systems is their very high volume-specific capacity C_{vol}, which can attain very high values: up to 0.1 to 0.2 kg H_2/L (or 3 to 6 kWh/L). This is more than the corresponding values for pure gaseous hydrogen compressed to 700 bar (0.03 kg H_2/L, not counting the weight of the tank) or pure liquid hydrogen (0.07 kg H_2/L). This apparent paradox can be resolved by realizing that in the crystal lattice of the metal host, hydrogen is present in an atomic state rather than in a molecular state. Because of the weight of the metal hosts, which is sometimes very considerable, the gravimetric figures are actually not as good as the volumetric figures. For this reason, prime interest in hydrogen storage in the form of metal hydrides is in the field of small portable power sources. Among the systems being discussed, hydrides of iron–titanium alloy ($FeTiH_x$), and of vanadium (VH_x) are of particular interest. The intercalated hydrogen atoms are quite mobile in these structures. In other structures, such as MgH_x, covalent interactions between hydrogen and the metal are predominant, and the mobility of the hydrogen atoms is restricted.

Basically, the "refueling" of the host, that is, hydrogen intercalation, is facilitated by elevated pressure of the gaseous hydrogen and lower temperature. The "discharge" or deintercalation of hydrogen and its delivery to the consumer are facilitated by elevated temperature of the storage system and reduced pressure of the hydrogen gas. In many cases, unfortunately, these processes are quite slow. Sometimes a certain acceleration is possible through special catalysts. Another problem is the performance degradation (decrease of the specific capacity) sometimes observed when intercalation and deintercalation have been repeated many times.

The work toward practical hydrogen storage systems using metal hydrides is very promising, but still in an early stage, and numerous problems associated with the kinetics of the processes and the stability of the systems remain to be solved. Many of these points have been discussed by Eberle et al. (2006).

11.4.4 Carbon Materials

In the mid-1980s, various carbon materials forming nano-sized structures were discovered: fullerenes, nanotubes, nanofibers, and so on. That activated carbons are excellent adsorbents for many different substances has long been known. Adsorption on carbon materials as a rule is physical in its character, occurring due to van der Waals forces. The energy of such interactions is very low, on the order of few kJ/mol. Physical adsorption of hydrogen can occur via the same mechanism on carbon materials. When the nanostructured carbon materials became available, there was hope that they could find practical use for temporary hydrogen storage. Toward the end of the twentieth century it was reported by a number of research groups [e.g., the group of N. M. Rodriguez (Chambers et al., 1998)] that extremely high values of the specific adsorption capacity (several tens of weight percent) could indeed be attained on such materials. These reports have aroused great interest and stimulated many investigations in this field.

Subsequent investigations have not confirmed the initial data. In fact, very remarkable physical adsorption of hydrogen does occur on the nanostructures, but does not go beyond 3 wt%.

A detailed analysis of hydrogen adsorption on various nanostructured carbon materials has been reported in a thorough review by Ströbel et al. (2006). These authors regard it as quite possible that in the future, carbon structures will be found on which hydrogen adsorption in amounts of or even above 6 wt% could be achieved.

REFERENCES

Ahmed S., M. Kumpelt, *Int. J. Hydrogen Energy*, **26**, 291 (2001).

Amendola S. C., S. L. Sharp-Goldman, M. S. Janjua, et al., *J. Power Sources*, **85**, 186 (2000).

Batista M. S., E. I. Santiago, E. M. Assaf, E. A. Ticianelli, *J. Power Sources*, **145**, 50 (2005).

Chambers A., C. Park, R. T. K. Baker, N. M. Rodriguez, *J. Phys. Chem. B*, **102**, 4253 (1998).

Chandra M., Q. Xu, *J. Power Sources*, **156**, 190 (2006).

Eberle U., G. Arnold, R. von Helmolt, *J. Power Sources*, **154**, 456 (2006).

Huang C., R. Jiang, M. Elbaccouch, et al., *J. Power Sources*, **162**, 563 (2006).

Kamarudin S. K., W. R. W. Daud, A. Md. Som, et al., *J. Power Sources*, **157**, 641 (2006).

Kojima Y., Y. Kawai, H. Nakanishi, S. Matsumoto, *J. Power Sources*, **135**, 36 (2004).
Lee H. K., H. Y. Choi, K. H. Choi, et al., *J. Power Sources*, **132**, 92 (2004).
Megede, D. zur, *J. Power Sources*, **106**, 35 (2002).
Onda K., T. Kyakuno, K. Hattori, K. Ito, *J. Power Sources*, **132**, 64 (2004).
Onda K., K. Ichihara, M. Nagahama, et al., *J. Power Sources*, **164**, 1 (2007).
Peters R., R. Dahl, K. Klüttgen, C. Palm, D. Stolten, *J. Power Sources*, **106**, 238 (2002).
Pinilla J. L., R. Utrilla, et al., *J. Power Sources*, **169**, 103 (2007).
Soler L., Macanas J., M. Munoz, J. Casado, *J. Power Sources*, **169**, 144 (2007).
Wild, P. J. de, R. G. Nyquist, F. A. de Bruijn, F. R. Sobbe, *J. Power Sources*, **159**, 995 (2006).

Reviews

Cheekatamarla P. K., C. M. Finnerty, Reforming catalysts for hydrogen generation in fuel cell applications, *J. Power Sources*, **160**, 490 (2006).
Frusteri F., S. Freni, Bio-ethanol, a suitable fuel to produce hydrogen for a molten carbonate fuel cell, *J. Power Sources*, **173**, 200 (2007).
Ströbel R., J. Garche, P. T. Moseley, et al., Hydrogen storage by carbon materials, *J. Power Sources*, **159**, 781 (2006).

CHAPTER 12

ELECTROCATALYSIS

12.1 FUNDAMENTALS OF ELECTROCATALYSIS

The science of electrocatalysis provides the connection between the rates of electrochemical reactions and the bulk and surface properties of the electrodes on which these reactions proceed. The degree to which an electrode will influence the reaction rates differs for different electrochemical reactions. In complex electrochemical reactions having parallel pathways, such as reactions involving organic substances, the electrode material might selectively influence the rates of certain individual steps and thus influence the selectivity of the reaction: that is, the overall direction of the reaction and the relative yields of primary and secondary reaction products.

Historically, electrocatalytic science developed from investigations into cathodic hydrogen evolution, a reaction that can be realized for many metals. It was found in a number of studies toward the end of the nineteenth century that at a given potential, the rate of this reaction differs by many orders of magnitude between metals.

Research in electrocatalysis was strongly stimulated in the early 1960s by efforts toward the development of various types of fuel cells. Studies were initiated on the various factors influencing the rates not only of hydrogen evolution but also of other reactions, particularly cathodic oxygen reduction and the complete oxidation of simple organic substances ("fuels") to carbon dioxide. The term *electrocatalysis* began to be used in a universal fashion

Fuel Cells: Problems and Solutions, By Vladimir S. Bagotsky

following publications of Thomas Grubb, in which he showed that the electrochemical oxidation of hydrocarbons (methane, ethane, ethylene, etc.) is possible at platinum electrodes at temperatures below 150°C, in contrast to gas-phase oxidations, which for the same metal catalysts would proceed with perceptible rates, only at temperatures above 250°C.

In 1965, synergistic (nonadditive) catalytic effects were discovered in electrochemical reactions. It was shown in particular that the electrochemical oxidation of methanol on a combined platinum–ruthenium catalyst will occur with rates two to three orders of magnitude higher than at pure platinum even though pure ruthenium is catalytically altogether inactive.

Appreciable interest was stirred by the successful use of nonmetallic catalysts such as oxides and organic metal complexes in electrochemical reactions. From 1968, work on the development of electrocatalysts on the basis of the mixed oxides of titanium and ruthenium led to the fabrication of active, low-wear electrodes for anodic chlorine evolution, which under the designation Dimensionally Stable Anodes (DSA) became a workhorse of the chlorine industry. All these achievements were of great practical as well as theoretical importance and brought electrocatalysis into being as a separate branch of science.

Electrocatalytic reactions have much in common with ordinary (chemical) heterogeneous catalytic reactions, but electrocatalysis has certain characteristic special features:

1. The rate of an electrochemical reaction depends not only on given system parameters (composition of the catalyst and electrolyte, temperature, state of the catalytic electrode surface) but also on electrode potential. Thus, in a given system, the potential can be varied by a few tenths of a volt; as a result, the reaction rate will change by several orders of magnitude.
2. In electrochemical reactions, the catalyst surface is in contact, not only with the reacting species but also with other species (i.e., the solvent molecules and the electrolyte ions), which in turn influence the properties of the surface and give rise to special reaction features.
3. In electrochemical systems, not only the reactant species but also electrons must be supplied to and/or withdrawn from the catalyst particles, since the electrons are involved directly in all electrochemical reactions.

Because of features 1 and 3, in particular, catalysts for electrochemical reactions must have a certain degree of electronic conductivity.

Like other heterogeneous chemical reactions, electrochemical reactions are always multistep reactions. Some intermediate steps may involve the adsorption or chemisorption of reactants, intermediates, or products. Adsorption processes as a rule have a decisive influence on the rates of electrochemical processes.

The reaction rate will be influenced not only by the surface concentration of reacting species but also by other factors, which may include the orientation of these species on the surface and their bonding to neighboring species. This implies that an analysis of electrocatalytic phenomena must include a full consideration of all features of adsorption.

Metal catalysts are most commonly used in fuel cells (and in many other areas of applied electrochemistry). The catalytic activity of metals is correlated significantly with their positions in the periodic table of the elements, since it depends on the electronic structure of the metal atoms.

High catalytic activity is found in the transition metals or *d*-metals (i.e., metals consisting of atoms with unfilled *d*-electron shells). They include the metals of subgroups IVB to VIIIB of the periodic table, most prominently the metals of the platinum and iron group but also such metals as vanadium, molybdenum, and manganese. The catalytic activity of *sp*-metals (i.e., nontransition metals from subgroups IB, IIB, IIIA, and IVA, including "mercurylike" metals such as cadmium, lead, and tin) is much lower, but it should be pointed out that such a distinction between active transition metals and inactive nontransition metals is conditional and remains somewhat ambiguous.

Platinum is the most universal catalyst for many electrochemical reactions. It is considered to be stable over a wide range of potentials in most solutions, including strongly acidic and strongly alkaline. Despite its high cost, it is widely used in fuel cells (and many other electrochemical devices). The large investments associated with platinum electrocatalysts are usually paid back by appreciably higher efficiencies.

The scientific literature abounds in attempted correlations between the catalytic activities of metals and some set of their bulk properties (electron work function, bond energy of intermediates, etc.). Such correlations would help in understanding the essence of catalytic action and enable a conscious selection of the most efficient catalysts for given electrochemical reactions.

When summarizing the results of these attempts it is only possible to say that all observed correlations are secondary. The primary factor influencing the catalytic activity of all metals and most of their physicochemical parameters are the special features of electron structure of their atoms, which in turn are related to their atomic number (i.e., their position in the periodic table). A very important criterion for this electron structure is the *percent d-character*, which characterizes the number of unpaired electrons in the *d*-orbitals of the individual metal atoms. Because of the vacancies existing in these orbitals, metals will interact with electron-donating species forming electron pairs. It is this interaction that determines the special features of adsorption of these species and, as a consequence, the catalytic activity of a given metal.

The catalytic activity of an electrode is determined not only by the nature of the electrode metal (its bulk properties) but also by the composition and structure of the surface on which the electrochemical reaction takes place. These parameters, in turn, depend on factors such as the method of electrode preparation, the methods of surface pretreatment, the conditions of storage, and other factors all having little effect on the bulk properties.

The major factors that characterize the structure and state of catalyst surfaces and may influence their activity are the crystallographic orientation of the surface and of its individual segments, the presence of structural defects of different types or of points (*active sites*) with special properties, and the presence of accidental or consciously added traces of foreign matter.

In view of the vast amount of experimental data in this field, it is very difficult to draw general conclusions that would hold for different, let alone all electrocatalytic systems. The crystallographic orientation of the surface undoubtedly has some specific influence on adsorption processes and on the electrochemical reaction rates, but this influence is rather small. It can merely be asserted that the presence of a particular surface orientation is not the decisive factor for high catalytic activity of a given catalyst surface.

Special activity is often attributed to lattice defects such as dislocations, kinks, vacancies, stacking faults, and intergrain boundaries emerging at the crystal surface. Experiments carried out with catalysts containing differerent numbers of defects have shown, however, that it will not be justified to identify crystallographic defects emerging at the electrode surface with the active sites responsible for catalytic activity of the electrode as a whole.

For higher efficiency in catalytic action and smaller quantitative needs, the catalysts often are used in a highly disperse state (or, to use modern terminology, as nanosized particles). These highly disperse metal catalysts most often are supported by an electronically conducting substrate or carrier that should provide for uniform supply or withdrawal of electrons (current) to/from all catalyst crystallites. The substrate should also serve to stabilize the disperse state of the catalyst and retard any spontaneous coarsening of the catalyst crystallites.

As a rule, the nanosized catalysts are polydisperse (i.e., their crystallites and/or crystalline aggregates come in different sizes and shapes). For particles of irregular shape, the concept of (linear) size is undefined. For such a particle, the diameter d of a sphere that has the same number of metal atoms or volume may serve as a measure of particle size.

The specific surface area of an ideal monodisperse catalyst per unit of its mass is related to its diameter d as $S = 6/\rho d$ (where ρ is the density). In the case of platinum, a particle diameter of 5 nm corresponds to a specific surface area of $56\,m^2/g$. With a particle size reduced to 2.2 nm, $S = 127\,m^2/g$, which is the largest specific surface area attainable with platinum. At this crystallite diameter all platinum atoms of a particle are surface atoms (i.e., the particle contains no "inert" bulk atoms). The specific surface area of disperse catalysts can be measured quite accurately by low-temperature nitrogen or helium adsorption (the BET method) or, in the case of platinum, in terms of the amount of charge consumed for the electrochemical adsorption or desorption of a monolayer of hydrogen atoms. In subsequent sections we consider specific questions that arise in the use of different catalysts when building and operating fuel cells.

In the Bacon cell, the first successful modern fuel cell, porous nickel was used as a material for the electrodes, doing double duty as conductor and catalyst for the current-producing electrochemical reactions (in the cathode the nickel was lithiated). A little later, Justi introduced Raney nickel and Raney silver as the catalytic electrodes: nickel for the hydrogen anode and silver for the oxygen cathode. It soon became evident, however, that to build sufficiently efficient and compact fuel cells, more active but also more expensive platinum catalysts supported by a noncatalytic electrode material had to be used. Today, practically all fuel cells operating at temperatures below 200°C use catalysts made of platinum and/or other platinum group metals. For this reason, all research efforts in the electrocatalysis of fuel cells have focused on two problems: (1) raising the efficiency of utilization of these catalysts to reduce the amounts needed, and (2) developing new fuel cell catalysts free of platinum metals.

12.2 PUTTING PLATINUM CATALYSTS ON THE ELECTRODES

In the first fuel cell models of the PEMFC and DMFC type developed in the 1960s, considerable amounts of platinum catalyst (4 to 10 mg/cm^2) that were applied directly to the current-collecting substrate (the electrode) were used. In view of these large noble-metal needs, broad development and commercialization of fuel cells was ruled out in those years. By the mid-1990s, wide research efforts had made it possible to reduce the amount of platinum in the electrodes to values below 1 mg/cm^2 by prior deposition on carbon-black supports and by improving the structure of the electrode's active layer with an added ionomeric solid electrolyte (Srinivasan et al., 1990; Gottesfeld and Zawodzinski, 1997). This new technology using less platinum did not detract from the performance figures, which to the contrary were improved significantly as well.

Yet in view of the prospects of future mass production and massive applications of fuel cells in new areas, such as the electric car and portable electronic devices, further reduction of the platinum content to below 0.1 mg/cm^2, along with new technologies for the manufacture of electrodes with platinum catalyst more amenable to mass production, were needed.

In this context, an efficiency criterion, often used in recent years for the utilization of platinum in electrodes, is formulated as γ_{eff}^{Pt}; it is the power of a fuel cell with electrodes using 1 mg of platinum catalyst and given in W/mg. This parameter is related to fuel cell production cost, while the parameter p_S (in mW/cm^2) is related to the size (total surface area of the electrodes) that is needed for a fuel cell of given output.

The traditional way of depositing highly disperse platinum onto the various supports is by reducing platinum ions Pt^{4+} from a solution of chloroplatinic acid (H_2PtCl_6), either electrochemically by passing a cathodic current, or chemically by adding reducing agents such as formaldehyde, sodium borohydride, and the like. These methods are very convenient in laboratory practice and are often useful as well for the industrial production of membrane-electrode assemblies for PEMFCs and DMFCs. However, difficulties arise when depositing catalyst layers containing very low quantities of platinum by these methods. In mass production of fuel cell electrodes, it proved very difficult to secure sufficient uniformity of the catalyst layers. Moreover, as a multistep process, these methods are too expensive for mass production, although it must be said that work toward improving them has continued until the present time (see Plyasova et al. 2006, and Zeis et al., 2007). A number of alternative methods for depositing small amounts of platinum or other, expensive metal catalysts on different substrates have been developed in recent years.

12.2.1 Sputter Deposition

As a way of depositing thin layers of metals onto substrates, sputter deposition (SD) is a technology used widely in the semiconductor industry. Its advantages are simplicity, relatively low cost, the possibility of precise control of the amount of metal deposited and of layer thickness, and the feasibility of using a variety of metal targets (sources) and substrates.

In the SD process, metal atoms are liberated from the target containing the metal(s) needed as a deposit, by ion bombardment, possibly aided by magnetron action, and then condense as a thin film on the substrate. Gruber et al. (2005) used SD to deposit a layer of platinum about 200 nm thick, containing 5 μg/cm^2, on commercial gas-diffusion electrodes from Eltech Systems. A laboratory model of a PEMFC having two such electrodes (with a total of 10 μg/cm^2) yielded a maximum energy density of 124 mW/cm^2 (at a current density of about 300 mA/cm^2), corresponding to $\gamma_{\text{eff}}^{\text{Pt}} = 12.4$ W/mg. For comparison, standard-type MEA containing 10 times more platinum (a total of 107.5 μg/cm^2) produced 203 mW/cm^2 under the same conditions, equivalent to a value of $\gamma_{\text{eff}}^{\text{Pt}}$ of about 2 W/mg. It is seen that with the thin platinum layer, the utilization efficiency increases by a factor of 10 while the specific energy density decreases slightly.

A more detailed description of equipment used to produce thin mixed Pt–Ru coatings on a gas-diffusion layer is given by Caillard et al. (2006) (Figure 12.1). Using an inductive plasma excited by a planar coil powered by a RF generator of 13.56 MHz in an argon atmosphere (pressure 0.5 Pa), high-energy argon ions were produced and used to bombard the target. The substrate was about 5.5 cm away from the target. The authors pointed out that the system is very flexible and that layers of different structure and Pt/Ru ratios can be deposited by changing individual parameters.

A number of workers noticed insufficient adhesion of the deposit to the substrate and insufficient long-term stability under operating conditions as defects of SD. It was pointed out, on the other hand, that the method, unlike in

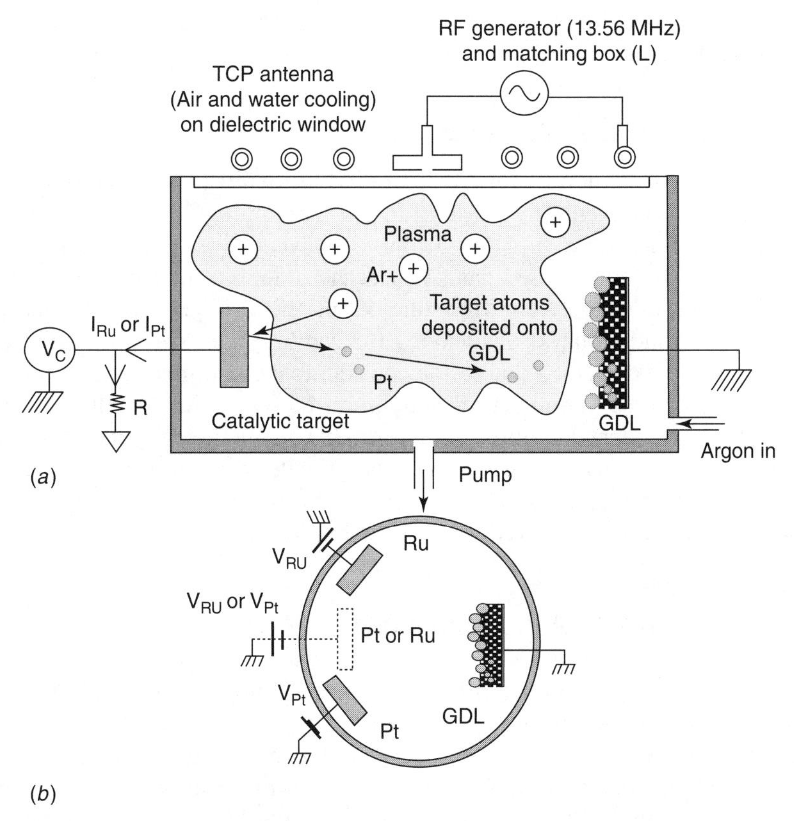

FIGURE 12.1 Schematic of a plasma sputtering reactor: (a) side view; (b) top view. (From Caillard et al., 2006, with permission from Elsevier.)

its applications in the semiconductor industry, is not required to yield continuous films when used to produce catalytic deposits; deposition in the form of individual, catalytically active surface segments will be sufficient.

12.2.2 Dual Ion Beam–Assisted Deposition

Dual ion beam–assisted deposition is a method based on physical vapor deposition (PVD). A plasma of ions to be deposited is generated by bombardment of a target with low-energy ions. The ions are then extracted from the plasma and accelerated to be included into the PVD layer growing on the substrate.

Saha et al. (2006) reported layers 25 to 75 μm thick containing 0.04 to 0.08 mg/cm^2 platinum that were obtained by this method. For MEA with two electrodes containing 0.04 mg/cm^2 each (i.e., a total of 0.08 mg/cm^2), an energy density of 270 mW/cm^2 was obtained at a voltage of 0.65 V, corresponding to a utilization efficiency of $\gamma_{eff}^{Pt} = 3.4\,W/mg$. With electrodes of the ordinary type containing $1\,mg/cm^2$ of Pt, this factor was only 0.9 W/mg.

12.2.3 Electrospray Technique

The electrospray method was used in the work of Benitez et al. (2005). Here an ultrasonically homogenized solution (an ink) containing 20 wt% of Pt/C and Nafion ionomer in an aqueous-alcoholic solvent was pushed through a capillary tube by compressed gas. An electric potential difference of 3300 to 4000 V was applied between the solution in the capillary and the carbon substrate. The individual drops leaving the capillary lose solvent by evaporation and become dispersed due to the considerable coulombic repulsion forces that develop from a significant accumulation of space charge. They form an aerosol mist that deposits onto the substrate. This method is very simple and may be used for a relatively economical large-scale manufacture of electrodes. The problems of preparing highly efficient electrodes that have an ultralow platinum content are discussed in a review by Wee et al. (2007).

12.3 SUPPORTS FOR PLATINUM CATALYSTS

An appreciable increase in the utilization efficiency of platinum catalysts in fuel cells was attained when the highly dispersed platinum was deposited not directly onto the conductive electrode base but onto carbon black or other carbon materials serving as an intermediate base for the nanodispersed catalyst. On carbon supports, the nanosized platinum crystallites were less subject to recrystallization and coarsening. In addition, new technical devices such as adding Nafion ionomer to the active mass have helped to considerably improve the contact between catalyst and solid electrolyte (a Nafion-type membrane). Carbon black was found to be a very convenient support for platinum catalysts. It is readily available and not expensive. Certain blacks (such as furnace black Vulcan XC-72) have special surface properties that have a favorable effect on catalyst activity.

It was soon found, however, that catalysts in the form of platinum deposited onto carbon black (Pt–C) have a rather important defect, inasmuch as during long-term use in a fuel cell, the surface of the carbon-black particles becomes oxidized, particularly so at the positive electrode, which upsets the contact between the platinum catalyst and the carbon black (in some cases the crystallites "slip away" from the carbon black). For this reason it became urgent to search for other supports for the highly disperse platinum catalysts.

12.3.1 Other Carbon Supports

In recent years, carbon nanotubes (CNTs) have received attention as supports for platinum catalysts. They have an internal diameter of 3 to 10 nm, an external diameter of 6 to 20 nm, and lengths of up to 100 nm. They have a large specific surface area, low resistivity, and are highly stable under the conditions of electrochemical experiments. Preliminary experience testifies to their promise for fuel cells. Wu and Xu (2007) studied two types: single-walled (SWCNTs) and multiwalled (MWCNTs). It was seen when examining the structure of platinum deposits on these nanotubes that the crystallites had the same size on the two types of tubes but differed substantially in their morphology. On SWCNTs they were close to spherical and were anchored mainly at the boundaries of the SWCNT bundles and the Nafion binder. This unique morphology contributes to an improved utilization of platinum. Therefore, SWCNTs offer advantages over MWCNTs, their benefits including an abundance of oxygen-containing surface groups and a highly mesoporous three-dimensional structure. In their reviews, Yang et al. (2006a,b) discuss the effects of ultrasound on the properties of Pt–Ru catalysts deposited on MWCNTs.

Platinum deposits on onionlike fullerenes (OLFs) were studied by Xu et al. (2006). The Pt particles dispersed uniformly on OLFs had an average diameter of 3.05 nm, smaller than the 4.10 nm on Pt/Vulcan XC-72 prepared by the same method, and exhibited a peak current for methanol electrooxidation 20% higher than that on Pt/Vulcan XC-72.

12.3.2 Transition Metal Oxides as Supports

It was shown as early as 1969 by Hobbs and Tseung that platinum particles supported by tungsten trioxide (WO_3) have special catalytic properties. On such deposits, electrochemical hydrogen oxidation in acidic solutions is markedly accelerated. This may be due to the fact that because of the difference in electron work functions between the Pt nanoparticles and the foreign support, electrons may in part be transferred between these phases. In view of their small size, the catalyst particles will experience a large change in the concentration of free electrons (and thus the electronic state), leading to changes in their adsorption and catalytic properties. Another possible reason is hydrogen spillover: a partial transfer of adsorbed hydrogen from the surface of the Pt particles to the oxide surface producing a tungsten brass (H_xWO_3; $0<x<1$).

The influence of different methods of platinum deposition on WO_3 was studied by Chen and Tseung (2000). A practical use of this support in fuel cells is difficult, owing to the partial dissolution of WO_3 in the acidic medium (i.e., in contact with the proton-conducting membrane). It was shown by Raghuveer and Viswanathan (2005) that some increase in stability of the WO_3 in an acidic medium is produced when bringing a certain amount of Ti^{4+} ions into the crystal lattice of WO_3.

Kulesza et al. (2001) suggested a bifunctional catalyst containing platinum nanoparticles and having the microstructure of polynuclear oxocyanoruthenium dispersed in a reactive matrix of nonstoichiometric tungsten(VI,V) oxides.

This composite catalyst proved to be more active than a simple catalyst of the Pt–WO_3 type having the same Pt content. The authors suggested as a possible explanation for this observation that hydroxoruthenium species formed within the polynuclear microstructure have an activating effect on neighboring Pt nanoparticles.

Zhang et al. (2006) developed a new, simple method for the deposition of palladium nanoparticles on vanadium oxide nanotubes. These tubes are readily produced by low-temperature hydrothermal synthesis, and in contrast to carbon nanotubes, are now available in sizable quantities (tens of grams). Palladium is deposited at room temperature from a 0.1% $PdCl_2$ solution (containing the nanotubes) by adding formaldehyde. The palladium particles produced had diameters of 7 to 13 nm. This catalyst exhibited good activity toward methanol electrooxidation in an alkaline medium (0.1 M NaOH). The activity of the catalyst was preserved after 2 weeks in water as well as after 400 voltammetric cycles. Simple preparation of this catalyst is regarded by the authors as a very important advantage.

12.3.3 Platinum Nanoparticles in a Matrix of Conductive Polymer

Toward the end of the twentieth century, research workers were attracted by a new type of electrode consisting of a film of conducting polymer (such as polyaniline) into which a metal catalyst is inserted by chemical or electrochemical deposition. The specific catalytic activity (per unit of working surface area) exhibited by platinum crystallites in polyaniline toward methanol oxidation was higher than that of ordinary disperse platinum (Mikhailova et al., 2001). This may be due both to special structural features of platinum crystallites formed in the polymer matrix and to their placement in a three-dimensional matrix, facilitating reactant access. The activity of the catalyst in polyaniline strongly depends on the deposition conditions. A similar situation is found for platinum crystallites incorporated into films of poly(vinylpyridine) (Maksimov et al., 1998). It was pointed out by Napporn et al. (1996) that the properties of platinum-containing catalysts dispersed within a polymer matrix decisively depend on their method of preparation.

The catalyst's incorporation into a polymer film provides a very convenient possibility for studying the properties of multimetal catalysts. Three-component catalysts of the Pt–Ru–X type (where X = Au, Co, Fe, Mo, Ni, Sn, or W) incorporated into polyaniline films were studied in detail by Lima et al. (2001). Very encouraging results were obtained for a system including molybdenum (about 5 at%), which produced current densities an order of magnitude higher than at the Pt–Ru catalyst under analogous conditions. Catalysts incorporated into polypyrrol films were studied (Kulesza et al., 1999) for a more detailed understanding of the properties of composite catalysts consisting of platinum and the polynuclear oxocyanoruthenium complexes mentioned in Section 12.3.2.

12.4 PLATINUM ALLOYS AND COMPOSITES AS CATALYSTS FOR ANODES

12.4.1 Work Toward Improved Pt–Ru Catalysts

The catalytic problems associated with the anodic oxidation of methanol are very similar to those associated with the anodic oxidation of CO-containing technical hydrogen. At present, the standard catalyst for both reactions is a mixed Pt–Ru catalyst (with about 50 at% of each metal) obtained by their joint chemical or electrochemical deposition from solutions of simple or complex compounds on carbon black.

In the meantime, the search for other modified platinum catalysts has continued on a large scale, for two reasons. First, in many instances the catalytic activity of said system is insufficient. For the oxidation of pure hydrogen (free of CO traces), it is markedly lower than that of pure platinum. In methanol oxidation, the activity is not high enough for fuel cells powering an electric car. Second, the content of ruthenium in platinum ores is very low, and for this reason it would be difficult to plan large-scale production of PEMFCs and DMFCs for electric transport applications on the basis of ruthenium-containing catalysts. One way to raise the efficiency of Pt–Ru catalysts is to raise their degree of dispersion. Steigerwalt et al. (2001) reported development of a nanocomposite containing Pt–Ru and graphite nanotubes. The metal clusters had a mean diameter of 6 to 7 nm. The activity of this composite in methanol electrooxidation was 50% higher than that of ordinary Pt–Ru deposits on carbon black.

A marked increase in specific activity was attained when applying the platinum catalysts not by ordinary chemical or electrochemical deposition but by sputter deposition (Gruber et al., 2005). Using this method, one can produce an active layer consisting of alternating thin layers of the catalyst and of a carbon–Nafion composition; one can also cover a layer of pure platinum with a thin layer of ruthenium, which by oxidizing CO impurities in technical hydrogen protects the platinum from poisoning (Haug et al., 2002).

It is a defect of the Pt–Ru catalysts that ruthenium partly dissolves from the alloy and leaves the catalyst with a purely platinum surface layer; this will happen under certain conditions when the supply of methanol is insufficient and the potential moves in the positive direction. According to the data of Wang et al. (2006), adding nickel so as to attain a Pt/Ru/Ni ratio of 3:3:1 not only improves the catalyst's activity but also slows ruthenium dissolution.

12.4.2 Alloys of Platinum with Other Metals

Apart from Pt–Ru-based catalysts, large interest has been displayed in recent years in the catalytic properties of alloys (or joint deposits) of platinum and tin. Like ruthenium, tin adsorbs oxygen readily and over a certain range of potentials yields surface oxides. The mechanism of the accelerating effect of

Pt–tin alloys on electrooxidation reactions is probably the same as with the Pt–Ru alloys and is related to the bifunctional properties of the alloy surface: adsorption of organic species occurring on platinum sites and that of oxygen-containing species on tin sites, followed by chemical interaction between the species on both sites. Evidence as to the relative merits of the Pt–Ru and Pt–Sn alloys is contradictory. Both alloys are more active than pure platinum in the electrochemical oxidation of methanol. However, work involving Pt–Sn alloys has focused primarily on ethanol electrooxidation and on the development of ethanol fuel cells (see Section 4.11.1). Important research in this direction is being conducted in France by the research group of Lamy et al. (2004). In contrast to methanol electrooxidation, added ruthenium not only fails to improve the platinum catalyst's activity in ethanol oxidation, but actually depresses this actitivy in certain cases. A slight accelerating effect is observed when adding rhodium or rhenium to platinum. With 10 to 20 at% of tin in Pt, there is a sharp increase in the rate of ethanol electrooxidation. Unfortunately, though, Pt–Sn catalysts are not able to resolve the basic problem in the electrocatalysis of ethanol oxidation: that of breaking the C–C bond as required for the complete (12-electron) oxidation of ethanol to CO_2.

Of great interest is the use of intermetallic compounds of platinum with rare-earth metals such as cerium and praseodymium for anodic methanol oxidation, known from the work of Lux and Cairns (2006). This combination is attractive inasmuch as it involves two metals that differ strongly in their own electrode potentials: Pt with $E^0 = +1.2\,V$ and Pr with $E^0 = -2.3\,V$(SHE), and thus in their electronic structure. However, for the same reason, traditional methods of preparing joint disperse deposits of these metals by chemical or electrochemical reduction in a solution of the corresponding salts fail in such a situation. Lux and Cairns developed a new technology for preparing disperse powders of such compounds: by thermal decomposition of complex cyanide salts of these metals. The catalyst obtained had some activity in ethanol oxidation (although somewhat lower than that of Pt–Ru catalyst). It is also true that even this catalyst has not been able to produce complete (12-electron) oxidation of ethanol. The use of catalysts containing rare-earth elements in low-temperature electrochemical reactions has not been studied much until now and may be a promising area of electrocatalysis.

12.4.3 Composite Platinum Catalysts

Nanodisperse platinum catalysts on transition-metal oxide supports such as WO_3 were considered in Section 12.3.2. It was pointed out there that the activity of these catalysts was markedly higher than that of the same catalysts supported by carbon black. In the present section we consider composite catalysts obtained when depositing oxides on catalysts of the Pt–C type. Shim et al. (2001) put nanosized particles of the oxides WO_3 and NiO_2 on the finished Pt–C catalyst by exposing it to solutions of the corresponding salts and by subsequent chemical deposition of the oxides. It was found that this produces a certain increase (by 10 to 50%) in the true surface area of the platinum catalyst (in cm^2 Pt/cm^2 electrode), and an increase (by a factor of 1.5 to 3) in the efficiency of platinum utilization (in mA/mg Pt). These effects are attended by a decrease in bond energies of the adsorbed hydrogen and oxygen with the surface of the platinum particles. The authors offered two explanations for the effects observed: hydrogen spillover from the platinum surface to the oxide surface, and a linear boundary between two phases serving as the catalytically active sites in this composite catalyst (adlineation model).

Ioroi et al. (2002) used another sequence for preparing a composite catalyst containing molybdenum oxide (MoO). The oxide was first deposited in an amorphous form on the carbon-black support; then platinum was deposited on top of the product. The catalyst was tested in the electrochemical oxidation of hydrogen. When working with pure hydrogen, the catalyst had practically the same activity as the control sample (without MoO). However, when working with technical hydrogen containing 100 ppm carbon monoxide, the catalyst still had its high performance figures, whereas those of the control sample had degraded drastically.

Composite platinum catalysts containing complex oxides of the perovskite type (ABO_3 where A = Sr, Ce, La, etc., and B = Fe, Ni, Pt, Ru, etc.) were studied by Deshpande et al. (2006). The perovskites were prepared by combustion synthesis, where an aqueous solution of the starting components was evaporated to dryness, then the residue ignited or self-ignited, quickly heating up to temperatures above 1500°C. It was found that when platinum was introduced into the composite during the first step (i.e., was added to the aqueous solution), the properties of the catalyst were markedly better than with platinum introduced into the finished perovskite matrix.

More data concerning the use of platinum catalysts for anodes in fuel cells are reported in the reviews by Liu et al. (2006), Wee and Lee (2006), and Antolini (2007).

12.5 NONPLATINUM CATALYSTS FOR FUEL CELL ANODES

Relatively coarse porous nickel was used as the material for the hydrogen anode in the first modern hydrogen–oxygen fuel battery, developed by Bacon in 1960 and working in alkaline solution. In the fuel cells developed during the same years by Justi, much more highly disperse *skeleton* Raney nickel, *nanosized* in current terminology, was used as catalyst and electrode material of the hydrogen anode. From then on, nickel catalysts were used in the anodes of practically all alkaline hydrogen–oxygen fuel cells. Moreover, highly disperse nickel proved to be a good catalyst for anodic methanol oxidation and was used in certain alkaline methanol–oxygen fuel cell types. It is remarkable that catalysts on the basis of disperse metallic nickel currently are used in the anodes of all high-temperature fuel cells with solid electrolyte.

A certain danger exists when using a highly disperse nickel catalyst in the anodes of fuel cells operated with aqueous electrolyte solutions, for the following reason. If in a battery built up from a number of fuel cells, the supply of fuel (hydrogen or methanol) suddenly stops or is strongly reduced in a particular cell while the battery supplies current into the external circuit, current is forced through this cell by the other cells that continue to work, and the disperse nickel at the anode is oxidized, forming the oxide NiO by reaction with water. Then the electrode and with it the entire cell has lost its ability to function as a fuel cell within the battery.

For the electrooxidation of alcohols in alkaline medium (see Section 6.4), Shen et al. (2006) suggested a platinum or palladium catalyst promoted with 25 wt% of nickel oxide NiO deposited on a carbon-black support. According to their data, this additive substantially accelerates the electrooxidation of methanol in an alkaline medium. Tarasevich et al. (2005) suggested a Ru–Ni catalyst deposited on carbon black for the electrooxidation of ethanol in an alkaline medium; it is considerably more active than pure ruthenium. Under the operating conditions of fuel cells in acidic media as well as in contact with proton-conducting membranes of the Nafion type, the use of nonplatinum catalysts is highly restricted, owing to corrosion problems.

It has been known for a long time that tungsten carbide (WC) is a good catalyst for hydrogen oxidation in acidic solutions. In contrast to platinum, tungsten carbide remains active even in CO-containing hydrogen (McIntyre et al., 2002). A practical use of this material is made difficult by the fact that an accidental shift of potential in the positive direction causes its surface to become oxidized (much more so than the surface of disperse nickel) and to lose its catalytic activity.

A special case of electrocatalysis is encountered in the enzyme and bacterial biofuel cells discussed in Section 9.2, where very restrictive conditions are imposed on the function of biocatalysts: a pH value of the medium of about 7 and a temperature close to room temperature.

An interesting piece of work was reported by Mondal et al. (2005), who deposited a film of polyaniline (about 35 mg/cm^2) by electrochemical means on an electrode of stainless steel. This electrode proved to be active for the electrooxidation of ascorbic acid in 0.5 M sulfuric acid solution. A fuel cell model with such an anode and a standard oxygen electrode of platinum had an open-circuit voltage (OCV) of 0.5 V. At a current density of 10 mA/cm^2, the cell's working voltage at room temperature was about 0.1 V; at 70°C it went up to 0.35 V. This is an indication for the promise of metal-free catalysts consisting of conductive polymers, in different varieties of fuel cells. Work related to the use of nonplatinum catalysts for electrochemical (cathodic) oxygen reduction is discussed in Section 12.6.

12.6 ELECTROCATALYSIS OF THE OXYGEN REDUCTION REACTION

Selecting a suitable catalyst for the oxygen electrodes is one of the key problems in the development of fuel cells operated at temperature below 200 to 250°C. The energy losses encountered in such fuel cells are actually associated, primarily with the oxygen electrodes. The polarization of the oxygen electrode is high, not only because the cathodic reduction of oxygen is relatively slow but also because even at open circuit, the potential of such an electrode is about 0.2 V more negative than its thermodynamic value (1.23 V at 25°C). A corresponding shift is included in its working potential when current flows. This is observed no matter which of the known catalysts is used in the electrode.

In acidic solutions, the most active catalyst is platinum, which is used in practically all PEMFCs and DMFCs with proton-exchange membranes. Since oxygen reduction is so slow, at the oxygen electrode of hydrogen–oxygen fuel cells a larger amount of platinum catalyst must be used than at the hydrogen electrode. In alkaline solutions, catalysts consisting of disperse silver or catalysts containing cobalt oxides and/or oxides of other metals are active.

The kinetics and mechanism of electrochemical oxygen reduction at different catalysts has been widely studied from the late 1950s onward, as can be seen in the monographs of Hoare (1968), Breiter (1969), and Kinoshita (1992).

One important question that up to now has not been answered conclusively is that of the nature of the open-circuit potential exhibited by an oxygen electrode (Section 12.6.1).

An important requirement that must be met by a catalyst for the oxygen electrode is that of accelerating the reduction reaction that follows the four-electron mechanism:

$$O_2 + 4H^+ + 4e^- \rightarrow 2H_2O \tag{12.1}$$

rather than that following the two-electron mechanism:

$$O_2 + 2H^+ + 2e^- \rightarrow H_2O_2 \tag{12.2}$$

which not only is less favorable energetically (just two electrons per oxygen molecule and a less positive potential) but also yields hydrogen peroxide as a product, which is harmful for the membrane and other components of the fuel cell. At platinum catalysts the four-electron mechanism prevails, but at carbon electrodes the two-electron mechanism prevails. At other electrodes (catalysts) the reaction may proceed in parallel along both mechanisms.

Recent work related to oxygen electrodes in fuel cells has mainly been as follows:

- Attempts to raise the activity of the platinum catalyst further and to find other active catalysts for the oxygen reduction reaction in acidic media (including operation in contact with proton-conducting membranes)

- The development of catalysts for the oxygen electrode that are fairly insensitive to methanol that has crossed through the membrane in DMFCs
- For acidic and alkaline media, the development of catalysts for the ORR that do not contain platinum or other noble metals
- The development of bifunctional catalysts that can be used for both cathodic reduction and anodic evolution of oxygen

12.6.1 Open-Circuit Potential of the Oxygen Electrode

The main reaction occurring at the oxygen electrode in fuel cells,

$$O_2 + 4H^+ + 4e^- \rightleftharpoons H_2O \tag{12.3}$$

is invertible, that is, may occur in both the cathodic direction (oxygen reduction to water in fuel cells) and the anodic direction (oxygen evolution from water in electrolyzers). Despite this, at temperatures below about 200 to 300°C, the reaction is not in equilibrium (is not reversible in the thermodynamic sense). At open circuit, when no current flows, a value of electrode potential E_0 is established at the electrode that differs from the thermodynamic value E_{thermod}, and has been called the open-circuit potential OCP. The value of E_{thermod} is 1.23 V at a temperature of 25°C. The value of E_0 depends on the way it is measured. When the electrode is simply in contact with the electrolyte while gaseous oxygen is present, its value is 1 ± 0.1 V. Figure 12.2 shows how the potential of this electrode depends on current density under the conditions of minor anodic and cathodic polarization. When the values of potential measured under cathodic polarization are extrapolated to zero current density, a value of 1 ± 0.1 V is obtained again. Doing the same with the values of potential measured under anodic polarization, a value of 1.5 ± 0.1 V is now obtained for E_0. This phenomenon is practically independent of the electrode metal at which reaction (12.1) takes place. It is also independent of the type of electrolyte and is seen in both acidic and alkaline solutions.

This "jump" in potential away from the thermodynamic value by $\Delta E = |E_{\text{thermod}} - E_0| \cong 0.3 - 0.4$ V has huge consequences for a number of fields of applied electrochemistry. For fuel cells it implies an irrevocable loss of about 30% of the electrical energy being generated, in addition to the ohmic and polarization losses. For commercial water electrolyzers (and also for the electrolyzers used in hydroelectrometallurgy), it implies additional consumption of electrical energy.

As the temperature is raised, the size of this potential jump ΔE decreases gradually. The thermodynamic value of E_{thermod} gradually becomes less positive with increasing temperature, and the value of E_0 becomes gradually more positive. At temperatures above about 400 to 600°C, this jump is no longer observed (detailed data are not available in the literature for this region of temperatures).

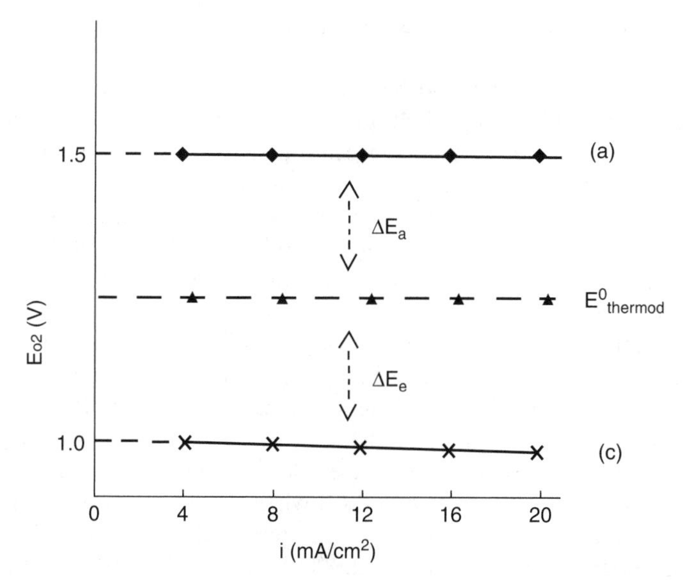

FIGURE 12.2 Dependence of the potential of an oxygen electrode on current density in the region of low values of polarization: (*a*) anodic, (*c*) cathodic. $E^0_{\text{thermod.}}$, thermodynamic value of the oxygen potential.

In view of the large importance of this value for applications, it was the subject of numerous studies in the 1950s and 1960s. Different reasons have been offered: the influence of oxide layers present on the electrode surface; the influence of traces of hydrogen peroxide accumulating in the electrolyte layer next to the electrode, and so on. The exact origin of the phenomenon has not yet been established.

12.6.2 Raising the Activity of Platinum Catalysts for the ORR

In a number of studies going back to the 1980s and relating to PAFC development, it was shown that polarization of the oxygen electrode will decrease by about 50 to 80 mV at intermediate current densities in acidic solutions when instead of pure platinum, alloys of platinum with iron, nickel, or cobalt (also chromium and analogous metals; such catalysts are often designated as Pt–M/C) are deposited on the carbon-black support. A number of explanations were offered. In one of the first of these studies (Jalan and Taylor, 1983), the activity increase was attributed to a change in interatomic distances on the catalyst surface. Paffett et al. (1988) attributed it to an increase in surface roughness (or true surface area) of the electrode that results from dissolution of the less noble component, chromium in their case, from the alloy surface. Later, Mukerjee et al. (1995) attributed it to a change in electronic state of the platinum alloys, more precisely an increase in the number of *d*-band vacancies.

Research activity in this field has been quite important and is continuing. Thus, the effect of alloys deposited on a support by chemical vapor deposition (Seo et al., 2006b) and by dual ion-beam-assisted deposition (IBAD; Gullá et al., 2006) has been studied. Another advantage of such platinum-alloy catalysts that was noted by many workers is their reduced sensitivity to methanol crossing over to the oxygen electrode in DMFCs and the associated improvement in the stability of the electrode's potential (Yuan et al., 2006; Baglio et al., 2007).

Some doubts exist as to the stability of such alloy catalysts during long-term operation in acidic media. On the one hand, slight selective dissolution of the less noble metal produces surface roughening, hence an increase in true surface area and in catalyst activity. On the other hand, two negative consequences cannot be ruled out: (1) the promoter action of the additive will be lost if all of the alloy partner disappears by dissolution; and (2) when the alloy partner then deposits on the membrane, this may cause the membrane to be clogged or destroyed. It was shown in specific studies (Yu et al., 2005; Seo et al., 2006a) that a marked deterioration of the properties of such alloys will not occur during relatively short-term tests (e.g., 500 hours). Aspects of the stability of such catalysts were discussed by Antolini et al. (2006). In a review, Bezerra et al. (2007) suggested that the loss of alloy catalyst properties during long-term operation will not go beyond the limits of short-term change; it is a phenomenon observed in all catalysts and has to do with the gradual recrystallization and coarsening of the particles. Unambiguous statistical data confirming the stability of platinum-alloy catalysts during long-term operation of fuel cells are not yet available in the literature.

Studies into the reasons for enhanced catalytic activity of platinum alloys for the ORR also continue. It is thought that changes in electron structure are the basic factor, although it is not ruled out that all the factors mentioned are actually operating simultaneously. A rather high catalytic activity for the ORR—approaching that of pure platinum—was found for alloys of cobalt with palladium, another platinum group metal (Sovadogo et al., 2004; Wang et al., 2007).

12.6.3 ORR Catalysts Without Platinum

Two types of catalysts without platinum are employed:

1. Catalysts that contain no platinum or other platinum group metals but may contain noble metals (gold, silver) or other rare and expensive metals
2. Catalysts that do not contain any noble or other expensive metals

In the present section we examine catalysts for the ORR not containing platinum that belong to the first of these two groups. Catalysts of the second group are considered in the following section.

Gold

Baker et al. (2006) have shown that small amounts of gold (9.4 wt%) deposited on powder of hydrated tin oxide (SnO) make a good catalyst for four-electron oxygen reduction in acidic and alkaline solutions.

El-Deab and Ohsaka (2006) found that a catalyst obtained by the electrochemical deposition of manganese dioxide (MnO_2) nanoparticles on glassy carbon, followed by the deposition of very small amounts of gold on these particles, will also be active for the oxygen reaction in alkaline solution. These authors believe that two-electron oxygen reduction occurs on gold, while at the MnO_2 particles the hydrogen peroxide produced on gold decomposes instantaneously to water and oxygen. In this way the overall reaction is equivalent to the four-electron process.

Silver

Silver is a good catalyst for four-electron oxygen reduction in alkaline solutions. In 1962, Justi and Winsel developed an oxygen electrode consisting of skeleton (Raney) silver. Alkaline fuel cells with such electrodes were actually manufactured in subsequent years, but later, in view of the higher demands placed on space-flight fuel cells, the silver catalysts were replaced by platinum catalysts.

Ruthenium-Based Chalcogenides

In 1986, Alonso-Vante and Tributsch communicated that semiconducting ruthenium–molybdenum chalcogenides having the general formula $Mo_xRu_yXO_2$ (with X = chalcogen: essentially, one of the elements O, S, Se, Te) and forming *Chevrel phases* exhibit good catalytic activity for oxygen reduction in acidic solutions. This discovery was followed by many studies in this direction using similar compounds synthesized in different ways. They were found to be highly stable chemically in acidic media. It was soon found that the catalytic activity is not restricted to Chevrel phases, but other varieties of such chalcogenides are active as well. The December 2007 issue of the *Journal of Applied Electrochemistry* contains a number of papers reporting studies performed with catalysts on the basis of ruthenium selenide. It was shown in particular (Kulesza et al., 2007) that the catalytic activity of this compound can be raised by adding tungsten trioxide (WO_3). It was pointed out in many of the papers that such chalcogenides are markedly less sensitive than platinum catalysts to methanol.

12.6.4 ORR Catalysts Without Noble Metals

Mercury that is no catalyst for practically any electrochemical reaction unexpectedly exhibits a rather high catalytic activity for the two-electron reduction of oxygen to hydrogen peroxide [reaction (12.2)]. Carbon materials, graphite, and more particularly, active carbon also display some catalytic activity for the two-electron reduction of oxygen in alkaline solutions. Carbon cathodes are found in the widely used zinc–air cells (the "batteries" for hearing aids) that do not have to meet high current loads. Transition metal oxides having a perovskite structure, such as cobalt oxide (Co_3O_4), are often used to raise the activity of carbon cathodes in zinc–air cells. These additives are very active catalytically, and raise the activity of the electrode as a whole, but in addition they catalyze the disproportionation of hydrogen peroxide to water and oxygen, thus effectively changing the two-electron reaction producing the hydrogen peroxide to a more efficient four-electron reaction. These oxides are unstable in acidic solutions.

In 1964, Jasinski discovered that certain macrocyclic compounds of transition metals (Fe, Co, etc.), N_4 compounds such as phthalocyanins and tetraazaannulenes, are very active catalysts for oxygen reduction in acidic solutions. However, their stability during prolonged work in acidic media was found to be very low. Sometime later it could be shown (Bagotsky et al., 1977–1978) that after thermal treatment at 700 to 800°C these compounds, despite their partial decomposition, not only retain their catalytic activity but acquire the chemical stability required for long-term operation. Sometimes their activity even increases. Heat-treated compounds of this type then became the subject of numerous studies.

When such catalysts were used, not in liquid acidic solutions but in contact with proton-conducting membranes, certain difficulties were seen, so the compounds noted above did not find practical applications in PEMFCs and DMFCs.

In a review by Wang (2005), detailed data can be found for different types of ORR catalysts without platinum, both for the variety with other noble metals and that without noble metals.

12.6.5 Bifunctional Oxygen Catalysts

Bifunctional catalysts are those that should be active both for the cathodic reduction of oxygen to water and for the anodic evolution of oxygen from water. This constitutes a problem, because upon anodic polarization (such as that required for oxygen evolution), the surface of most catalysts becomes covered with oxide layers and hence loses its ability for cathodic oxygen reduction. As reported in Section 10.1.4, according to data of Swette et al. (1994), bifunctional properties occur in a catalyst consisting of a mixture of highly disperse platinum particles and iridium oxide (IrO_2) in a 1:1 weight ratio. This mixed catalyst is rather active for both oxygen evolution and oxygen reduction. The mechanism of its action has not yet been established. Possibly, anodic oxygen evolution occurs basically on the iridium oxide particles, while cathodic oxygen reduction occurs on the platinum particles, here found to be less strongly oxidized. This catalyst can be used in both acidic and alkaline electrolyte solutions.

In alkaline solutions, bifunctional properties are exhibited by catalysts having the pyrochlore structure $A_2B_2O_7$, where A = Pb, Bi and B = Ru, Ir (Horowitz et al., 1983), and by oxide catalysts having the perovskite structure (e.g., $La_{0.6}Ca_{0.4}CoO_3$) (Wu et al., 2003). The properties of bifunctional oxygen electrodes are discussed in greater detail in a paper by Jörissen (2006).

12.7 THE STABILITY OF ELECTROCATALYSTS

As a rule, the activity of catalysts for electrochemical reactions (like that of catalysts for chemical reactions in general) falls off with time. The degree and rate of this decline depend on a large number of factors: the catalyst type, its method of preparation, its working conditions (i.e., composition of the electrolyte solution, temperature, current density, working potential, length of the time of use) as well as the conditions of storage prior to initiating the work. The main reason for a drop in activity are various side reactions that occur both while the catalyst is stored in an idle condition and while it is working under current flow:

1. Corrosion (spontaneous dissolution) of the catalytically active material, and hence a decrease in the quantity present. Experience shows that contrary to widespread belief, marked corrosion occurs even with platinum metals. For smooth platinum in sulfuric acid solutions at potentials of 0.9 to 1.0 V (SHE), the steady rate of self-dissolution corresponds to a current density of about 10^{-9} A/cm^2. Also, because of enhanced dissolution of ruthenium from the surface layer of Pt–Ru catalysts, their exceptional properties are gradually lost and they are converted to ordinary, less active platinum catalysts.
2. In fuel cells, where platinum catalysts deposited onto carbon black or other carbon supports are often used, corrosion of the carbon support is a very troublesome problem. This corrosion not only leads to a deterioration of the contact between catalyst and substrate that is needed for it to function, but may also cause the catalyst to drop away from the electrode.
3. Spontaneous recrystallization of the catalyst, leading to a decrease in its specific surface area (sintering of catalysts) and possibly also to a change in the chemical and/or phase composition of the surface layer. The recrystallization of disperse catalysts that occurs as a result of surface diffusion of its atoms or of the permanent dissolution and redeposition of its surface atoms is caused by the general tendency of materials to attain a lower surface area and lower excess surface energy. The rate of this process is higher the higher the degree of dispersion of a catalyst.
4. The adsorption and accumulation of various impurities from the electrolyte or surrounding atmosphere on the catalyst surface. The rate of accumulation of impurities on the catalyst surface depends on its activity for adsorption, which often is parallel to its catalytic activity.

A number of problems associated with the stability of platinum catalysts operating in PEMFCs have been discussed in a review by Shao et al. (2007).

REFERENCES

Alonso-Vante N., H. Tributsch, *Nature*, **323**, 431 (1986).

Antolini E., J. R. C. Salgado, E. R. Gonzalez, *J. Power Sources*, **160**, 957 (2006).

Baglio V., A. Stassi, A. Di Blasi, et al., *Electrochim. Acta*, **53**, 1360 (2007).

Bagotsky V. S., M. R. Tarasevich, K. A. Radyushkina, et al., *J. Power Sources*, **2**, 233 (1977–1978).

Baker W. S., J. J. Pietron, M. E. Teliska, et al., *J. Electrochem. Soc.*, **153**, A1702 (2006).

Benitez R., J. Soler, L. Daza, *J. Power Sources*, **151**, 108 (2005).

Caillard A., C. Coutanceau, P. Brault, et al., *J. Power Sorces*, **162**, 66 (2006).

Chen K. Y., Z. Sun, A. C. C. Tseung, *Electrochem. Solid-State Lett.*, **3**, 10 (2000).

Deshpande K., A. Mukasyan, A. Varma, *J. Power Sources*, **158**, 60 (2006).

El-Deab M. S., T. Ohsaka, *J. Electrochem. Soc.*, **153**, A1365 (2006).

Gruber D., N. Ponath, J. Müller, F. Lindstaedt, *J. Power Sources*, **150**, 67 (2005).

Gullá A. E., M. S. Saha, R. J. Allen, S. Mukerjee, *J. Electrochem. Soc.*, **153**, A366 (2006).

Haug A. T., R. E. White, J. W. Weidner, et al., *J. Electrochem. Soc.*, **149**, A868 (2002).

Hobbs B. S., A. C. C. Tseung, *Nature*, **222**, 556 (1969).

Horowitz H. S., J. M. Longo, H. H. Horowitz, *J. Electrochem. Soc.*, **130**, 1851 (1983).

Ioroi T., N. Fujiwara, Z. Siroma, et al., *Electrochem. Commun.*, **4**, 442 (2002).

Jalan V., E. J. Taylor, *J. Electrochem. Soc.*, **130**, 2299 (1983).

Jasinski R., *Nature*, **201**, 1212 (1964).

Jörissen L., *J. Power Sources*, **155**, 23 (2006).

Kulesza P. J., M. Matczak, A. Wolkiewicz, et al., *Electrochim. Acta*, **44**, 2131 (1999).

Kulesza P. J., B. Grzibowska, M. A. Malik, et al., *J. Electroanal. Chem.*, **512**, 210 (2001).

Kulesza P. J., R. Miecznikowski, B. Baranowska, et al., *J. Appl. Electrochem.*, **37**, 1439 (2007).

Lamy C., S. Rousseau, E. M. Belgsir, et al., *Electrochim. Acta*, **49**, 3901 (2004).

Lima A., C. Coutanceau, J.-M. Léger, *J. Appl. Electrochem.*, **31**, 379 (2001).

Lux K. W., E. J. Cairns, *J. Electrochem. Soc.*, **153**, pt. I, A1132; pt. II, A1147 (2006).

Maksimov Yu. M., B. I. Podlovchenko, T. L. Azarchenko, *Electrochim. Acta*, **43**, 1053 (1998).

McIntyre D. R., G. T. Burstein, A. Vossen, *J. Power Sources*, **107**, 67 (2002).

Mikhailova A. A., E. B. Volodkina, O. A. Khazova, V. S. Bagotsky, *J. Electroanal. Chem.*, **509**, 119 (2001).

Mondal S. K., R. R. Raman, A. K. Shukla, N. Munichandraiah, *J. Power Sources*, **145**, 16 (2005).

Mukerjee S., S. Srinivasan, J. McBreen, *J. Electrochem. Soc.*, **142**, 1409 (1995).

Napporn W. T., H. Laborde, J.-M. Léger, C. Lamy, *J. Electroanal. Chem.*, **404**, 153 (1996).

Paffett M. T., J. G. Beery, S. Gottesfeld, *J. Electrochem. Soc.*, **135**, 1431 (1988).

Plyasova L. M., I. Yu. Molina, A. N. Gavrilov, et al., *Electrochim. Acta*, **51**, 4477 (2006).

Raghuveer V., B. Viswanathan, *J. Power Sources*, **144**, 1 (2005).

Saha M. S., A. F. Gullá, R. J. Allen, S. Mukerjee, *Electrochim. Acta*, **51**, 4680 (2006).

Seo A., J. Lee, K. Han, H. Kim, *Electrochim. Acta*, **52**, 1603 (2006a).

Seo J. S., H.-I. Joh, H. T. Kim, S. H. Moon, *Electrochim Acta*, **52**, 1676 (2006b).

Shen P. K., C. Xu, R. Zeng, Y. Liu, *Electrochem. Solid-State Lett.*, **9**, A39 (2006).

Shim J., C.-R. Lee, H. K. Lee, J. S. Lee, E. J. Cairns, *J. Power Sources*, **102**, 172 (2001).

Sovadogo O., K. Lee, K. Oishi, et al., *Electrochem. Commun.*, **6**, 105 (2004).

Srinivasan S., D. J. Manko, H. Koch, et al., *J. Power Sources*, **29**, 367 (1990).

Steigerwalt E. S., G. A. Deluga, D. E. Cliffel, C. M. Lukehart, *J. Phys Chem.*, **105**, 8097 (2001).

Swette L. L., A. B. LaConti, S. A. McCatty, *J. Power Sources*, **47**, 343 (1994).

Tarasevich M. R., Z. R. Karichev, V. A. Bogdanovskaya, et al., *Electrochem. Commun.*, **7**, 141 (2005).

Wang Z. B., G. P. Yin, P. F. Shi, Y. C. Sun, *Electrochem. Solid-State Lett.*, **9**, A13 (2006).

Wang W., D. Zheng, C. Du, et al., *J. Power Sources*, **167**, 243 (2007).

Wu G., B.-Q. Xu, *J. Power Sources*, **174**, 148 (2007).

Wu N.-L., W.-R. Liu, S.-J. Su, *Electrochim. Acta*, **48**, 1567 (2003).

Xu B., X. Yang, X. Wang, et al., *J. Power Sources*, **162**, 160 (2006).

Yu P., M. Pemberton, P. Plasse, *J. Power Sources*, **144**, 11 (2005).

Yuan W., K. Scott, H. Cheng, *J. Power Sources*, **163**, 323 (2006).

Zeis R., A. Mathur, G. Fritz, et al., *J. Power Sources*, **165**, 65 (2007).

Zhang K.-F., D.-J. Guo, X. Liu, et al., *J. Power Sources*, **162**, 1077 (2006).

Reviews and Monographs

Antolini E., Catalysts for direct ethanol fuel cells, *J. Power Sources*, **170**, 1 (2007).

Bagotsky V. S., *Fundamentals of Electrochemistry*, 2nd ed., Wiley, Hoboken, NJ, 2005, Chap. 28.

Bezzera C. W. B., L. Zhang, K. Zhang, H. Liu, et al., A review of heat-treatment effects on activity and stability of PEM fuel cell catalysts for oxygen reduction reaction, *J. Power Sources*, **173**, 891 (2007).

Breiter M. W., *Electrochemical Processes in Fuel Cells*, Springer-Verlag, Berlin, 1969.

Demirci U. B., Theoretical means for searching bimetallic alloys as anode electrocatalysts for direct liquid-fed fuel cells, *J. Power Sources*, **173**, 11 (2007).

Gottesfeld S., T. A. Zawodzinski, Polymer electrolyte fuel cells, in: R. C. Alkire, H. Gerischer, D. M. Kolb, C. W. Tobias (eds.), *Advances in Electrochemical Science and Engineering*, Vol. 4, Wiley-VCH, Weinheim, Germany, 1997, p. 195.

Hoare J. P., *The Electrochemistry of Oxygen*, Wiley, NewYork, 1968.

Jörissen L., Bifunctional oxygen/air electrodes, *J. Power Sources* **155**, 23 (2006).

Justi E. W., A. Winsel, *Fuel Cells Kalte Verbrennung*, Steiner, Wiesbaden, Germany, 1962.

Kinoshita K., *Electrochemical Oxygen Technology*, Wiley, Hoboken, NJ, 1992.

Lamy C., S. Rousseau, E.M. Belgsir, et al., Recent progress in direct ethanol fuel cell: development of new platinum-tin electrocatalysts, *Electrochimica Acta*, **49**, 3908 (2004).

Lipkowski J., P. N. Ross (eds.), *Electrocatalysis*, Wiley, New York, 1998.

Liu H., C. Song, L. Zhang, J. Zhang, H. Wang, D. P. Wilkinson, A review of anode catalysis in the direct methanol fuel cell, *J. Power Sources*, **155**, 95 (2006).

Shao Y., G. Yin, Y. Gao, Understanding and approaches for the durability issues of Pt-based catalysts for PEM fuel cell, *J. Power Sources*, **171**, 558 (2007).

Wang B., Recent development of non-platinum catalysts for oxygen reduction reaction, *J. Power Sources*, **152**, 1 (2005).

Wee J.-H., K. Y. Lee, Overview of the development of CO-tolerant electrocatalysts for proton-exchange membrane fuel cells, *J. Power Sources*, **157**, 128 (2006).

Wee J.-H., K. Y. Lee, S. H. Kim, Fabrication methods for low-Pt-loading electrocatalysts in proton exchange fuel cell systems, *J. Power Sources*, **165**, 667 (2007).

Yang C., X. Hu, C. Dai, L. Zhang, H. Jin, S. Agathopoulos, Ultrasonically treated multi-walled carbon nanotubes (MWCNTs), *J. Power Sources*, **160**, 187 (2006a).

Yang C., X. Hu, D. Wang, et al., Ultrasonically treated multi-walled carbon nanotubes (MWCNTs) as PtRu supports for methanol electrooxidation, *J. Power Sources*, **169**, 187 (2006b).

Yu X., S. Ye, Recent advances in activity and durability enhancements of Pt/C catalytic cathode in PEMFC, *J. Power Sources*, Part 1, **172**, 133 (2007); Part 2, **172**, 145 (2007).

CHAPTER 13

MEMBRANES

Membranes have two functions in fuel cells. On the one hand, they are separators preventing any direct electronic contact between the electrodes that would constitute an internal short circuit in the cell. On the other hand, they have the very important function of a solid electrolyte, which makes the electrode processes at both electrodes possible and provides ionic contact between these electrodes so that current can pass when outside contact is made across a load between electrodes.

In view of these functions, membranes in fuel cells must satisfy the following basic requirements:

- High ionic conductivity
- Lack of electronic conduction
- High mechanical stability
- High chemical stability and heat resistance under the operating conditions of a fuel cell
- Lack of permeability for reactants used in the fuel cell
- Availability and acceptable cost
- Convenient handling during fuel cell assembly

Fuel Cells: Problems and Solutions, By Vladimir S. Bagotsky

13.1 FUEL CELL–RELATED MEMBRANE PROBLEMS

At present, in almost all work on fuel cells of the PEMFC and DMFC type, Nafion-type proton-conducting membranes based on perfluorinated sulfonic acids (PFSAs) have been used (Figure 13.1). Some properties of these membranes had been reported in Section 3.1.1. They have a high chemical stability and fully adequate protonic conductivity. Fuel cells built with such membranes offer rather good performance and relatively long life. It is not an exaggeration to say that the broad development of fuel cells and the general interest displayed in fuel cells today would have been impossible without the development and availability of these membranes.

Yet Nafion-type membranes do exhibit certain defects. The chief defect responsible for lack of a really broad commercialization of fuel cells containing such membranes is their high cost (about $\$700/m^2$). This figure is unlikely to come down soon, since it is founded in a highly complex manufacturing technology.

Important further shortcomings of Nafion-type membranes are the following:

1. The protonic conductivity and mechanical properties are highly sensitive to the membrane's moisture content, which in turn is related to that of the ambient atmosphere, which implies that complex systems must be used to maintain the proper water balance in a membrane-type fuel cell.
2. In view of point (1), these membranes cannot be used at temperatures above 130 to 150°C.
3. These membranes are highly permeable to methanol and certain other substances likely to be used in fuel cells.
4. These membranes are very sensitive to ions of heavy metals (e.g., iron), which when getting into the membrane because of the corrosion of metal parts within the fuel cell drastically lowers their electrical and mechanical performance.
5. These membranes exhibit a certain degree of degradation during long-term operation in fuel cells (see Section 13.2).

It is for these reasons that rather broad efforts are made in many laboratories (a) to modify membranes based on PFSA-type polymers so that the foregoing defects may be overcome, (b) to design new types of membranes from different polymers or other materials, and (c) to design alkaline membranes exhibiting hydroxyl ion conduction.

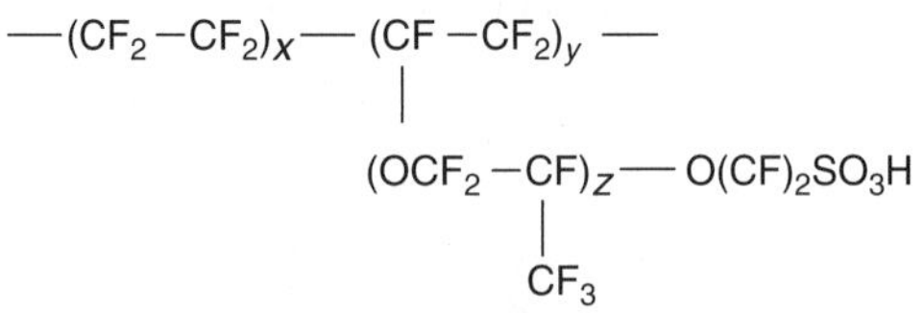

$$—(CF_2-CF_2)_x—(CF-CF_2)_y—$$

$$(OCF_2-CF)_z—O(CF)_2SO_3H$$

$$CF_3$$

FIGURE 13.1 Structure of Nafion.

13.2 WORK TO OVERCOME DEGRADATION OF NAFION MEMBRANES

It was believed for a long time that Nafion-type membranes were extremely stable chemically. This opinion was derived from the fact that the chemical C–F bonds of the carbon skeleton constituting the backbone of this polymer are exceptionally strong, requiring a very high energy (about 460 kJ/mol) to be broken. Long-term experiments soon showed, however, that a gradual degradation will occur under the operating conditions of fuel cells and is revealed by fluoride ions (F^-) being set free. The number of F^- ions set free can serve as a certain quantitative indicator of the degree of degradation.

The following events are regarded as the major membrane degradation mechanism. In a fuel cell, a certain minor diffusion of oxygen occurs from the cathode chamber through the membrane to the anode chamber, where by reaction with hydrogen, traces of hydrogen peroxide (H_2O_2) are formed. The catalytic decomposition of this substance produced by traces of iron ions leads to the formation of free oxygen-containing radicals $^{\bullet}OH$ or $^{\bullet}OOH$. These radicals react with membrane end groups of the type of $–CF_2COOH$ that have been formed in small amounts during membrane synthesis, which yields CO_2 and HF.

It was seen in detailed studies involving x-ray photoelectron spectroscopy and FTIR spectroscopy (Chen et al., 2007) that apart from the end groups named above, the membranes contain further "weak spots" that may become sites where degradation processes start. These are C–H groups (which are always formed during membrane synthesis) and even the sulfonic acid groups $–SO_3H$ themselves. Reacting with the oxygen-containing free radicals, the latter may be transformed to S–O–S bonds undergoing further oxidation to SO_2.

DuPont Fuel Cells, the company that was the first and major developer of Nafion-type membranes for fuel cells, continues to improve them and eliminate some of the defects mentioned (Curtin et al., 2004). Apart from purely technical work aiming, among other things, at a cost reduction, work is done to reduce the number of vulnerable groups in a membrane. A leading engineer at the company believes that this work should raise the working life of the membrane–electrode assemblies (MEAs) by a factor of 10 (Okine, 2006). Work is also underway in various places to find different ways of modifying Nafion-type membranes.

13.3 MODIFICATION OF NAFION MEMBRANES

13.3.1 Improvement in Moisture Content

Repeated attempts have been made to modify Nafion-type membranes so as to prevent their drying out at temperatures above 100°C by introducing inorganic materials that have hydrophilic properties. In such *composite membranes*, the polymer base secures protonic conduction of the membrane, while the inorganic filler raises the membrane's affinity for water. Metal oxides such as SiO_2, TiO_2, and ZrO_2 can be used.

In one of the first studies done in this field, workers at Princeton University (Yang et al., 2001; Costamagna et al., 2002) obtained very promising results with fillers consisting of zirconium phosphate and certain oxo derivatives of this salt. A fuel cell of the PEMFC type containing such a membrane was operated without problems for several hours at a temperature of more than 130°C, the voltage produced was 0.45 V at a current density of 1.5 A/cm^2. A cell with a membrane not containing such a filler degraded rapidly and irreversibly under the same conditions. The authors believe that the filler partly occupies large pores (measuring more than 10 nm) in the polymer, favoring capillary condensation of water in them. Similar data obtained with PEMFCs having a Nafion membrane containing TiO_2 powder (3 wt%) that had been calcined at 400°C were reported by Saccà et al. (2005).

Bauer and Willert-Porada (2005) tested membranes containing phosphates of zirconium and titanium in DMFC-type cells. They found that adding the filler produced a certain decrease in protonic conductivity, but also an almost twofold decrease in methanol permeation through the membrane, so that the overall efficiency of cells with the composite membrane was somewhat higher than that of cells without fillers in the membrane. The authors also pointed out that a considerable improvement in the mechanical stability of the membrane was produced by the filler.

Of great interest for composite membranes is the use of heteropoly acids (HPAs) such as silicotungstic acid (STA; $H_4SiW_{12}O_{40}$) and the analogous phosphotungstic acid (PTA) and phosphomolybdic acid (PMA). It was shown in one of the first studies in this direction (Tazi and Savadogo, 2000) that the water content of a Nafion membrane increased from 27% to 60% when STA was added to it. The membrane's conductivity increased accordingly. Composite membranes of this type have good chemical stability even at higher temperatures. Additional doping of the membrane with thiophene produced a substantial increase in fuel cell performance.

Zheng et al. (2007) developed a method for preparing composite membranes with SiO_2 that secured a high degree of homogeneity and flexibility. Instead of the usual aqueous-alcoholic solvent for membrane preparation, they used dimethylformamide as the solvent for the PFSA ionomer. In this way they were able to prepare a combined solution of PFSA and silica gel yielding a highly homogeneous film. The protonic conductivity was almost double that of a membrane without a filler.

13.3.2 Reduced Methanol Crossover

PFSA-based membranes have a combination structure built up from hydrophilic segments associated with the sulfonic acid groups and a backbone of hydrophobic, perfluorinated hydrocarbon within which the sulfonic acid groups are distributed uniformly. The branched network of interconnected hydrophilic segments provides the protonic conduction path of the membrane but at the same time offers methanol molecules a chance to diffuse from the anodic to the cathodic side of a DMFC. Therefore, anything done to reduce methanol crossover will cause some decrease in protonic conductivity.

Methanol crossover was reduced markedly when a barrier layer of benzimidazole was applied to the surface of a Nafion membrane while the protonic conductivity of the original Nafion membrane was preserved (Hobson et al., 2002). A similar effect was attained when sulfonated poly(vinyl alcohol) was introduced into the membrane (Shao and Hsing, 2002).

Choi et al. (2001) applied plasma etching and plasma palladium sputtering to Nafion membranes. This treatment depressed methanol crossover by about 35%. The current–voltage curves for fuel cells with treated membrane were better than those for fuel cells with an untreated membrane (data concerning the conductivity change were not reported in the paper).

Kim et al. (2003) suggested impregnating the Nafion membrane with palladium nanoparticles. According to their data, this membrane modification produced only a small conductivity decrease (by about 35%) but a very strong drop in methanol crossover (by a factor of 7).

13.4 MEMBRANES MADE FROM POLYMERS WITHOUT FLUORINE

Repeated attempts have been made to use materials less highly fluorinated than Nafion, to reduce the costs. However, such membranes have a markedly lower chemical stability, which inevitably would cut the expected membrane lifetime in fuel cells. The search continues for suitable membrane materials having a completely different chemistry.

13.4.1 Membranes on the Basis of Polybenzimidazole

Polybenzimidazole (the exact chemical name is poly[2,2′-(*m*-phenylene)-5,5′-bibenzimidazole]) is a polymer of the alkaline type with high thermal resistance (Figure 13.2). In a pure form it has a very low conductivity, on the order of 10^{-12} S/cm. Because of its alkaline nature, this polymer readily interacts with many acids, particularly so with phosphoric acid. This yields hydrogen bonds producing proton transitions. Workers at Case Western Reserve University,

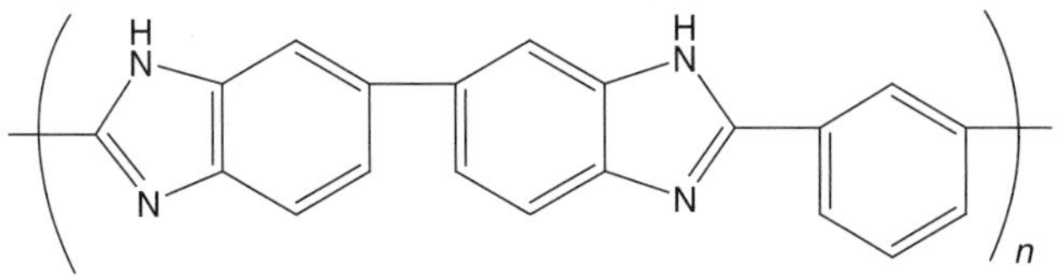

FIGURE 13.2 Structure of PBI.

Cleveland (Wainright et al., 1995; Ma et al., 2004) suggested films of PBI doped with concentrated (11 M) phosphoric acid for fuel cells. Such films have a very high water content (two molecules of water for every polymer repeat unit) and high protonic conductivity (up to 0.04 S/cm at a temperature of 200°C). A methanol fuel cell built with such membranes (0.1 mm thick) gave very good performance at 200°C: a voltage of about 0.45 V at a current density of 200 mA/cm^2. The high thermal resistance of PBI membranes doped with concentrated phosphoric acid was confirmed by Samms et al. (1996). Following these early publications, many investigations into PEMFC-type fuel cells were reported [see, e.g., Seland et al. (2006)].

Results of long-term tests (600 hours at 150°C) of PEMFCs with a PBI membrane doped with 85% phosphoric acid were reported in 2006 by Liu et al. In these tests, the membrane was not additionally humidified. Under these conditions the current density dropped from an initial value of 714 mA/cm^2 to 300 mA/cm^2. After 130 hours, a gradual performance decay of 82 μV/h began. This decay had accelerated to 270 μV/h toward the end of the test. Apart from catalyst recrystallization, the voltage loss was attributed to an obvious gradual attack of the membrane assumed to be due to free radicals of the type $^\bullet$OH and $^\bullet$OOH, as well as to a gradual leaching of phosphoric acid from the membrane.

13.4.2 Membranes on the Basis of Polyether Ether Ketone

Membranes made from polymers of the polyether ether ketone (PEEK) type are fluorine-free, too. These polymers contain phenyl groups linked by groups –O– and –(C=O)– (Figure 13.3). They can be sulfonated relatively readily, yielding proton-conducting sulfonated polyether ether ketone (SPEEK). The conductivity of such membranes is about 0.05 S/cm at 100°C and a relative humidity of 100%, and increases with increasing temperature. However, at temperatures above 150°C a considerable moisture loss is seen, and gradual degradation of the membrane begins.

Several attempts to make composite membranes on the basis of SPEEK have been reported. A marked increase in conductivity at temperatures of around 100°C can be attained by adding fillers such as SiO_2 or molybdophosphoric acid. Jones et al. (2005) reported building DMFCs with a SPEEK composite membrane containing zirconium phosphate. A battery consisting of such cells exhibited stable operation at 130°C during one month when working eight hours daily. No other data have been found for work with such membranes at temperatures of 130°C or more.

Jörissen et al. (2002) studied a composite membrane which in addition to SPEEK also contained PBI as well as a basically substituted polysulfone (bPSU). When operating a DMFC with such a membrane at a working temperature of 110°C, they saw a strong blocking of methanol crossover, but the fuel cell performance was poor.

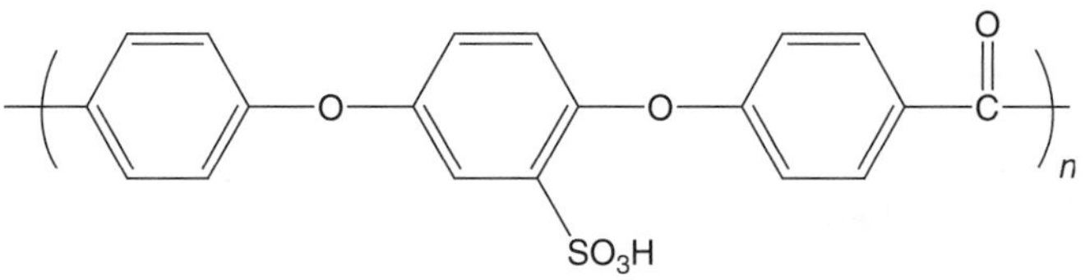

FIGURE 13.3 Structure of PEEK.

13.5 MEMBRANES MADE FROM OTHER MATERIALS

Membranes with promising properties have been prepared from polyphosphazenes in combination with sulfonimide (polyphosphazene-based sulfonimide; Chalkova et al., 2002) or with polyacrylonitrile (blended polyphosphazene/polyacrylnitrile; Carter et al., 2002).

Low methanol crossover was also seen with membranes prepared from poly(vinyl alcohol) that contained mordenite (a zeolite variety; Libby et al., 2001).

Various aspects of the work on composite membranes prepared from different polymers have been discussed in detail in a review by Savadogo (2004).

13.6 MATRIX-TYPE MEMBRANES

For a number of years, workers at Tel Aviv University have developed membranes very different from those derived from proton-conducting polymers that were described above. Peled et al. (2000) described matrix-type membranes made from nano-sized powders (<15 nm) of nonconducting materials having good water retention. These include the oxides of silicon, zirconium, and aluminum. With a suitable binder [poly(vinylidene fluoride) (PVDF)], highly porous films (porosity 50 to 90%) having a thickness between 30 and 1000 μm were cast from these powders. The films were impregnated with acids having a low vapor pressure, such as sulfuric, phosphoric, or polyfluoroaryl acids. These films have good mechanical properties and a high protonic conductivity (0.45 S/cm at room temperature). According to these authors, the most important advantage of these membranes over Nafion is their much lower cost and a lack of sensitivity to the detrimental effects of heavy-metal ions that are formed in fuel cells.

Workers of the same group suggested using polytetrafluoroethylene (PTFE) in membranes of this type to raise the thermal resistance (Reichman et al., 2007). Nano-sized ceramic particules produced from tetraethyl orthosilicate [TEOS; $Si(OC_2H_5)_4$] by sol–gel synthesis were introduced into a porous PTFE layer to give the material some degree of hydrophilicity and to secure a sufficiently high moisture content. In a matrix material thus obtained, the volume ratio of hydrophilic and hydrophobic pores was approximately 1:1. A membrane 137 μm thick was impregnated with an aqueous 3 M H_2SO_4 solution. At room temperature the membrane had a conductivity of 0.11 S/cm. It had a good thermal resistance (up to 300°C).

A matrix-type membrane in which cesium dihydrogen phosphate (CsH_2PO_4) was used as a proton-conducting material was reported by Matsui et al. (2005). It is known that because of a phase transition, the conductivity of this material rises drastically at a temperature of 250°C (Haile et al., 2001). However, because of dehydration, the conductivity gradually decreases with time at such temperatures. Silicon pyrophosphate (SiP_2O_7) was suggested as a matrix material. The conductivity of such a matrix membrane was 44 mS/cm at 266°C. When the temperature was dropped below 250°C, the expected sudden drop in conductivity was not observed; instead, the conductivity decreased smoothly to a value of about 1 mS/cm at 100°C. Thus, this membrane retained useful properties throughout the temperature range examined (110 to 287°C). Lack of a sudden drop in conductivity was attributed to formation of a new phase, $CsH_5(PO_4)_2$.

13.7 MEMBRANES WITH HYDROXYL ION CONDUCTION

The problems associated with the preparation of fuel cell anion-exchange membranes with hydroxyl ion conduction were noted in Section 6.3. Fauvarque et al. (1995) in the CNAM Laboratory for Industrial Electrochemistry in Paris reported preparing an anion-exchange membrane by mixing methanolic solutions containing equivalent amounts of polyethylene oxide (PEO) and KOH, evaporating part of the solvent, and casting an anhydrous film. At room temperature, such a membrane had a conductivity of about 10^{-3} S/cm. They discussed using it in alkaline nickel–cadmium storage batteries. In the same laboratory, Agel et al. (2001) tested a composite membrane prepared from a polymer of diazabicyclooctane (DABCO) and triethylamine (TEA) that was treated with 1 M KOH in order to be converted to the OH form. The transport number of the OH^- ions in this membrane was close to unity. A marked improvement in the performance of hydrogen–oxygen fuel cells containing such a membrane was attained when the space between the membrane and the electrodes was filled with a 1 M KOH solution saturated with polyacrylic acid.

In work done at the Royal Military College of Canada (Wan et al., 2006), the use of chitosan, a natural biopolymer and derivative of chitin that contains both hydroxyl and amino groups, was suggested. Hydrated membranes prepared from this polymer are chemically and thermally stable at temperatures of up to 200°C after cross-linking with glutaraldehyde (GA). They have a conductivity of about 10^{-3} S/cm. Since this is too low for fuel cells, the authors used a multilayer membrane with a porous matrix filled with KOH solution between two chitosan–GA layers. Hydrogen–air fuel cells having a composite membrane of this type gave a current density of 30 mA/cm^2 at a voltage of 0.2 V when the temperature was 50°C.

REFERENCES

Agel E., J. Bouet, J. F. Fauvarque, *J. Power Sources*, **101**, 267 (2001).

Bauer E., M. Willert-Porada, *J. Power Sources*, **145**, 101 (2005).

Carter R., R. Wycisk, H. Yoo, N. Pintauro, *Electrochem. Solid-State Lett.*, **5**, A195 (2002).

Chalkova E., X. Zhou, C. Ambler, M. A. Hofmann, et al., *Electrochem. Solid-State Lett.*, **5**, A221 (2002).

Chen C., G. Levitin, D. W. Hess, T. F. Fuller, *J. Power Sources*, **169**, 288 (2007).

Choi W. C., J. D. Kim, S. I. Woo, *J. Power Sources*, **96**, 411 (2001).

Costamagna P., C. Yang, A. B. Bocarsli, S. Srinivasan, *Electrochim. Acta*, **47**, 1023 (2002).

Curtin D. E., R. T. Lousenberg, T. J. Henry, et al., *J. Power Sources*, **131**, 41 (2004).

Fauvarque J. F., S. Guinot, N. Bouzir, et al., *Electrochim. Acta*, **40**, 2449 (1995).

Haile S. M., D. A. Boysen, C. R. I. Chisolm, R. B. Merle, *Nature*, No. 401, 910 (2001).

Hobson L. J., Y. Nakano, H. Ozu, H. Hayase, *J. Power Sources*, **104**, 79 (2002).

Jones D. J., J. Rozière, M. Marrony, C. Lamy, et al., *Fuel Cells Bull.*, No. 10, 12 (2005).

Jörissen L., V. Gogel, J. Kerres, J. Garche, *J. Power Sources*, **105**, 267 (2002).

Kim Y. M., K. W. Park, J. H. Choi, et al., *Electrochem. Commun.*, **5**, 571 (2003).

Libby B., W. H. Smyrl, E. L. Cussler, *Electrochem. Solid-State Lett.*, **4**, A197 (2001).

Liu G., H. Zhang, J. Hu, et al., *J. Power Sources*, **162**, 547 (2006).

Ma Y.-L., J. S. Wainright, M. H. Litt, R. F. Savinell, *J. Electrochem. Soc.*, **151**, A8 (2004).

Matsui T., T. Kukino, R. Kiguchi, K. Eguchi, *Electrochem. Solid-State Lett.*, **8**, A256 (2005).

Okine R., cited from *Fuel Cell Today*, Nov., 20, 2006.

Peled E., T. Duvdevani, A. Ahron, M. Melman, *Electrochem. Solid-State Lett.*, **1**, 210 (1998); **3**, 525 (2000).

Reichman S., A. Ulus, E. Peled, *J. Electrochem. Soc.*, **154**, B327 (2007).

Saccà A., A. Carbone, E. Passalacqua, et al., *J. Power Sources*, **152**, 16 (2005).

Samms S. R., S. Wasmus, R. F. Savinell, *J. Electrochem. Soc.*, **143**, 1225 (1996).

Seland F., T. Berning, B. Børresen, R. Tunold, *J. Power Sources*, **160**, 27 (2006).

Shao Z.-G., I-M. Hsing, *Electrochem. Solid-State Lett.*, **5**, A185 (2002).

Tazi B., O. Savadogo, *Electrochim. Acta*, **45**, 4329 (2000).

Wainright J. S., J.-T. Wang, D. Weng, et al., *J. Electrochem. Soc.*, **142**, L121 (1995).

Wan Y., B. Pepply, A. M. Creber, et al., *J. Power Sources*, **162**, 105 (2006).

Yang C., P. Costamagna, S. Srinivasan, *et al.*, *J. Power Sources*, **103**, 1 (2001).

Zheng R., Y. Wang, S. Wang, O. K. Shen, *Electrochim. Acta*, **52**, 3895 (2007).

Reviews

Inzelt G., M. Pineri, J. W. Schultze, M. A. Vorotyntsev, Electron and proton conducting polymers: recent developments and prospects, *Electrochim. Acta*, **45**, 2403 (2000).

Iojoiu C., F. Chabert, M. Maréchal, et al., From polymer chemistry to membrane elaboration: a global approach of fuel cell polymeric electrolytes, *J. Power Sources*, **153**, 198 (2006).

Li X., E. P. L. Roberts, S. M. Holmes, Evaluation of composite membranes for direct methanol fuel cells, *J. Power Sources*, **154**, 115 (2006).

Neburchilov V., J. Martin, H. Wang, J. Zhang, A review of polymer electrolyte membranes for direct methanol fuel cells, *J. Power Sources*, **169**, 221 (2007).

Savadogo O. Emerging membranes for electrochemical systems: II. High temperature composite membranes for polymer electrolyte fuel cell (PEFC) applications, *J. Power Sources*, **127**, 135 (2004).

CHAPTER 14

SMALL FUEL CELLS FOR PORTABLE DEVICES

Early in the 1990s, a variety of new portable devices started to become mass consumer goods: notebook-type computers, videocameras, digital still cameras, cellular phones, camcorders, audio and video players, medical appliances for individual use, and others. For their power supply, various storage batteries were used: nickel–cadmium and nickel–hydride at first, and later, lithium-ion batteries. Such batteries sustain their electronic devices for no longer than several hours, and when empty, require many hours of recharging, which considerably detracts from the convenience offered by the devices.

It was in this context that attention turned to the development of new types of small fuel cells and to power plants of small size assembled from such fuel cells. The introduction of these small power plants to domestic technical applications opened ways for a broad commercialization of fuel cells, and with it, a flow of new sources of financing that would support further fuel cell development.

Apart from domestic applications, a need for new, improved power sources of small size also arose in different portable devices to be used in the military sector, such as night-vision devices and individual communication equipment.

Apart from their longer uninterrupted function, fuel cells have the important advantage over classical storage batteries that after exhausting their fuel supply, one may practically instantaneously recharge them simply by replacing an empty cartridge with a new one filled with fresh fuel. The special operating

Fuel Cells: Problems and Solutions, By Vladimir S. Bagotsky

features of different types of power plants for portable devices are considered in more detail in Section 15.4.

Portable electronic devices need very low power for their operation, often milliwatts or at most a few watts and no more than 10 W. Small low-power fuel cells designed as power supplies for portable devices have been named *micro-fuel cells* or *mini-fuel cells* (mini-FC), the latter term being preferred. The concept of a miniaturization of fuel cells is also current.

Mini-fuel cells are developed most often on the basis of hydrogen–oxygen PEMFCs and liquid-fuel DMFCs (DLFCs). A special case of mini-fuel cells on the basis of SOFC is discussed in Section 14.5. A solid electrolyte is used in all these systems. Attempts to build mini-fuel cells using a liquid electrolyte do not appear in the literature.

14.1 SPECIAL OPERATING FEATURES OF MINI-FUEL CELLS

One of the most important jobs of ordinary (large) fuel cells is the highly efficient conversion of chemical fuel energy to electrical energy. For a higher efficiency of fuel utilization, power plants involving such fuel cells include many peripherals to monitor and regulate the rate of fuel supply and external parameters such as temperature and pressure to influence the efficiency.

In mini-fuel cells to be used in portable devices, the basic need is a lower volume and weight of the power plant, that is, higher values of the specific energy per unit volume (kWh/L) and per unit mass (kWh/kg). To achieve this, one must first reduce the number of auxiliary devices considerably and develop a fuel cell structure that is able to work with a minimum set of these peripherals. This can be attained in different ways. One may, for exemple, design a fuel cell operating with a defined constant current density, that is, a constant rate of reactant use and product formation that is independent of the current load. In the system, the fuel cell will then have included in parallel a small buffer storage battery absorbing the excess current at low loads and supplementing the fuel cell at high loads when a device giving rise to variable loads is attached.

An important factor for a simplified structure of the fuel cell and fuel power plant is that of supplying fuel as the only reactant while drawing the other (oxidizing) reactant directly from the ambient air without using pumps or ventilators for its intake (i.e., using air-breathing cells). In exactly the same way, the heat evolved should be transferred directly to the ambient medium without heat-transfer fluids or forced airstreams. This task is easier inasmuch as heat evolution is insignificant at the low operating power of mini-fuel cells.

Often, the mini-power plants do not come as a separate unit or module but are integrated into the devices powered by them, and share their housing. This, however, considerably interferes with an easy communication of the fuel cell with the ambient atmosphere supplying its oxygen needs and serving as a heat and product sink.

Another important requirement for power plants serving portable devices is convenient handling. In particular, these plants should admit frequent on–off cycles. When turning power on, the power source should attain its nominal performance within a very short time. In this context the fuel cell's working temperature should not differ much from ambient temperature and would, for instance, be situated within limits of 15 and 45°C. Often, a small storage battery will be included in the power plant for an initial powering of peripherals and, if needed, some heating of the fuel stack. This could be the same battery as the load-leveling buffer battery mentioned above.

14.2 FLAT MINIATURE FUEL BATTERIES

The traditional structure of a fuel battery is that of a vertical stack of alternating bipolar plates, membrane–electrode assemblies, and heat-exchange plates that is compressed with the aid of massive end plates and tie bolts. This structure is poorly adapted for building mini-fuel plants. Particularly for the applications mentioned, special small fuel cells are needed that have design principles different from those of their large cousins.

In a conventional fuel cell, the electrochemical process securing energy conversion is concentrated in the membrane–electrode assembly. This assembly takes up a negligible fraction of the volume (and mass) of the fuel cell. By far the largest volume fraction (over 90%) is taken up by end plates, bipolar plates, heat exchangers, gas-distribution devices, and the like. One might, of course, attempt to minimize the volume and weight of all these design elements that are peripheral to the electrochemical process. However, the possibilities of doing so are limited and will not make any substantial difference.

A flat (two-dimensional) fuel–battery structure where one of the dimensions—the height—is much smaller than the others is much better suited to the design of a small power block for portable devices. Free access of air to the cathode and the elimination of heat and reaction products are greatly facilitated in such a structure, owing to the large ratio of exposed surface area to volume of the fuel cell. It will be seen in the following that with a flat battery structure the structure of its component fuel cells can also be simplified significantly. In addition, existing new technologies admitting mass production and an important cost reduction can be used to make small flat, fuel batteries. A last argument for a flat fuel–battery structure is its convenient fit in the equipment being powered: for instance, a notebook, where the space set aside for the power source will also be flat.

The basic design of a flat fuel battery and its differences from the conventional battery design are shown in Figure 14.1. In a conventional design the individual cells of the battery are vertically stacked (Figure 14.1a). The positive electrode of one cell is in contact with the negative electrode of the neighboring cell via a bipolar plate (acting as the junction between cells). In a flat battery design, the individual cells of the battery are arranged next to each other in a

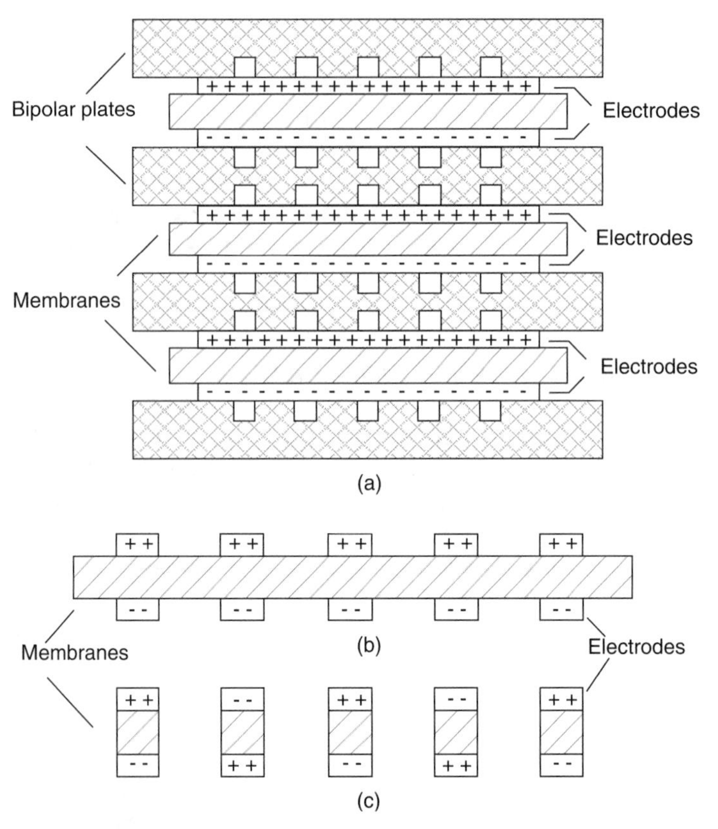

FIGURE 14.1 Basic designs of fuel cell stacks: (a) conventional design of vertical stacks; (b) band design of flat stacks; (c) flip-flop design of flat stacks.

single plane. Two such planar arrangements are possible. In one of them (Figure 14.1b) the negative electrodes of all individual cells are situated on one side of the electrolyte membrane, the positive electrodes of all cells are situated on the other side of this membrane. This arrangement is called the *band design*. In the other arrangement (Figure 14.1c), negative and positive electrodes alternate on each side of the membrane in such a way that each segment of the membrane is provided with electrodes of opposite polarity on the two sides. Such an arrangement is called the *flip-flop design*.

Each of these two versions has advantages and shortcomings. In the band design, the reactant gases are supplied simultaneously to all electrodes: the fuel gas on one side of the membrane, simultaneously to all negative electrodes, and air on the other side of the membrane, simultaneously to all positive electrodes in the battery. This eliminates the need for the complex gas-distribution systems (manifolds) that must be used if each gas is to be fed to each cell individually. The side of the battery with the positive electrodes may be left entirely open so as to freely admit air oxygen. Apart from a simpler gas supply, sealing the battery is also substantially simplified in this version, in that only that part of the battery requires sealing where the fuel gas supply is located. A considerable defect of this version is the complex electrical commutation of the individual cells. For a series connection of the cells to a single electrical circuit, the connections between cells must either go through the membrane, which will upset its integrity, or cross over beyond the edges of the membrane, which will give rise to sealing problems.

In the flip-flop design, the connections between cells link the positive electrode of one cell by the shortest path with the negative electrode of the neighboring cell. This version has the very important defect that a single membrane cannot be used for the entire battery. In any battery consisting of cells connected in series, the electrolytes of two neighboring cells must not be in mutual contact, as this is equivalent to an internal short of the battery. In fact, in each cell, the electrical potential of the electrolyte is different from that in another cell, and upon contact, marked currents will arise that cause self-discharge of the battery and constitute useless reactant consumption. For this reason the membrane electrolyte in a battery of the flip-flop type must be cut up into individual pieces so that each pair of electrodes will have its own electrolyte. Apart from this problem, a battery of the flip-flop type again requires an individual reactant supply to the electrodes in each cell.

All component parts of the battery (electrodes, electrolyte, the connections between cells) should be arranged on a flat support that provides mechanical stability and current collection. The reactant is supplied to the electrodes via channels in this support. Different materials—metals, plastics, and composites—may be used to make the support. Two types of support are most commonly found: wafers of doped (semiconducting) silicon and printed-circuit boards (PCBs). Both types have long been known and are widely used in the electronics industry. Different processing technologies have been developed, such as the microelectromechanical system (MEMS) technology, photolithographic patterning, and various kinds of etching. All these technologies were introduced successfully to the production of flat fuel batteries. The availability of equipment for these technologies has been essential for fast progress in this direction of fuel cell development.

14.3 SILICON-BASED MINI-FUEL CELLS

A version of a methanol–air fuel cell has been described in one of the first papers on mini-fuel cells on silica support (Kelley et al., 2000). A thin layer of silicon nitride was laid down by low-pressure chemical vapor deposition on the surface of a polished silicon wafer oriented in the <100> plane. Then membranelike segments of thickness 150 μm were formed by photolithographic patterning and chemical etching with KOH solution, and small holes of diameter less than 40 μm serving for gas supply were formed within the segments. On one of the wafers forming the fuel cell anode, Pt–Ru catalyst was deposited. The cathodic catalyst, consisting of platinum, was deposited directly onto a Nafion-type membrane. When assembling the cell, the anodic and cathodic silicon wafers were placed onto the membrane in such a way that the gas holes in the two plates were aligned. The entire cell (anode wafer, cathode wafer, and electrolyte membrane) was subjected to hot pressing. The active working surface area of this mini-cell was 0.25 cm^2. At a temperature of 70°C and a voltage of 0.2 V, the current density was about 0.3 mA/cm^2, almost as high as in an analogous methanol–air cell having a working surface area 100 times larger.

Meyers and Maynard (2002) proposed two versions of hydrogen–oxygen fuel cells with doped silicon as the supports: a bilayer cell and a monolithic cell. In the bilayer version, anodes and cathodes were prepared on two different wafers. A layer of porous silicon was deposited on the silicon wafer. Then the top part of this porous layer was subjected to electropolishing and electrochemical etching. These processes were programmed in such a way that surface-layer segments closer to the surface and deeper segments were removed in alternation, so that a network of channels or tunnels was created on the wafer surface, to be used for a reactant gas supply. On the surface of these channels, a layer of the corresponding catalyst was then deposited, and finally, a membrane was cast over the entire wafer surface from a Nafion solution. A cell was made by superimposing two such wafers having different electrodes. The series connection of the individual cells in the battery was provided according to the band design. In the monolithic version, all these operations were conducted on two different sides of the same silicon wafer.

Hayase et al. (2004) used a polished *n*-type silicon wafer having a thickness of 100 μm and oriented in the <100> plane that was covered with a 500-nm layer of silicon oxide. Gas channels 80 to 90 μm deep (see Figure 14.2) were formed by photolithographic patterning and chemical etching in a 15%

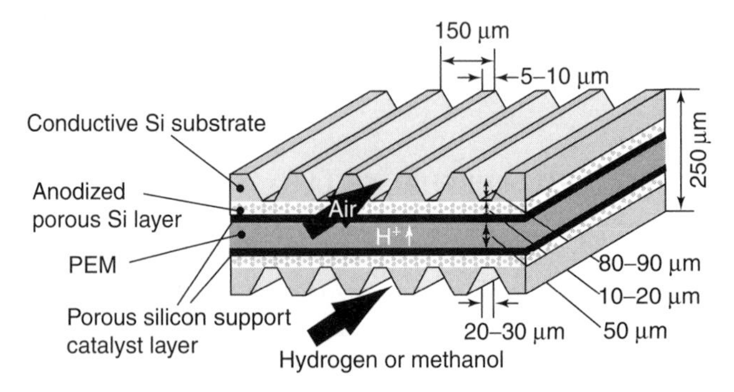

FIGURE 14.2 Schematic view of the miniature fuel cell design. A PEM sheet is hot-pressed with monolithically fabricated Si electrodes. Catalyst metals are supported by porous Si formed by anodization. (From Hayase et al., 2004, with permission from The Electrochemical Society.)

solution of tetramethylammonium hydroxide (TMAH). The oxide layer was removed from both sides by etching in dilute HF solution. A layer of porous silicon was then formed by anodizing in concentrated HF solution. For a uniform deposition of the platinum catalyst within the porous layer, 50 mM HF was added to the platinizing bath to remove the oxide layer from the pore walls and enable electroless deposition of platinum metal on the walls. Two such electrodes were hot-pressed with the Nafion in between. The total thickness of the cell was less than 300 μm.

Chu et al. (2006) described a mini-fuel cell on a silicon support in which 5 M formic acid solution was used as a fuel. These workers believe that design versions where a membrane of the Nafion type is sandwiched between two silicon chips will not exhibit stable operation, since depending on the ambient conditions, the membrane will shrink and swell periodically. They suggested that a layer (a membrane) of nanoporous silicon with a thickness of 50 to 150 μm that is filled with 0.5 M sulfuric acid solution should be used as the electrolyte. On one side of this membrane, nanodisperse palladium was applied to serve as the anodic catalyst. Platinum as the cathodic catalyst was applied to the other side of the porous silicon membrane. With a 100-μm membrane at room temperature, such a mini-cell having a diameter of the working zone of 0.53 cm and working at a current density of 300 mA/cm^2 had a voltage of about 0.3 V (maximum specific power: 94 mW/cm^2).

14.4 PCB-BASED MINI-FUEL CELLS

Just like the production of items supported by silicon chips, printed-circuit-board technology today has attained mass-production maturity. The boards are a laminated support of glass fiber/epoxy resin or other, similar composites. The support is about 1.5 mm thick. A thin layer of copper of about 35 μm is plated onto this support, and after photolithographic patterning, the electrical circuit needed is etched out of it. The copper layer serves as current collector when such a plate is used as the support for a mini-fuel cell.

A detailed description of a PCB-based mini-fuel cell was given in a paper by Schmitz et al. (2003). A serpentine flow field was machined in one of the plates, to be used as the anode. A series of parallel rectangular holes were cut from the cathode plate to secure free access of air to the cathode. A membrane–electrode assembly of commercial origin 35 μm thick was glued with adhesive tape to the anode plate. The two plates and the MEA sandwiched between them were pressed together by six screws with a controlled torque. The active working surface area of the cell measured $20 \times 50\,mm^2$. An insulating gasket with a notch having the diameter of the active region was inserted into the cell to prevent direct contact between the copper layers of the two plates.

The cell had a thickness of about 3.5 mm. It worked at room temperature in an air-breathing mode. Hydrogen was supplied to the anodic region at

30 cm^3/min. In the point of maximum specific power of 110 mW/cm^2, the current density was 275 mA/cm^2, the voltage 0.4 V.

Using PCB technology, one gains additional advantages. It is possible to use several independent plates simultaneously, stacked and separated by insulating gaskets. Within such a multilayer structure, the electrical connections between the layers are accomplished via special holes in the plates. However, an important problem associated with the use of PCBs for fuel cells was also pointed out by the authors. When such a cell has been operated for more than 100 hours, corrosion of the copper coatings due to their contact with the electrolyte leads to pronounced degradation of the cell. A protective coating of the copper layer may be used to prevent this attack. The best results were obtained with a combined electrolytic coating consisting of 10 μm nickel and 1 μm gold. Gold plating technology is well known and well developed in the electronics industry.

A multicell battery on the basis of a PCB, producing voltages of up to 15 V, has been described by O'Hayre et al. (2003). In a cell of volume 3.5 cm^3 and maximum voltage about 1 V, a maximum specific power of 400 W/L was attained, while in a battery consisting of 16 cells that had a volume of 46.9 cm^3, the maximum voltage was 14.8 V and the maximum specific power was 53 W/L.

14.5 MINI-SOLID OXIDE FUEL CELLS

Beckel et al. (2007) reported on a new development involving mini-fuel cells supported by micro-hot plates. The plates are thin membranes (of about 1 μm), consisting of a dielectric and suspended over an opening in a silicon substrate. A heater is laid out on the membrane so that its temperature may be raised to above 500°C. Such hot plates have been used in recent years to produce various devices, such as sensors and microreactors. Recent success toward achieving a lower working temperature for solid-oxide fuel cells creates the hope that it may be possible to use such hot plates for building mini-SOFCs. The authors studied one of the first steps toward making such a mini-SOFC: by applying to the membrane a cathode layer and examining the structural properties of this layer. A 1-μm membrane measuring 1×1 mm^2 consisted of two layers of silicon nitride (SiN) deposited by low-pressure chemical vapor deposition (LPCVD). An electric heater made of platinum was laid out on the membrane so that at a power of 120 mW the membrane temperature could be raised to 600°C. The heater was covered with a thin layer (225 μm) of tantalum (Figure 14.3).

A layer of cathode material of the LSCF type ($La_{0.6}Sr_{0.4}Co_{0.2}Fe_{0.8}O_3$) was deposited by air-pressurized spray pyrolysis of a mixed solution containing the components. After 60 minutes of deposition, the layer had a thickness of 0.5 to 1 μm. Initially, the layer was amorphous, but after annealing at 600°C using the built-in heater, it acquired a crystalline structure. In its physical properties (electrical conductivity), the layer did not differ from analogous layers deposited onto a compact substrate. This indicates that it would basically

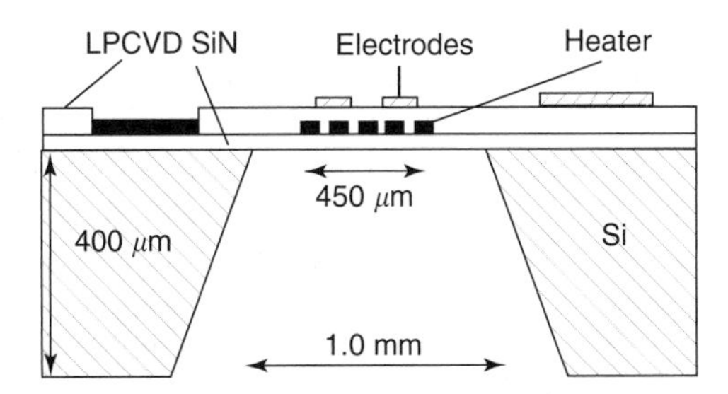

FIGURE 14.3 Schematic cross-sectional view of a micro-hot plate made of a Pt heater between two SiN films, covered by electrodes. (From Beckel et al., 2007, with permission from Elsevier.)

be possible to use hot plates for building mini-SOFCs. The next steps in this direction are the development of technologies to deposit an electrolyte layer and a layer with anodic catalyst onto the same hot plate.

14.6 THE PROBLEM OF AIR-BREATHING CATHODES

In all types of fuel cells, but particularly in small cells, it is very important to be able to use ambient air directly (passively) as the oxidizing agent without additional equipment for its compression or for its artificial injection. Often, the share of this additional equipment (compressors, pumps, systems to regulate their operation) in the total volume and total mass of a fuel cell power plant is rather high. In addition, all this equipment consumes a significant fraction of the electrical energy produced. Not rarely, it is precisely operational failures of these systems that are the reason for premature termination of an entire power plant's operation. Cells using air directly (passively) have been termed *air-breathing* or *self-breathing cells.*

The task of building such "freely breathing" cells is complicated by the fact that in most versions of fuel cells, water as a reaction product is evolved on the cathode, and a circulating stream of air is needed to remove it. Often, this airflow is also used to eliminate the heat of reaction. For plants of large power output (and large dimensions), it generally is more profitable to use actively injected air rather than solving all the problems associated with a passive air supply. When building plants of low power and small size, however, it will be practically unavoidable to use a passive air supply.

Despite the importance of this problem, little research has been done to examine passive air supply in detail. The main condition for the passive use of air is a sufficiently large contact surface area between the cathode and the surrounding atmosphere. In fuel batteries of flat design, one of the sides may be left completely open. In fuel batteries designed for notebooks, the back side of the monitor in the lid housing the battery could be left open almost completely.

A large contact surface area between the fuel cell and the surrounding atmosphere has a large defect, however, inasmuch as the ambient conditions (temperature, relative humidity, the presence of pollutants) are unstable and variable in time. These changes cannot remain without an effect on the internal state of the fuel cell when the contact surface area is large.

Certain aspects of the operation of mini-fuel cells with a large contact surface area between cathode and ambient air were examined by Schmitz et al. (2004) in the instance of a methanol–air mini-fuel cell. They studied the elimination of product water when the open contact surface area was in different spatial orientations: horizontally pointing up, and vertical. The tests were conducted while ambient temperature varied between 25 and 31°C. It was found that cell operation was more stable with a vertical orientation of the contact surface area. In this orientation, the values of relative humidity at points close to the contact surface area were lower than in the horizontal orientation. As an explanation, the authors suggested that the water vapor produced will partly condense yielding liquid droplets, which, under the effect of gravity, will slip down from the vertical surface.

Pan (2006) noted that in a fuel cell with a passive air supply, much more stable results were obtained when the cathodic catalyst was deposited directly onto the electrolyte membrane rather than onto the gas-diffusion layer.

There have been other publications mentioning an air-breathing fuel cell in the title (e.g., Jaouen et al., 2005), but it must be pointed out that on the whole, the important problem of fuel cell operation with a passive air supply from the ambient atmosphere and a passive elimination of reaction products and heat has not been investigated sufficiently.

14.7 PROTOTYPES OF POWER UNITS WITH MINI-FUEL CELLS

14.7.1 Specific Performance Indicators

The most important performance indicators for power supplies designed for portable equipment are the specific energy per unit mass (weight), γ_m (J/kg) and/or per unit volume, γ_v (J/L). Often, the miniplants with fuel cells are used in portable equipment as a replacement for lithium-ion batteries, which have specific performance indicators of 150 Wh/kg and 350 Wh/L).

A power plant with mini-fuel cells comprises both the fuel battery itself with all peripherals and the containers with the reactant (fuel). The weight of the battery together with the ancillary equipment, $G_{\mathrm{Bat+Anc}}$, to a first approximation is proportional to the electric power, P_e, generated; that is, $G_{\mathrm{Bat+Anc}} = \alpha P_e$. The weight of the containers with the fuel depends on the required length of time, τ, of uninterrupted operation of the power plant (i.e., without container replacement), and to a first approximation is proportional to the electrical energy, W_e, produced during this time (when working at a constant power, $W_e = P_e\tau$; i.e., $G_{cont} = \beta W_e = \beta P_e \tau$). The total weight, G_{PU}, of the power plant is the sum of weights of the individual components: $G_{PU} = G_{cont} + G_{Bat+Anc}$. The specific energy of the power plant per unit weight, γ_m, is given by W_e/G_{PU}. From this it can be seen that

$$\gamma_m = \frac{P_e\tau}{\beta P_e\tau + \alpha P_e} = \frac{\tau}{\alpha + \beta\tau}, \tag{14.1}$$

which means that the specific energy is larger the longer the required time of uninterrupted operation, and hence, the larger the share of containers with fuel in the total mass of the power plant (or, the smaller the share of the mass of the fuel battery itself). The same holds true for the specific energy per unit volume.

In papers concerning the different prototypes of small fuel cell plants, unfortunately, the data reported are often not sufficiently detailed for an unambiguous estimate of the contributions of the various components to the specific performance indicators (the values of the constant coefficients α and β).

14.7.2 Power Plants with DMFCs

Xie et al. (2004 at Motorola, Inc.) reported building a mini-power plant of 2 W using a six-cell battery of DMFCs of the conventional type with bipolar plates of graphite. In the plant, methanol and air were actively supplied with the aid of pumps. The working surface area of each cell was 6 cm^2. The battery operated with a current density of 20 mA/cm^2 and a voltage of 3 V (0.5 V per cell). The ancillary equipment (three pumps, sensors, and the like) consumed 0.375 W of the electrical energy produced. During a 5-hour test, the battery consumed 10.5 mL of pure methanol from a container with a specific energy supply of 956 Wh/L and 1080 Wh/kg (the weight of the container was not given) and produced about 10 Wh of electrical energy. The design values for one week of uninterrupted operation (168 hours) were: weight of the system, 0.685 kg, volume, 0.913 L; specific energy, 490 Wh/kg and 368 Wh/L.

Shimizu et al. (2004) provided data for an individual cell having an active surface area of 35 cm^2, a passive supply of methanol, and direct access of air from the ambient atmosphere (air breathing). A tank for methanol holding 28 mL was built into the cell. Using 4 M methanol solution, the cell worked with a current density of 36 mA/cm^2 and a voltage of 0.3 V.

Miesse et al. (2006, Korea Institute of Science and Technology) reported the development of a system for laptop-type computers. The dimensions of the system were $20.5 \times 8.5 \times 8.25$ cm^3 (1438 cm^3); the weight without fuel was 1.5 kg and with fuel it was 1.8 kg. The system contained a 15-cell battery with bipolar graphite plates weighing 0.65 kg. A formic acid solution (2.1 M) was the fuel for the battery. The system also contained a miniature pump for the fuel supply, a miniature air compressor, fans for cooling, and a power conditioning system. The system included a battery for startup, which was recharged during operation. The system produced 30 W of electrical power (with a specific power of 60 mW/cm^2). From this output, about 6 W was used for the internal needs of the ancillary equipment listed.

14.7.3 Power Plants with PEMFCs

A flat battery consisting of six hydrogen–air cells of the PEMFC type supported by two silicon wafers and connected in series was described by Zhang et al. (2007). The active surface area of the battery measured $1.2 \times 1.2\,cm^2$ ($1.44\,cm^2$). Hydrogen was supplied from a container with a metal hydride material having a large hydrogen uptake: the compound $MmNi_5$ (Mm = mischmetall, a mixture of rare-earth metals). The container was placed in a water bath at about 60°C to desorb the absorbed hydrogen. The hydrogen flow was 40 mL/min. Direct access of air was provided from the ambient atmosphere. The battery was operated at a temperature of 20°C and a relative humidity of the air of 50%. An almost linear relation between battery voltage and current was obtained which began with a value of 5.3 V at zero current and arrived at about 1 V at a current of 0.5 A. The maximum power of 0.9 W was attained at an overall current density of 250 mA/cm^2. When drawing a power of more than 0.5 W, deterioration of the parameters was seen.

Urbani et al. (2007) reported building a 15-W power plant with a battery consisting of 10 hydrogen-air fuel cells of the PEMFC type with bipolar graphite plates 5 mm thick. The plant measured $20.5 \times 13 \times 28$ cm^3 (746 cm^3). Hydrogen was supplied to the battery from a container with metal hydride via a special pressure regulator. A dc–dc converter was added to the battery, which provided a stable output voltage of 11.3 V. A small fan consuming 1.5 W stabilized the temperature inside the plant. It turned on automatically when the battery voltage was above 4 V. With the fan turned on, the maximum energy density of 82 mW/cm^2 was attained at a current density of 80 mA/cm^2 and a voltage of the individual cells of 0.43 V. The plant worked at a temperature of 25°C and a relative humidity of 45 to 55%. It was used as an uninterrupted power supply for 3 hours to a portable DVD player. It worked for more than 100 hours without additional servicing (except for periodic hydride container replacement).

14.7.4 Mini-Reformed Hydrogen Fuel cells

Hydrogen–air fuel cells of the PEMFC type have higher performance indicators than those of methanol–air fuel cells of the DMFC type. However, manipulating hydrogen that is stored either as a gas in small pressurized tanks or in the absorbed state as a metal hydride is more difficult than manipulating methanol or any other type of liquid fuel. The question has therefore been discussed of building mini-power plants on the basis of PEMFC-type cells, including a mini-reformer for the liquid fuel (methanol or a borohydride solution; see the review of Kundu et al., 2007). Such systems are of interest from the point of view of user convenience and their potential performance parameters. It must be taken into account, however, that when using methanol in such systems, there may be a small delay during startup because the reformer must first be heated up.

In a number of papers, the development of mini-plants for reforming was reported. Seo et al. (2004) developed such a mini-reformer for the steam conversion of methanol. This reformer contained a methanol evaporator and the reactor proper. The unit was built from stainless steel plates. These plates contained 20 parallel channels that were 0.5 mm wide, 0.6 mm deep, and 30 mm long. A $Cu/ZnO/Al_2O_3$ steam-conversion catalyst was deposited in these channels. A mixture of methanol and water was added to the evaporator. The conversion process occurred at a temperature of 200 to 260°C. The full unit measured $7 \times 4 \times 3$ cm^3 (84 cm^3). The unit generated hydrogen sufficiently fast for a 10-W fuel cell.

Kundu et al. (2006) reported building a mini-reformer based on a silicon wafer having a structure similar to that described in Seo et al.'s paper. The unit's total volume was 0.9 cm^3; its productivity matched a fuel cell power of 2.4 W.

14.8 CONCLUDING REMARKS

Construction of the small fuel cells discussed above is a relatively new field of development in the history of fuel cells. Basic work in this field did not begin until after the beginning of the twenty-first century. It can be seen from the review given above that the results obtained so far are fragmentary. A relatively large number of the papers published refer to new types of flat mini-fuel cells on supports consisting of silicon wafers or of printed circuit boards, but so far the full potential of these supports is little exploited. Most of the work concerning complete power plants with small fuel cells is still based on conventional fuel cell structures with bipolar plates.

Work in this field is exploding. There can be no doubt that considerable progress will be achieved, and in the future such power plants will find the same mass-consumer acceptance as those consumer electronic devices that they are supposed to power.

REFERENCES

Beckel D., D. Briand, A. Biberle-Hütter, et al., *J. Power Sources*, **166**, 143 (2007).

Chu K.-L., M. A. Shannon, R. J. Masel, *J. Electrochem. Soc.*, **153**, A1562 (2006).

Hayase M., T. Kawase, T. Hatsuwasa, *Electrochem. Solid-State Lett.*, **7**, A231 (2004).

Jaouen F., S. Haasl, W. van der Wijngaart, et al., *J. Power Sources*, **144**, 113 (2005).

Kelley S. C., G. A. Deluga, W. H. Smyrl, *Electrochem. Solid-State Lett.*, **3**, 407 (2000).

Kundu A., J. H. Jang, H. R. Lee, et al., *J. Power Sources*, **162**, 572 (2006).
Meyers J. P., H. L. Maynard, *J. Power Sources*, **109**, 76 (2002).
Miesse C. M., W. S. Jung, K.-J. Jeong, et al., *J. Power Sources*, **162**, 532 (2006).
O'Hayre R., D. Braithwaite, W. Hermann, et al., *J. Power Sources*, **124**, 459 (2003).
Pan Y. Y., *J. Power Sources*, **161**, 282 (2006).
Schmitz A., M. Tranitz, S. Wagner, et al., *J. Power Sources*, **118**, 162 (2003).
Schmitz A., S. Wagner, R. Hahn, et al., *J. Power Sources*, **127**, 197 (2004).
Seo D. J., W.-L. Yoon, Y.-G. Yoon, et al., *Electrochim. Acta*, **50**, 719 (2004).
Shimuzu T., N. Momma, M. Mohamedi, et al., *J. Power Sources*, **137**, 277 (2004).
Urbani F., S. Squadrito, G. Giacoppo, et al., *J. Power Sources*, **169**, 334 (2007).
Xie C., J. Bostaph, J. Pavio, *J. Power Sources*, **136**, 55 (2004).
Zhang X., D. Zheng, T. Wang, et al., *J. Power Sources*, **166**, 441 (2007).

Reviews

Kamarudin S. K., W. R. W. Daud, S. L. Ho, A. Hasran, Overview of the challenges and developments of micro-direct methanol fuel cells, *J. Power Sources*, **163**, 743 (2007).
Kundu A., J. H. Jang, C. R. Jung, et al., Micro-fuel cells: current developments and applications, *J. Power Sources*, **170**, 67 (2007).
Qian W., D. P. Wilkinson, J. Shen, H. Wang, J. Zhang, Architecture for portable direct liquid fuel cells, *J. Power Sources*, **154**, 202 (2006).

CHAPTER 15

MATHEMATICAL MODELING OF FUEL CELLS

FELIX N. BÜCHI

Electrochemistry Laboratory, Paul Scherrer Institut, Villigen, Switzerland

Over the past two decades, modeling has become a highly important tool for fuel cell technology development. The mathematical simulation work has two main goals:

1. *Understanding*. Fuel cells, as electrochemical reactors with gradients of species, temperature, and potential in all dimensions and over a broad range of scales from nanometers to meters, are complex devices. Changing a single parameter (e.g., the gas humidity in a PEMFC) can result in effects on the scale of reaction kinetics and other parameters all the way to temperature distribution in a fuel cell stack. This multitude of effects, as well as their consequences felt in important parameters such as efficiency or power density, are difficult or impossible to comprehend without modeling approaches.
2. *Design*. In the interlinked processes of the transport of species, heat, and electrons in the various phases, one of the processes is often found to be limiting. On a small scale, physical properties such as the current density distribution in a catalyst layer are extremely difficult to measure. Therefore, modeling is an alternative to the trial and error of experimental work when developing and designing such structures. On a larger scale, such as that of gas distributors in the flow field or a manifold in stacks, design of the structures using an experimental approach is difficult and needs to be supplemented by sophisticated modeling tools. In many engineering disciplines, the use of sophisticated models for design

Fuel Cells: Problems and Solutions, By Vladimir S. Bagotsky

purposes has become a standard tool (e.g., in predicting the mechanical properties of complex parts and structures).

The theories developed in the 1960s for gas-diffusion electrodes (Newman and Tobias, 1962) were the basis for the first fuel cell models published in the 1970s in the field of phosphoric acid fuel cells. A fast acceleration of model development was observed in the 1990s, along with increased general interest in fuel cell technology and certainly also in the availability of cheap and powerful computing resources. The importance of modeling in recent years is highlighted by the fact that during the past five years, more than 1000 papers have been published on the topic, and that articles dealing with mathematical simulation are among the most cited papers in the fuel cell literature.

A distinction between the different modeling approaches found in the literature can be provided in various ways. To give an overview, in this chapter the dimensionality of the models is used as the criterion. Zero- up to fully three-dimensional approaches are known. These dimensions are illustrated in Figure 15.1. Whereas zero-dimensional models are single equations and one-dimensional approaches are a description of processes orthogonal to the electrolyte, simulations in two and more dimensions also include the mass, heat, and charge transport in the plane of the flow field.

With the exponential increase in computing power over the past 40 years, sometimes referred to as Moore's law (Moore, 1965), and the associated decrease in the cost of computing, the models have become more and more complex. Therefore, to some extent the dimensionality also reflects the development in time.

The basic physical phenomena are the same for all types of fuel cells. They are described by the conservation equations for mass, energy, momentum, and

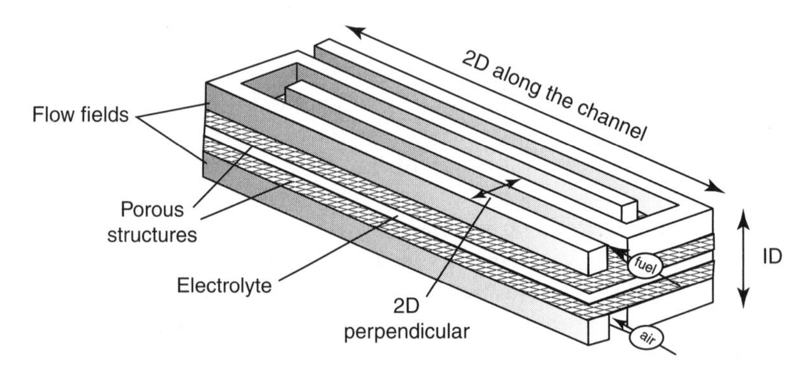

FIGURE 15.1 Schematic of a slice through a fuel cell. Dimensions indicated are: 1D, orthogonal to the electrolyte; 2D along the channel, combining the 1D and the direction along the channel; 2D perpendicular, combining the 1D and the direction perpendicular to the main flow.

charge. In addition, specific equations are used that deal with the electrochemical processes:

- Faraday's law relating electric current and mass turnover
- Butler–Volmer or Tafel equations, relating electric current and potential

and with the very important processes of mass and energy transport in the cell:

- Navier–Stokes equations for the motion of the fluids
- Darcy's law for fluid flow in porous media
- Stefan–Maxwell equation for multispecies diffusion
- Ohm's law for electric current conduction
- Fourier's law for heat conduction

The correct description of energy and mass transport in the various phases of the fuel cell is a key for the mathematical description and simulation of cells.

15.1 ZERO-DIMENSIONAL MODELS

Equations describing global properties of fuel cells are denoted as zero-dimensional models. An example is the description of the current–voltage characteristics of a fuel cell where the different overvoltages are considered: for example, in the form (Ticianelli et al., 1988)

$$E = E_{\mathrm{oc}} - b\log(i) - Ri \tag{15.1}$$

where E denotes the cell voltage, E_{oc} the open-circuit voltage, b the Tafel slope, and R the cell resistance. This approach is rarely used anymore on the level of single fuel cells, because little can be learned from it. However, it is still used in models simulating entire fuel cell systems or power trains, including fuel cells, where one does not need details describing the behavior of the fuel cell stack and where one wishes to have a low computational effort.

15.2 ONE-DIMENSIONAL MODELS

One-dimensional models include the dimension orthogonal to the electrolyte encompassing the catalyst and porous layers and the gas volumes on either side. The transport of species and energy is only considered normal to the various layers.

The modeling approach is thus reduced to the most important dimension of the problem. Whereas in many cases and even in the early models, the entire sandwich was considered (Springer et al., 1991), the modeling domain is often restricted to a single component, such as the catalyst layer (Eikerling and Kornyshev, 1998), the electrolyte, the porous electrodes, or transport layers. In many cases the domain is also limited to the cathode (Svensson et al., 1996), considered to be the limiting interface of the fuel cell.

Apart from the electrochemical kinetics, a major issue for one-dimensional models in the transverse direction is transport in porous structures, most often the transport of mass, but also that of heat and of charge. While in high-temperature systems this includes the transport of species in gas mixtures in the voids and the transport of charge in the solid and of heat in both phases, in low-temperature fuel cells containing liquid water, the description in the voids extends to the high complexity of a two-phase system, where phase changes of water provide complex coupling to the temperature of the solid phase.

Charge transport in the electrolyte and mass transport in the gas phase are the dominant transport phenomena in fuel cells. Mass transport in gas mixtures is generally described by the Stefan–Maxwell equations:

$$\frac{dx_i}{dz} = \sum_{j \neq i} \frac{x_i N_j - x_j N_i}{p D_{ij}^{\text{eff}}} \tag{15.2}$$

where z is the one-dimensional direction of the model normal to the electrolyte, x_i denotes the mole fraction of species i, p is the total pressure, and D_{ij}^{eff} is the effective binary diffusion coefficient of species i and j. For porous materials the effective diffusion coefficient is defined by $D_{ij}^{\text{eff}} = D_{ij}(\varepsilon/\tau)$, where ε and τ are the porosity and tortuosity of the porous material.

Evidently, all important processes that occur in a fuel cell and involve mass transport and charge transfer are temperature dependent, but many of the earlier one-dimensional models have been formulated isothermally: that is, assuming that temperature differences in the various phases are small (Springer et al., 1991). In recent years, however, the inclusion of the temperature field has become common (Shah et al., 2007). In low-temperature fuel cells, this is highly important, since only in this way will it be possible to capture the critical effects of the local phase change of water.

15.3 TWO-DIMENSIONAL MODELS

When considering two dimensions, in addition to the transverse coordinate, one includes either the dimension along the channel or the dimension perpendicular to the channel. Both are of technical relevance, so let us consider first the more general case along the channel. In this case, especially when using models for cells of a technical size with active areas of several hundred square centimeters, the model domain reveals an aspect ratio of about 1000: tens of centimeters along the channels but only several hundred micrometers in the transverse direction. Such high aspect ratios pose severe problems for the definition of adequate grids for modeling. For this reason, dimensionally reduced grids are formulated based on the reasoning that transport phenomena in the direction along the channel are dominated by convection in the flow channels. An inspection of the general steady-transport equation,

$$\mathrm{div}(\rho v \Phi) - \mathrm{div}(\Gamma\, \mathrm{grad}\, \Phi) = S \tag{15.3}$$

for the property Φ and for the orders of magnitudes of the strength of convection ρv and diffusion conductance $\Gamma/\partial x$ lead to the dimensionless Péclet number $P_e = uL/\Gamma$. Here ρ, v, Γ, and S are the density, convection velocity, transport coefficient, and the source term, respectively; L represents the characteristic length. A formal application of Eq. (15.3) to the in-plane transport of reactants in complete cells leads to a Péclet number on the order of 10^{10}. This indicates that convective transport inside the flow channel is far more important than parallel diffusive fluxes in the gas-diffusion media or the membrane.

This finding leads to the development of pseudo-two-dimensional models for the along-the-channel case, where transport in the solid, liquid, and porous phases along the channel is neglected (Berg et al., 2004; Freunberger et al., 2006). This reduces computational costs considerably and allows extended parameter studies to be performed with only a minor compromise in accuracy.

In the case of two dimensions represented by the directions through the plane and transverse to the main fluid flow, the aspect ratio given by the width of the flow channel over the thickness of the electrochemical components is close to 1. Therefore, common two-dimensional modeling approaches are feasible here (Jeng et al., 2004). These models focus on an understanding of the limiting processes in pore transport and in the catalyst layers. These simulations may also be used to optimize the cross-sectional design of the gas-distribution structures. Again, the transport properties of the porous media for mass and energy become important parameters.

15.4 THREE-DIMENSIONAL MODELS

Whereas lower-dimensional modeling approaches are primarily used to foster an understanding of the influence of material properties and operating parameters, three-dimensional modeling of the fluid or heat flows or full-physics simulations of the cell can be used for design purposes. Layout of the structures of bipolar plates with respect to mass flow, pressure drop and temperature distribution, the sizing of manifolds in a stack, or design of the cooling structures requires support by sophisticated mathematical simulation. For these three-dimensional problems, commercial solvers and finite volume or finite difference methods are generally employed to solve the multiscale problems (Nguyen et al., 2004; Shimpalee et al., 2004). Accurate modeling of the fluid properties is based on the Navier–Stokes equation. In its form, generally used for incompressible fluids but which holds well even when dealing with "compressible" fluids such as gases near room temperature, it reads

$$\rho\left(\frac{\partial v}{\partial t} + v \cdot \nabla v\right) = -\nabla p + \mu \nabla^2 v + f \tag{15.4}$$

where ρ is the density, v the velocity, p the pressure, and μ the dynamic viscosity of the fluid, while f denotes other forces, such as gravity.

The general fluid properties can be described quite accurately in this way, but even the most sophisticated models are not able to describe or predict all cell properties precisely. This is because not all parameters are known well enough and because the porous materials, in particular, often have anisotropic properties with respect to their structure and therefore also for their effective mass and energy transport properties.

The full three-dimensional description of cells of technical size having active areas of several hundred square centimeters, requires a huge computational effort, such that even with today's computing resources, compromises may be required with respect to resolution of the small structures (Roos et al., 2003).

15.5 CONCLUDING REMARKS

Up to this point, all the description given has applied to models for the steady state. However, in real-life applications the power drawn from a fuel cell will be modulated according to the needs of the consumer. Whereas in some applications, such as stationary CHP, the modulation may be slow and thus not so important, in other applications, such as automotive power trains, a highly dynamic operation is required. The models, including all dimensions in space, must then be extended into the time domain. For complex models describing cells of technical size, the computational effort is very significant and requires adequate resources. Therefore, transient modeling has been taken up only in recent years.

It is important at this point to acknowledge that mathematical simulations still depend on experimental work. The complexity of fuel cells is such that results obtained from models must be validated in detail against experimental data. It should be acknowledged that to date, the majority of innovations were based on experimental work. It is quite probable that in the future, with increasing maturity of the technology, mathematical modeling will become a more and more important tool.

REFERENCES

Berg P., K. Promislow, J. St.-Pierre, et al., *J. Electrochem. Soc.*, **151,** A341 (2004).

Eikerling M., A. A. Kornyshev, *J. Electroanal. Chem.*, **453,** 98 (1998).

Freunberger S. A., M. Santis, I. A. Schneider, et al., *J. Electrochem. Soc.*, **153,** A396 (2006).

Jeng K. T., S. F. Lee, G. F. Tsai, C. H. Wang, *J. Power Sources*, **138**, 41 (2004).

Moore G., *Electronics*, **38,** 114 (1965).

Newman J. S., C. W. Tobias, *J. Electrochem. Soc.*, **109,** 1183 (1962).

Nguyen, P. T., T. Berning, N. Djilali, *J. Power Sources*, **130**, 149 (2004).

Roos M., E. Batawi, U. Harnisch, Th. Hocker, *J. Power Sources*, **118**, 86 (2003).

Shah A. A., G.-S. Kim, P. C. Sui, D. Harvey, *J. Power Sources*, **163**, 793 (2007).

Shimpalee, S., S. Greenway, D. Spuckler, J. W. Van Zee, *J. Power Sources*, **135**, 79 (2004).

Springer, T. E., T. A. Zawodzinski, S. Gottesfeld, *J. Electrochem. Soc.*, **138**, 2334 (1991).

Svensson A. M., S. Sunde, K. Nisancioglu, *Solid State Ionics*, **86–88,** 1211 (1996).

Ticianelli E. A., C. R. Derouin, A. Redondo, et al., *J. Electrochem. Soc.*, **135,** 2209 (1988).

Reviews

Barbir F., Fuel cell modeling, Chapt. 7 in: *PEM Fuel Cells: Theory and Practice*, Academic Press, New York, 2005.

Cheddie D., N. Munroe, Review and comparison of approaches to membrane fuel cell modeling, *J. Power Sources*, **147**, 72 (2005).

Weber A. Z., J. Newman, Modeling transport in polymer–electrolyte fuel cells, *Chem. Rev.*, **104**, 4679 (2004).

PART IV

COMMERCIALIZATION OF FUEL CELLS

CHAPTER 16

APPLICATIONS

16.1 LARGE STATIONARY POWER PLANTS

According to a terminological distinction that is sometimes adopted, stationary fuel cell–based power plants may be divided into those of large power (more than 10 kW) and those of small power (less than 10 kW). Basically, the former are intended primarily to generate grid power, that is, provide power to settlements of various sizes or to relatively large industrial sites. In their intended functions, they would not differ from thermal, hydraulic, or nuclear power plants. Fuel cell–based power plants of small power are built to supply electric power to individual residential or administrative quarters or to remote individual power customers. The dividing line is quite arbitrary, of course.

In this section, large stationary power plants are considered. Small stationary power plants are considered in Section 16.2.

16.1.1 Basic Development Trends

It had been shown in earlier chapters that in quite a few countries, large power plants based on different fuel cell systems have been built. Using PAFCs, the U.S. company UTC produced a large number of 200-kW power units. In Japan, multimegawatt power plants were built. Using MCFCs, the U.S. company Fuel Cell Energy built a power plant with an output of up to

Fuel Cells: Problems and Solutions, By Vladimir S. Bagotsky

2 MW in Santa Clara, California. Using SOFCs, the company Siemens–Westinghouse built a 100-kW power plant in the Netherlands.

This shows that all these systems have already left the experimental R&D stage and became an economic reality. They have already demonstrated lower consumption of natural fossil fuels per unit of power generated and lower emission of greenhouse gases and other harmful products.

Other advantages over thermal power plants are associated with fuel cell–based power plants. They can be produced in modules from which power plants of different size and output can be put together. This attribute, known as *scalability*, facilitates the design and construction of plants adapted to specific local requirements.

There has been a steady increase in the number of fuel cell–based power plants built and put in service. Adamson's 2007 review published in the online information newsletter *Fuel Cell Today* provides detailed data for these developments in the economy.

In Figure 16.1 we show the number of such plants constructed every year between 1996 and 2007, as well as the electric power added every year (megawatts per year). It can be seen that in individual years, this figure was 20 to 25 MW. The average power output of these plants increased from 0.2 MW in the years between 1997 and 2003 to 0.3 MW in 2006, and to 0.57 MW in 2007. Figure 16.2 shows the percentage figures for the various fuel cell systems used year by year to build these plants. It can be seen that over a number of years, more than 40% of the new plants were MCFC-based high-temperature plants. It is notable that in many cases, PAFCs with concentrated phosphoric acid were used, a system that for a long time had been regarded as not promising and for which R&D efforts have ceased long ago.

A marked fraction of the plants were built using PEMFCs, which cost more than other fuel cell systems. A wider use of such fuel cells is probably justified in those cases where a cheap source of highly pure hydrogen is available, such as close to chlor-alkali industries, where pure hydrogen is a free by-product.

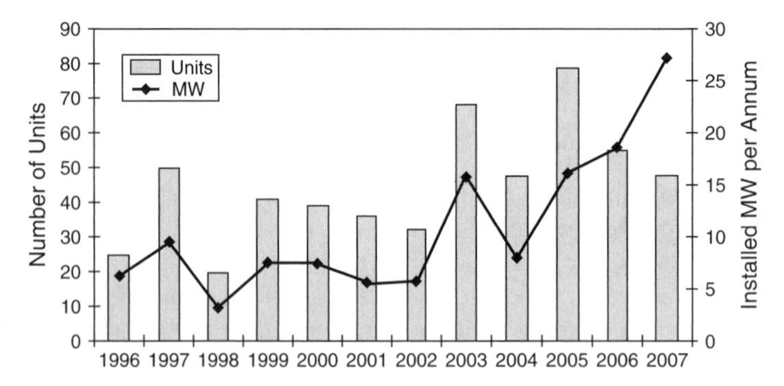

FIGURE 16.1 Annual number of large stationary fuel cell units and megawatts installed. (From Adamson, 2007, with permission from Fuel Cell Today, Inc.)

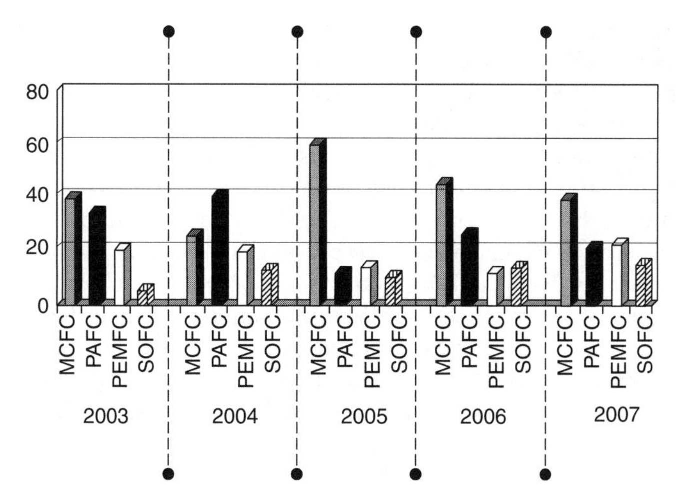

FIGURE 16.2 Technology type of large fuel cell units, by percentage. (From Adamson, 2007, with permission from Fuel Cell Today, Inc.)

Today, most fuel cell–based power plants are produced and set up in the United States due to active government support. A similar tendency can be seen in South Korea, where the government plans to have 300 such plants operating by 2012 (Bishoff, 2006). In Japan, where in earlier years many large FC-based power plants had been built and operated, attention is focused at present on the development of small stationary power plants and on power units for electric vehicles.

In most cases, large fuel cell–based stationary power plants are used for power production and at the same time for heat supply to customers in nearby locations [combined heat and power (CHP) systems]. This combined use of two different types of energy implies a very considerable increase in the total economic and energetic efficiency of such plants.

16.1.2 Hybrid Stations

Recent years have seen large efforts to build hybrid plants combining high-temperature fuel cells and gas turbines. The basic idea is that of highly efficient coordination of the gas and heat flows entering and leaving the two components of such combined power plants. This coordination serves to lower the energy losses and raises the overall plant efficiency dramatically.

Ghezel-Ayagh et al. (2005) describe work done in this direction by the company FuelCell Energy. The MCFCs with direct internal reforming of the hydrocarbon fuel (natural gas) developed by this company are used for this work. The main source of electric power (over 80%) in these hybrid plants are the fuel cells. The gas turbine produces an additional amount of electrical energy from the heat evolved by the operating fuel cells. The turbine also compresses the air fed into the fuel cells. Preliminary tests of such a hybrid power plant of 250 kW gave very promising results. The plant feeding power into the grid was operated for more than 6000 hours. Nitrogen oxide emissions into the atmosphere were minimal (less than 0.25 ppm), since the turbine was operated with heat from the fuel cells. These results are now used as a basis for building hybrid power plants of megawatt size.

Calculations show that such plants would have an overall efficiency for the direct conversion of chemical energy (natural gas) to electrical energy [referred to the lower heating value (LHV)] of 62% in the near term, 67% in the medium term, and possibly as much as 74.6% in the long term, when the gas turbines will have been further improved. Disregarding the turbine, the efficiency of the fuel cells alone would be 50% in the near term and 57% in the medium term.

Various schemes for a combination of SOFCs with gas turbines have been discussed by Winkler and Lorenz (2002). These authors also concluded that for large-scale hybrid plants, it is realistic in the long term to expect conversion efficiencies of up to 80%.* For power units of small size, such as those intended for electric vehicles, this conversion efficiency may attain values of 55%.

FIGURE 16.3 Overall view of a 3-MW SOFC-based Siemens–Westinghouse power station. (Courtesy of Siemens.)

*For methane oxidation, the highest energy conversion efficiency that is thermodynamically possible is close to 100%, according to Eq. (1.6).

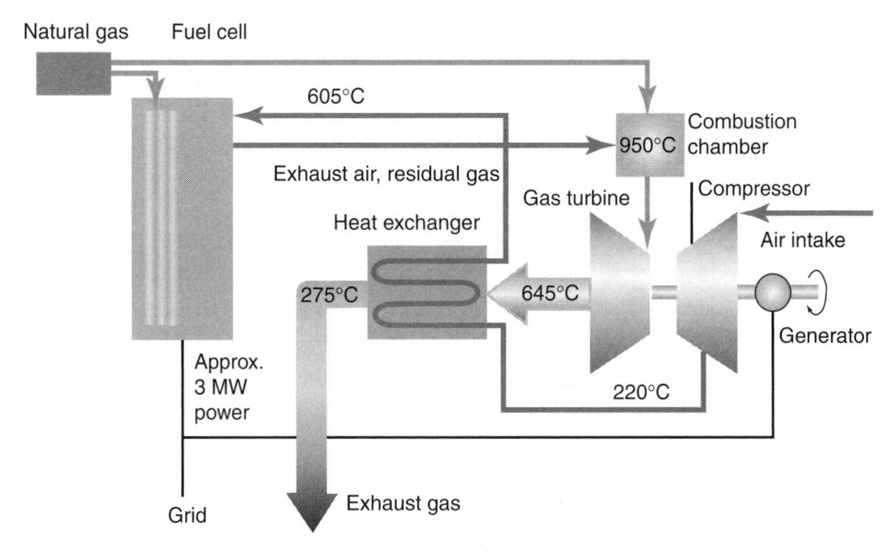

FIGURE 16.4 Scheduled to go onstream in 2012, the hybrid power plant shown in Figure 16.3 achieves an efficiency of around 70%, thanks to the combustion of residual hydrogen in a gas turbine as well as to other improvements. (Courtesy of Siemens.)

Figure 16.3 presents an overall view of a Siemens–Westinghouse 3-MW SOFC-based hybrid power station, and Figure 16.4, a scheme of the plant layout.

16.2 SMALL STATIONARY POWER UNITS

Fuel cell–based power plants that have an output of up to 10 kW are under vigorous development as well and find ever-wider practical uses. Figure 16.5 shows the number of such plants produced every year between 2001 and 2006. It can be seen that this number has increased each year. In 2007 the figure was expected to attain 3000 units. Approximately half of the units produced in 2006 had a power of about 1 kW, the other half an output of 1.5 to 10 kW, the numbers being distributed approximately evenly over this interval. The overwhelming number (more than 50%) of these plants were produced and set up in Japan, with the United States in second place. Figure 16.6 shows the fractions of the various types of fuel cells used over the years for these units. Most of the low-power units were built with PEMFC. The fraction of SOFCs has decreased gradually.

The Japanese company Ebara Ballard, a subsidiary of the well-known Canadian company Ballard, is the most important maker of PEMFCs. They developed a 1-kW power plant for combined heat and power production. It is

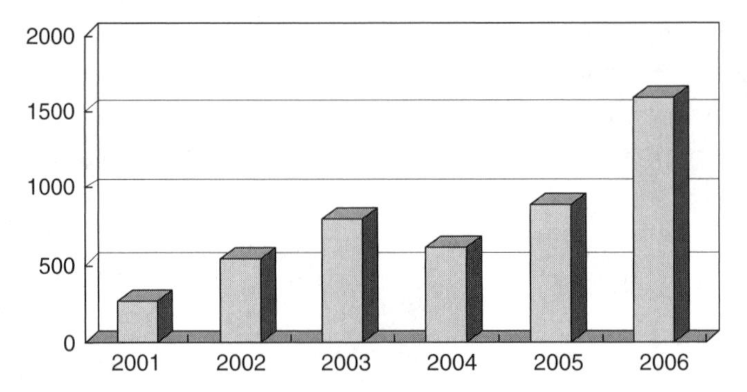

FIGURE 16.5 Annual numbers of small stationary fuel cell units installed. (From Adamson, 2006, with permission from Fuel Cell Today, Inc.)

remarkable that this unit is designed to be operated for a period of 10 years, in accord with requirements set by the Japanese government. A similar unit also designed for an operating time of 10 years was developed by the Japanese company Fuji Electric. These units cost $12,000 to $16,000 (Adamson, 2006).

The range of applications of small power plants is very large. They are designed primarily for the needs of communities for a combined supply of electric power and heat (hot water and heating) to individual structures of different sizes: individual cottages, administrative and office buildings, hospitals. At present, the cost of these units as a rule is not competitive (it should be no higher than $1500/kW). Therefore, part of the costs are supported by municipal or federal powers. In Japan, Prime Minister Junichiro Koizumi set

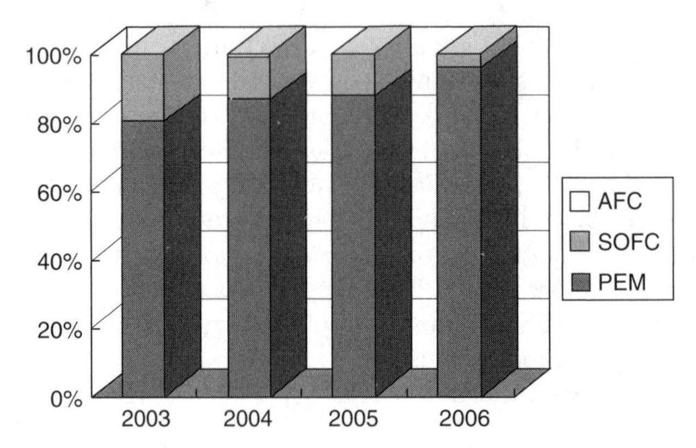

FIGURE 16.6 Technology type of small fuel cell units, 2003–2006, by percentage. (From Adamson, 2006, with permission from Fuel Cell Today, Inc.)

an example for the use of the most ecological system satisfying the needs for electrical energy and heat in a residence when he equipped his house with a 1-kW cogeneration unit [*Adv. Fuel Cell Technol.*, **8**, 18 (2004)].

Numerous examples of small fuel cell–based power units produced and used for domestic applications could be cited. Thus, the Vaillant Group, which has production facilities in a number of European countries, has developed a combined heat and power unit (CHP unit) designed for an electric power output of 4.6 kW and a heat output of 11 kW. Such units were installed in several countries (Germany, the Netherlands, Spain, Portugal). The European Commission has supported 30% of the full project costs, estimated at €8.6 billion [*Kommunalwirtschaft*, No. 2, p. 107 (2004)].

Another important area of application is the use of small power units as backup power in situations of sudden loss of grid power due to natural or technical problems. Such backup units are extremely important to those consumers who cannot tolerate power interruptions. This is the case for the various stationary telecommunications installations (e.g., receivers, transmitters, relay stations, signal amplifiers). In their 2006 paper, Perry and Strayer provided a detailed discussion of the problems associated with the use of fuel cells in backup applications.

Other users who can use units of this type are the surgery wards in hospitals, the computer units in traffic control and financial institutions, and the systems of emergency lighting in public spaces. Uninterruptible power is needed here to prevent loss of lives, severe accidents, or loss of valuable data that sometimes may not be recoverable.

Another important application of small power units is in remote-area power supply (RAPS) (Moseley, 2006). According to World Bank estimates, currently some 2 billion people around the world have no access to central power supply systems. For many of them, diesel electric generators are the only power source. These generators are cheap, but the cost of diesel fuel, including its transport to remote places, may sometimes be very high. Also, diesel generators operate with very large emissions of greenhouse gases and other harmful substances. A natural way out of this situation would be the use of natural energy sources such as solar energy via photovoltaics or wind energy via wind turbines. Using these natural resources, which are available only intermittently, one must have additional devices for temporary energy storage. Lead–acid batteries have been used for some time for such energy storage in the Amazonas river region in Peru. Another possibility is that of generating hydrogen in electrolyzers, and then using it in fuel cells (or using fuel cells of a regenerative type, described in Chapter 10).

An interesting project concerns plans to use fuel cells in the Antarctic base on Béchervaise Island. This project, supported by a grant of $600,000 from the Australian government, involves the production of hydrogen by electrolysis with wind energy and its use in low-power fuel cell units producing electric power and heat (Adamson, 2006).

16.3 FUEL CELLS FOR TRANSPORT APPLICATIONS

For more than three decades, the possible use of fuel cells as a power source for automotive transport applications has been explored widely. This scenario arose in connection with fuel supply problems during the energy crisis of the 1970s. Since then, it has been raised repeatedly in connection with air pollution in cities with rapidly growing motor car populations. This has caused a number of countries (and notably the U.S. state of California) to make it mandatory for carmakers to produce and sell a certain number of low, even "zero-emission" vehicles. Carmakers were thus forced to see how they could replace the internal combustion engines (ICEs) by other power units.

16.3.1 Alternatives to Traditional Vehicles with ICEs

Early in the twentieth century, cars powered by storage batteries were quite common. About 35% of all cars registered at that time in the United States were battery powered. They were much more convenient than vehicles with ICEs that had to be cranked to start the engine. The situation took a turn when an electric starter was patented in 1912 and was then widely introduced. This meant the practically complete displacement of electric cars by today's traditional ICE vehicles. Today, battery-powered vehicles are used only for transport within production plants, and for certain home deliveries, such as milk products and mail.

When electric cars are used, one has practically no local emissions (other than hydrogen and oxygen while charging the batteries). The only ecological consequence of this transport mode is the emission of greenhouse gases (CO_2) in remote power generation needed for the battery-charging requirements. Electric cars are user friendly. They have two major advantages: (1) ease of changing the operating mode (e.g., the speed), and (2) potential energy recovery while braking. These characteristics are particularly appropriate for their use in city (stop-and-go) traffic. Yet battery-driven cars have two large defects that have prevented their wide introduction so far: (1) a limited driving range (of only 60 to 200 km between battery-charging stops), and (2) the long time (several hours) needed to recharge the battery ("refuel" the vehicle).

The most important alternative to traditional cars with ICEs are hybrid cars, in which an ICE is combined with an electric drive and a battery. The electric drive is used primarily in city traffic, the ICE basically on the highway under constant driving conditions while simultaneously recharging the batteries. This type of car combines the advantages of the traditional car with those of electric vehicles while minimizing the defects of each.

16.3.2 Problems in the Use of Fuel Cell Vehicles

The advances made during the 1990s in the development of various types of fuel cells (mainly PEMFCs) have, of course, come to the attention of people developing passenger vehicles and other transport means, and were seen as a promise that an electric car could be developed with all its user convenience, wide driving range, rapid startup, and minimal emissions. Almost all the large carmakers embraced the problem and began design and experimental testing. Numerous joint projects were undertaken between car companies and companies developing and producing various types of fuel cells. Many demonstration vehicles having fuel cell power plants were developed in these projects. The additional financial resources of the carmakers stimulated further work on fuel cells and contributed to improving their electrical and operating characteristics.

It soon became evident that building an electric car with fuel cells is a task associated with numerous basic problems. These were first openly formulated in a review published in 2001 by McNicol et al. entitled: "Fuel cells for road transportation purposes—yes or no?" In the opinion of these authors, the major problem is associated with the selection of a suitable reactant (fuel) for road-transport fuel cells. The only fuel that is acceptable ecologically (lack of harmful emissions) and by its technical characteristics is hydrogen. However, the use of hydrogen is associated with the problems of onboard storage (a tank with highly compressed gas or a cryogenic vessel with liquified hydrogen, or a hydride). Moreover, an infrastructure for hydrogen transport and distribution is almost nonexistent. There is, of course, the possibility of decentralized production of electrolytic hydrogen (e.g., in special service stations), but the cost of this hydrogen would be many times higher than the current cost of gasoline. Hydrogen could be produced from gasoline or other petroleum products (for which the distribution infrastructure exists), by reforming directly onboard, but this approach was judged to be unrealistic by the authors, due to the complexity of the reforming equipment needed in light of the sulfur contaminants present in these fuels, and because of the ecologically harmful emissions that attend the reforming. Thus, the answer to the question formulated in the title of the McNicol et al. article was "no" for the routes just mentioned, as well as for routes involving other indirect uses of fuel (i.e., uses requiring prior processing). In the opinion of the authors, the only promising approach is that of using fuel cell power plants working with straightforward reactant supply to the fuel cells. Such reactants could be hydrogen (when problems of its transport and storage have been solved), or methanol and similar reducing agents (when the corresponding catalytic problems have been solved).

Attention was also called to the fact that a definite changeover from traditional cars to electric cars with fuel cells will be associated with the need for a fundamental restructuring, not only of the car industry and of oil production and refining operations, but also of many other sectors of the economy related to them.

It must be acknowledged that even today, no solution is in sight for many of the questions raised in McNicol et al.'s 2001 review. In a review entitled "Fuel cell vehicles: status 2007," published in 2007, Helmolt and Eberle discussed basically only two problems: onboard storage of hydrogen, and possibilities for lowering the cost of a fuel cell power plant. Concerning the first problem, the authors concluded that cryogenic hydrogen storage is not appropriate in transport applications, owing to the large, unavoidable losses of hydrogen while parking and even while driving. Practically the only acceptable method of hydrogen storage would be in lighter cylindrical tanks under a pressure of 700 bar. Even then, difficulties arise when trying to pack a number of tanks of sufficiently high capacity into an electric car. Under the criterion of volume taken up, hydrogen storage in a bound form as metal hydride would be advantageous, but for electric cars this method of storage is not practical, since the refueling (metal hydride formation) and subsequent liberation of hydrogen are associated—in systems known at this time—with the uptake and evolution of large amounts of thermal energy that would need heat-exchange equipment of unrealistically high capacity in an electric car.

As to the cost of the power plant, the final goal would be \$50/kW (about \$5000 for one electric car). The cost of the power plant depends not only on the cost of the fuel cell stack as such, but on that of all auxiliary units and systems, including hydrogen storage and supply. The authors do not rule out that the goal could realistically be achieved when changing over to mass production of such power plants (many millions of units).

The authors believe that the further development of electric cars with fuel cells will occur in three stages. In the first stage (until the year 2010), the basic technical and economic problems will be studied with a limited sample (on the order of 100). In the second stage (until the year 2015), these studies will be continued with a sample of several thousand cars and with a certain number of hydrogen service stations. In the third stage (beginning in about 2015), production technology should be developed and the corresponding production capacities built up. In addition, the infrastructure required for the use of these electric cars should be developed to a sufficiently large extent.

In a 2005 paper, Ahluwalia et al. analyzed the possibility of raising the efficiency of fuel cell–powered vehicles by connecting to the fuel cell power plant a device (a battery) that accumulates electrical energy during braking. On the basis of this energy recovery, fuel savings in such a hybrid car could amount to 27% in stop-and-go traffic, and to 15% under conditions of combined city and highway driving.

Colella et al. (2005) studied the ecological consequences of a possible changeover in U.S. road traffic from traditional cars with ICEs to electric cars with fuel cells. They considered several possibilities: hydrogen being obtained by an ecologically clean process (electrolysis with electric energy produced by wind power generators), hydrogen being produced by the reforming of natural gas, and hydrogen being produced by coal gasification. It was shown in this analysis that a considerable decrease in air pollution by nitrogen oxides, carbon monoxide, volatile organic substances, and particulate matter occurs in all these cases. Even with hydrogen obtained by the reforming of natural gas, one sees a considerable decrease in the emission of greenhouse gases (CO_2). Similar conclusions were reached by Granowskii et al. (2006).

In the most recent review on this topic (by Ahluwalia and Wang, 2008), entitled "Fuel cell systems for transportation: status and trends," which was published while work on this book was just about complete, two questions are examined: possibilities for further cost reductions, and possibilities for a further performance increase of power plants with PEMFC-type fuel cells and a direct hydrogen supply. According to the U.S. program DOE/FreedomCAR, the cost of said plants, which in 2005 was $125 per kilowatt of electricity, should come down to $45 per kilowatt of electricity by 2010 and to $30 by 2016. By comparison, ICEs cost $25 to 35 per kilowatt of electricity equivalent. The specific power of the plants should increase from 500 to 650 W/kg during the same period. In the review, various technical solutions are discussed that might make it possible to approach these goals if not to actually reach them. It is interesting to point out that according to data given in this review, in 2005 the cost of the fuel cell stacks was 63% of the entire power plant. By 2010 this share should come down to 48%. In this review, as in many others, the question of hydrogen storage in the vehicles was left out of consideration completely.

16.3.3 Perspectives for the Further Development of Fuel Cell Vehicles

Attention must be called to the fact that in practically all work concerning the use of fuel cells for road transport vehicles, only fuel cells using hydrogen as a fuel were considered. There can be no doubt that hydrogen–oxygen fuel cells (and in particular those of the PEMFC type) at present have been developed to such a degree that in all their technical parameters, they are fit for power plants of electric cars. Tests of different types of electric cars with such power plants, which have already been performed for almost 10 years, will undoubtedly be extended and broadened (probably involving hundreds of cars). It is only in this way that the data can be gathered that are needed concerning all special technical and operational features of this kind of transport. Fundamental problems in the broad development of these electric vehicles have been and continue to be the supply of hydrogen and the cost of their fuel cell power plants. Local electrolytic hydrogen generation would require a considerable power grid enhancement.

Probably the first instances of wider practical commercial use of fuel cell power plants will be in connection with heavier transport equipment used within cities and near cities: buses and trucks supplying merchandise to the local trade and materials to construction sites. This heavy transport equipment is responsible for large part of the air pollution in large cities. They can be much more readily equipped with compressed hydrogen tanks. Creating an infrastructure for the distribution of and refueling with hydrogen is much easier within a city than all over the country.

The first demonstration-type buses having a fuel cell power plant were introduced toward the end of the twentieth century. In London, such buses started operating in suburban commuter transit soon after 2000. During the years from 2001 to 2006, a program of the European Commission called Clean Urban Transport for Europe (CUTE) was run involving regular use of 27 fuel cell–powered buses in nine European cities. The success achieved by this initiative led to introduction of this type of bus transport in other cities of Europe, Asia, and North and South America (Crawley and Adamson, 2006).

As to the multimillion fleet of individual cars driven in the cities and outside, a massive changeover to fuel cell–powered electric cars will only become possible when, by further scientific and technical developments, new *highly efficient means* have become available for (1) onboard hydrogen storage and/or (2) the direct use of methanol or bioethanol in fuel cells, and (3) hydrocarbon fuel reforming to hydrogen. The changeover requires, first, that the technical and economic performance figures of fuel cell vehicles become comparable with or superior to those of traditional vehicles with ICEs or to hybrid cars (and these also undergo constant improvement); and second, that as a result of the global dwindling of reserves, the production of petroleum products drops off.

All these conditions will probably not become effective before the second half of the twenty-first century. The changeover from petroleum-based automobile transport to fuel cell–based electric-car transport will at any rate be gradual and stretch out over many years. As we said above, this changeover requires a radical restructuring of many sectors of the economy.

16.3.4 Fuel Cell–Based Power Plants for Other Transport Means

As to the use of fuel cell–based power plants in other transport means, we have repeatedly spoken in the present book about practical applications of such plants in manned spacecraft. The first examples were 1-kW PEMFC plants used in the 1960s in *Gemini* spacecraft, now outdated; then three 1.5-kW AFC plants each used in *Apollo*-type spacecraft in the 1970s, and finally, three 12-kW AFC plants each in the *Orbiter* space shuttles used until the present.

In these applications, where the power plant did not have to operate for more than a few days, the only possible alternative for powering the electrical equipment of the spacecraft would be batteries consisting of the usual galvanic cells, or storage batteries. Under these conditions, fuel cell–based power plants have much higher energy contents per unit volume and unit weight (mass) than batteries. Another great advantage is constant water production for the needs of the crew.

It is quite realistic to use fuel cell–based power plants for river, marine, and submarine vessels. In water transport, difficulties as such existing in the placement of devices for the storage or production of hydrogen in electric cars, are much more readily overcome. The problems of heat exchange (removal of heat from an operating power plant) are easier to solve on such vessels. In the case of large vessels and ferries, the diesel engines are provided not only for propulsion of the vessel but also for the generation of electrical energy needed for various operations (loading, unloading, lighting, communications, air conditioning, etc.). This part of the energy must also be generated while in port, which produces appreciable air pollution in the port area. Therefore, even here, considerable ecological effects would be gained when using fuel cells.

In a paper by Psoma and Sattler (2002), development work done in Germany on submarines with power generators based on PEMFC-type fuel cells was discussed in detail. Preliminary work in this direction had begun in the 1970s. By 2002, type 212A submarines were under construction, to be placed in service in the German and Italian navies by 2003. Modules with power between 30 and 120 kW and total power of 300 kW are used in these submarines. Oxygen for the fuel cells comes from cryogenic vessels kept outside the submarine's pressure hull. These vessels resist potential impact loads as well as the outside pressure changes occurring when the vessel dives to different depths. Hydrogen is stored in bound form as metal hydrides. This storage method provides the largest amount of hydrogen per unit volume (even larger than cryogenic storage). The equipment holding the metal hydrides is also kept outside the submarine's pressure hull. The cooling needed during charging (introduction of hydrogen into the metal alloy) is handled by land-based equipment. During discharge (liberation of hydrogen from the hydride), thermal energy set free during fuel cell operation is introduced into the storage system. This system of reactant storage secures longer underwater runs of the submarines than does traditional equipment based on storage batteries.

However, a further increase in the underwater range is not possible, owing to the large weight (and high cost) of the metal hydrides. For this reason, work on equipment for onboard steam conversion of methanol was begun in 1995. The CO_2 produced in the reforming process may be kept onboard the vessel in a liquid state and would be ejected periodically into the seawater. A plant of this type capable of supplying 240 kW worth of fuel cells with hydrogen has passed all necessary tests successfully. Production of type 214 submarines including such equipment was begun in 2001 for Greece, South Korea, and other countries. In the opinion of the authors, this indicates that power plants with fuel cells are the ultimate selection when building nonnuclear submarines.

For rail transport, the use of fuel cell–based power plants is technically possible without any doubt. However, whether it will be appropriate to replace the diesel engines used today by new types of power plants will depend on various economic factors, including the situation in the oil markets.

16.4 PORTABLES

In all likelihood, what for the sake of brevity we shall designate as "portables" will become the most important field of large-scale use of fuel cells over the next decade. This field covers the most diverse portable electronic and electrotechnical devices for daily life as well as for the many portable manufacturing aids that need electric power to function. Historically, probably the first such devices were pocket flashlights, which need power in the milliwatt range. The first walkie-talkie (portable transceiver or radio-communication equipment) was developed in 1940, and such devices became important in World War II. The first electronic quartz wristwatches appeared toward the end of the 1960s; they consume power in the microwatt range. At about the same time, various hearing aids and portable radios appeared. It was typically for all these devices that almost immediately after they appeared, they went into mass production, with millions of pieces made. The list of devices used as "portables" has become considerably longer during the last two decades, and their versatility and functions have grown tremendously: mobile phones, notebook-type portable computers, digital still and movie cameras, camcorders, and function sensors for communications, out-of-office work, mobile Internet connections, personal navigation, and sophisticated medical monitoring. Production volume has risen together with the levels of sophistication and utility.

16.4.1 Power Sources for Portables

The only power sources feasible for all these portables are electrochemical power sources, which today are the galvanic cells we call *batteries*. Disposable batteries were the classical power source of flashlights and still hold a very strong position (neutral Leclanché; later, alkaline "dry cells" of the zinc–manganese dioxide type). Rechargeable batteries became ever more important: first, nickel–cadmium, and more recently, nickel–hydride and lithium-ion batteries.

For convenient handling, a power source is usually placed somewhere inside the device, so it should respect certain limitations as to weight and volume. As a rule of thumb, a power source should not exceed 30 to 40% by mass and volume of the device powered by it. A similar upper limit of 30 to 40% applies to the cost.

It has become obvious in view of newer high-performance electronic devices powered by batteries that even the most perfect rechargeable lithium-ion batteries hardly meet desirable targets of operating time between recharges. Thus, a notebook may need recharging of its state-of-the-art lithium-ion batteries after as little as 2 to 3 hours. For a media player of the MP3 type, less than 10 hours is reached as a rule, while a mobile phone will require recharging after 3 to 4 hours of conversation. A major defect of rechargeable batteries is the time required for recharging, which may easily be more than 5 to 6 hours, and people tend to keep a second battery pack in reserve for that reason, thus doubling the weight of the power source.

It is in a rhythm of approximately 18 months that new generations of electronic equipment offering strongly enhanced capabilities are introduced. This is usually attended by distinctly higher power requirements. A particularly strong rise in power demand is expected in devices combining the functions of a mobile phone with those of a PC handling wireless broadband data transfer and processing (Dyer, 2002).

It is for these reasons that the mini-fuel cells described in Chapter 14 are looked upon as a superior power source able to supply higher currents for much longer times. It should be pointed out at once that from the outside, fuel cells (including their fuel and oxidizing agent) work by exactly the same principles of current generation as do ordinary batteries.

It is quite clear that we do not expect complete substitution of ordinary batteries by fuel cells in the portable field. Ordinary batteries will continue to maintain their leading market position as power sources for a large number of devices. Thus, disposable batteries are expected to maintain their importance for pocket flashlights and various medical devices, for example. Simple electronic devices such as portable radios, audio players, digital cameras, and so on, will continue to be powered by rechargeable nickel–cadmium and nickel–hydride batteries. Lithium-ion rechargeables are likely to continue in simpler mobile phones. Fuel cells will be attractive for more complex equipment, such as notebooks used for more than 2 hours at a time, for instance, where even lithium-ion batteries have insufficient energy density.

Small volume (V) and/or small weight (mass m) are prime requirements to be met by the power sources of portables. With an ordinary battery, these parameters increase in direct proportion to the length of time, t, it will function before disposal and replacement or recharge: $V_t = V_0 \cdot t$ (or $m_t = m_0 \cdot t$, where V_0 and m_0 are the volume and mass needed in unit time). With fuel cells, the volume (or mass) is the sum of a constant fuel cell volume V_{fc} and of volume V_f of the fuel tank + fuel, and it is only V_f that must be multiplied by the intended time of operation:

$$V_{fc(\text{total})} = V_{fc} + V_f t \tag{16.1}$$

(It is assumed that air is the oxidizing agent, so its volume or mass contribution is zero.)

As an example, consider power sources for a notebook consuming 10 W of electrical power. In Figure 16.7, the total volume of three types of power source is plotted against their operating times. Curve 1 refers to lithium-ion batteries common today, having a volume of about 250 cm^3 and a typical energy density of 130 Wh/L. Curve 2 refers to a DMFC version with $V_{fc} = 10\,cm^3$, a specific power of 5 W/L, and $V_f = 70\,cm^3/h$. Curve 3 refers to a DMFC version with $V_{fc} = 200\,cm^3$, a specific power of 200 W/L, and $V_f = 25\,cm^3/h$. Assuming that for the application considered, a volume of 300 to 400 cm^3 is the limit, lithium-ion batteries (curve 1) would provide an operating time of about 3 hours. Fuel cells of curve 2 taking up the same volume would provide approximately the same operating time. The basic advantage of using such a fuel cell would be that the long time required for recharging the battery will be replaced by the very short time required for replacing the fuel cartridge. With the fuel cell of curve 3, uninterrupted operation is possible for 7 to 8 hours without any large volume penalty.

It can be seen from the examples that long operating times of fuel cells depend mainly on a low value of V_f, which is the specific consumption of fuel, including the water needed to keep the membrane moist. It is worth mentioning that for

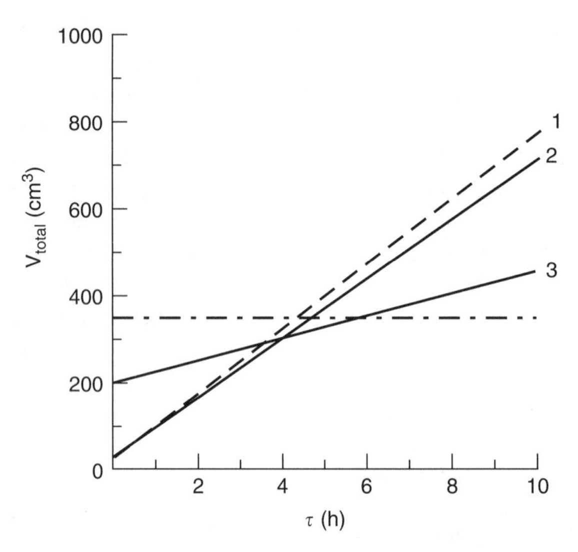

FIGURE 16.7 Volumes of three types of power sources: 1, lithium-ion battery having an energy density of 130 Wh/L; 2, DMFC with parameters of $V_{fc} = 10\,\text{cm}^3$ and $V_{fc} = 70\,\text{cm}^3/\text{h}$; 3, DMFC with parameters of $V_{fc} = 200\,\text{cm}^3$ and $V_{fc} = 25\,\text{cm}^3/\text{h}$. Dashed-dotted line; maximum admissible volume.

the 10-W DMFC, with 0.6 V output voltage of the individual cell, the consumption of methanol is a theoretical 4.2 cm^3/h when disregarding the water.

A similar situation is seen when comparing other battery and fuel cell parameters, such as their weight (mass) and even their cost. Thus, the cost of battery operation is mainly a function of their initial cost, which for lithium-ion batteries is about $1.2/Wh plus the cost of the battery charger (about $25), the grid power required for charging—as a variable depending on battery use—remaining essentially unimportant. In fuel cells, the variable component consisting of the cost of the fuel cartridges and fuel depends on operating time [basically, as in equation (16.1)], and makes a noticeable contribution to total operating cost.

16.4.2 Development Work on Fuel Cells for Portables

The development work on fuel cells for portables evolves quite differently from that on fuel cells for the other applications. The fuel cells intended for large and small stationary power plants and those intended for electric cars (considered in Sections 16.1 to 16.3) are to be seen in the context of global problems: lower consumption of hydrocarbon fuels (for which the resources are limited) and lower air pollution (greenhouse gases and other contaminants); hence society at large and all the governments take an active interest. Work on such systems has often been initiated and/or financed by government programs (see Chapter 17). More perfect power sources for portables are, to the contrary, strictly in the interest of individual makers and consumers. Development work in this area has been mainly market-driven in the past and did not benefit of any general programs involving government, with the important exception, however, of work toward power sources satisfying the needs of portable equipment for the military (see Section 16.5).

Today, almost half of all efforts worldwide to develop fuel cells for portable equipment come from North America, which has to do with the relatively large resources made available by the U.S. military (see Crawley's review, 2006). Japan carries a share of about 20%, the support coming from the many large companies making portables: Sony, Casio, Fujitsu, Hitachi, NEC, Toshiba, and others. Considerable contributions come from other countries, such as South Korea (Samsung), France, Italy, Germany, and some others. In almost 90% of all such developments, the fuel cells involved are those of the PEMFC and DMFC type (at approximately equal shares). Mini-fuel cells based on SOFC constitute a minor part; they are of interest primarily for military purposes.

All this work has led to numerous prototypes of fuel cell–based power sources. So far, prevalent practical applications are in the military field, while commercial production and uses in civil applications are still reluctant. Considering the broad introduction of multifunctional portable electronic equipment that is bound to happen in the years until 2010, one may expect fuel cells for portables in the civil sector to become competitive during that period and to reach a multimillion production level between 2015 and 2020.

Fuel cell–based power plants are widely useful not only as built-in power sources of portable devices but also as mobile charging stations for ordinary rechargeable batteries. It has been estimated that the number of devices depending on rechargeable batteries will be in the multibillions by 2010. In many situations (on moving installations, under field conditions, camping sites, huts, etc.) where grid power is not available for battery recharging, or is unreliable, and diesel generators are impractical, an appropriately sized fuel cell power plant is the only alternative possible.

Since government bodies have not been heavily involved in the development of fuel cells for portables (apart from the closed military sector), a question arises as to national or international regulations in this field. This is true in particular for aspects of standardization, testing, safety, and handling. As an example, consider a ruling of the International Civil Aviation Organization (ICAO) of January 1, 2007, authorizing cartridges with methanol or similar liquids to be used onboard commercial passenger aircraft for the operation of fuel cell–powered PCs. According to more recent general rules on liquids carried by passengers, it seems likely that such a use is ruled out at present.

16.5 MILITARY APPLICATIONS

From the very beginning of modern fuel cell development, potential applications for military purposes were an important driving force and source of financing of this R&D work. In the early 1960s, for example, the work of General Electric on membrane-type fuel cells that led to the power plants for *Gemini* spacecraft was financed in part by the US Navy's Bureau of Ships (Electronic Division) and by the U.S. Army Signal Corps (Crawley, 2007).

Speaking of military fuel cell applications, we must first point out the following. Most versions of fuel cell–based power plants are of ambivalent utility: They are just as good for civil as for military purposes. Thus, the stationary power plants of different sizes used for uninterruptible and emergency power supplies for military objects such as forts, command centers, radar stations, and the like do not differ in any way from similar power plants for civil use in hospitals, telecommunications installations, computer centers of banks, and so on. Power plants for automotive land and waterbound means of transport are equally good for civil and military vehicles. This is true even more directly for power sources intended to supply portable equipment. A lower volume and weight of all equipment carried by soldiers in combat (e.g., means of communication, means of orientation, night-vision devices) is a very important point for land forces.

One of the few examples of fuel cell use exclusively for military purposes which (so far) is without its civil analog is that for submarine propulsion described in Section 16.3.

It is out of considerations such as those recorded above that when speaking about work on fuel cells for military goals, we actually think of work financed by various military agencies.

In the Crawley review of 2007 mentioned earlier, data are given from which one can see the distribution of military support for the development, improvement, and production of fuel cells for various military uses: 45% for portable equipment, 22% for all kinds of ships, 11% for land vehicles, 10% for flying vehicles (unmanned aerial vehicles), traffic controllers, etc.), 6% for various weapons, and about 5% for stationary installations.

Fuel cell power plants for military applications differ from their civil analogs primarily in higher demands on reliability and trouble-free operation. They should keep working under the real conditions of military action, day and night, at any time of the year and whatever the weather. They should be simple to attend to and as insensitive as possible to faulty manipulation. They should admit all types of transport and shipping, including parachute dropping to destinations.

In 2004, Patil et al. reported data concerning work performed at the U.S. Army Communications and Electronics Research, Development and Engineering Center (CERDEC) on fuel cells for military uses. Their work is in three areas: (1) low-power plants (less than 20 W) for individual soldier and for various sensor devices; (2) medium-power plants (200 W to 2 kW) for silent observation posts and for the recharging of various types of storage batteries, and (3) high-power mobile and stationary plants (over 2 kW) for use as auxiliary power units.

An increasingly important vehicle application is a tactical mode of operation termed *silent watch*. These units are intended for uninterrupted battlefield observation. To this end they are equipped with many different electronic devices. An important requirement is the complete absence of acoustic and infrared emissions, as these would reveal their location. For this reason, a diesel generator could not be used to power the equipment. Ordinary batteries are not able to provide enough power. Fuel cells are the only possibility as the power supply.

A point of great importance in CERDEC work is that of users being able to work with the logistic fuel known as JP-8 kerosene or jet propellant, for which regular supply logistics exist, at least in all power plants of a certain capacity. To this end, new and improved processes for the reforming of this fuel are being developed to derive hydrogen from it that can be used in fuel cells.

Patil et al. also pointed out that using fuel cells will provide the army not only with longer operating times of equipment (which means more efficient use) but also with considerable savings on expenditures: 1 gallon of methanol costing less than $0.50 is the energy equivalent of disposable (primary) batteries woth $2240.

REFERENCES

Ahluwalia R. K., X. Wang, A. Rousseau, *J. Power Sources*, **152**, 381 (2005).

Bishoff M., *J. Power Sources*, **154**, 461 (2006).

Colella W. G., M. Z. Jacobson, D. M. Golden, *J. Power Sources*, **150**, 150 (2005).

Dyer C. K., *J. Power Sources*, **106**, 31 (2002).

Ghezel-Ayagh H., J. Walzak, D. Patel, et al., *J. Power Sources*, **152**, 219 (2005).

Granowskii M., I. Dincer, M. A. Rosen, *J. Power Sources*, **157**, 411 (2006).

Moseley P. T., *J. Power Sources*, **155**, 83 (2006).

Patil A. S., T. G. Dubois, N. Sifer, et al., *J. Power Sources*, **136**, 220 (2004).

Perry M. L., E. Strayer, *Intelec 2006 Proc.*, p. 230 (2006).

Psoma A., G. Sattler, *J. Power Sources*, **106**, 381 (2002).

Winkler W., H. Lorenz, *J. Power Sources*, **105**, 222 (2002).

Reviews

Adamson K.-A., Small stationary applications 2006, *Fuel Cell Today*, Dec. 2006.

Adamson K.-A., Large stationary survey 2007, *Fuel Cell Today*, Sept. 2007.

Ahluwalia R., X. Wang, Fuel cell systems for transportation: status and trends, *J. Power Sources*, **177**, 167 (2008).

Butler J., Transport-2008, *Fuel Cell Today*, July 2008.

Crawley G., Military survey 2006, *Fuel Cell Today*, June 2006.

Crawley G., Military survey 2007, *Fuel Cell Today*, May 2007.

Crawley G., K.-A. Adamson, *Market survey: buses*, *Fuel Cell Today*, Dec. 2006.

Helmolt R. von, U. Eberle, Fuel cell vehicles: status 2007, *J. Power Sources*, **165**, 833 (2007).

Huleatt-James N., Hydrogen Infrastructure, *Fuel Cell Today*, February 2008.

McNicol B. D., D. A. J. Rand, K. R. Williams, Fuel cells for road transportation purposes—yes or no? *J. Power Sources*, **100**, 47 (2001).

Wee J.-H., A feasibility study on direct methanol fuel cells for laptop computers based on a cost comparison with lithium-ion batteries, *J. Power Sources*, **173**, 424 (2007).

CHAPTER 17

FUEL CELL WORK IN VARIOUS COUNTRIES

At present, work on fuel cells is conducted in practically all developed countries of the world. This work proceeds in many research organizations (such as university laboratories, academic research centers, national laboratories) as well as in many large and small private companies and organizations. Various government services and public associations are involved as well. It is rather difficult to keep track of the total number of such entities. Data from various sources diverge: The 7th edition of Fuel Cell 2000s *Fuel Cell Directory* (2008) lists more than 985 organizations. The *Fuel Cell Manufacturer Directory* (from Energy Business Reports, 2006) lists 2802 fuel cell manufacturers.

17.1 DRIVING FORCES FOR FUEL CELL WORK

Many different driving forces are behind this broad growth of the fuel cell field. Scientific institutions are interested in further scientific and technical progress. The commercial organizations wish to see a larger production volume and increased sales volume of their market products, and hence are interested in improved engineering processes and performance figures.

Certain categories of fuel cell work touch the interests of countries as a whole, inasmuch as they promise solutions to global ecological and geopolitical problems, and for this reason are sustained and financed by government bodies. Some of these global problems are listed below.

Fuel Cells: Problems and Solutions, By Vladimir S. Bagotsky

Dwindling World Resources of Fossil Fuels

The oil crisis of the 1970s erupted chiefly as a result of technical and political difficulties that arose in the supply of oil from oil-producing countries to oil-consuming countries. At present, facing the first decades of the twenty-first century, another problem is more and more strongly evident. A large part (estimates say somewhat less than one-half) of the world's proven oil reserves has already been used. Each year, oil demand considerably exceeds the amount of oil held in deposits newly opened, and demand increases steadily. This is true, of course, for American and European countries, but even more so for countries such as Brazil, Russia, India, and China (the *BRIC countries*) and other emerging economies now included in the term. Realistic reasons exist to expect world oil resources to be almost completely exhausted by 2050 at present growth rates. The problem is much less acute when considering coal reserves. Various estimates predict that at current consumption levels, world coal resources will last more than another 150 years. For natural gas, the situation is also much less critical than it is with oil. Tapped resources are much larger than oil resources, and every year large new deposits are discovered. The deposits of *methane hydrates* ($CH_4{\cdot}nH_2O$) recently discovered deep under the oceans in various places feed additional optimism. Crystals of this material are stable at temperatures below 15°C and at high pressures such as those existing in depths of 1000 to 3000 m (Zegers, 2006). This would indicate that a depletion of fossil fuel resources is to be expected during the next decades for oil and oil products, creating problems for practically all types of transport.

Release of Greenhouse Gases and Global Warming

Most countries of the world, except for the United States and a few others, have signed and ratified the *Kyoto Protocol*, a treaty that was negotiated in December 1997 in the city of Kyoto, Japan, and came into force on February 16th, 2005. The Kyoto Protocol is a legally binding agreement under which industrialized countries will reduce their collective emissions of greenhouse gases by 5.2% relative to the output in 1990. The limitations imposed by the protocol have induced signatory countries (and others, too) to start implementing considerable improvements in all processes using coal, oil products, and other carbon-containing fuels and to look for new or alternative energy sources. We should point out here that the role of methane as a more potent greenhouse gas than CO_2 has been largely ignored so far.

Air Pollution in Large Cities

Incessant growth of the number of commercial and private vehicles circulating in large cities has led to such a level of air pollution that a real menace has developed for the health and even the life of the population, primarily children, older people, and people with respiratory illnesses or impairment and allergies. Operating ICEs of cars and trucks emit not only carbon dioxide (contributing to the greenhouse effect) but also substances harmful or toxic to humans, such as carbon monoxide (CO), different nitrogen oxides (NO_x), various sulfur compounds, certain organic substances (some carcinogenic), and various particulates containing carbon and other elements. In many cities the municipal powers have been forced to take administrative measures (such as limitations by vehicle type or time of day). Government action has also been introduced (such as the well-known law in California demanding a certain fraction of "zero-emission" vehicles, or German laws demanding filters for diesel particulates). The ill-famed "London smog" was a severe air-pollution problem caused by inappropriate fuels and heating systems early in the twentieth century, and a very severe episode in 1952 prompted the U.K. Clean Air Act of 1956.

Work related to the foregoing global problems is supported as a rule by government funds in the context of national projects or programs conducted in individual countries or groups of countries (such as the European Union). Progress in fuel cell work has been helped decisively worldwide by the scenarios described above.

Parallel to the work on fuel cells, another type of energy-related development work was begun in the 1970s in developed countries, variously characterized as work toward a "hydrogen economy" or a "hydrogen energy scenario." Progress in this new field was soon found to be intimately related to progress in fuel cells, so a separate appreciation of work in these two fields is hardly possible.

17.2 FUEL CELLS AND THE HYDROGEN ECONOMY

The idea that some sort of a hydrogen economy ought to be set up arose in the early 1970s, when in the context of the oil crisis, most countries of the world started to experience difficulties in the supply of oil and oil products, and a feeling developed that world oil resources would soon be depleted, and that world resources of other fossil energy sources would also be depleted in the more distant future. Global warming and rising city air pollution provided additional stimuli. Thus, the problems that were important for the fuel cell field have stimulated discussion of a desirable *cleaner* hydrogen economy.

The volume of model calculations and of diverse theoretical and experimental studies addressing the problems of a hydrogen economy soon became very significant. Many workers in different fields became involved, and large international meetings were organized at frequent intervals. Specialists get together annually for progress review and exchange. Elsevier Publishers began distributing a special *International Journal of Hydrogen Energy* in 1976, which beginning in 2008 is being published as 24 issues per year. An International Association for Hydrogen Energy has its headquater in Miami, Florida. There has been discussion of an approaching "hydrogen civilization" [Editorial, *Int. J. Hydrogen Energy*, **33**, 1, (2008)]. Wikipedia, the free online encyclopedia, provides this definition: "A *hydrogen economy* is a hypothetical economy in which the energy needed for motive power (for automobiles and other vehicle types) or electricity (for stationary applications) is derived from reacting hydrogen with oxygen."

Things actually go further than that. To be the basic source of energy for the consumer, hydrogen in such an economy would have to be a major energy vector supplementing or replacing electric power. Hydrogen is readily produced from water by electrolysis with electrical energy that would be derived from nuclear, solar, wind, and hydro power stations in an ecologically clean way. Potential world hydrogen resources made available according to that scheme could hardly dry up. Combustion of hydrogen (its oxidation by oxygen) yields water vapor as the only product. Air pollutants are normally not generated except in those cases where the combustion temperature is high, and toxic nitric oxides may be produced when air nitrogen reacts with air oxygen. In this way, two global problems are solved simultaneously: that of an everlasting energy supply, and that of a clean atmosphere.

It can be concluded from what we have said that the final aims to be achieved by a hydrogen economy are very similar to those to be achieved by fuel cells. For this reason, progress in the two fields is very closely related. As the universal energy vector in a hydrogen economy, hydrogen is also the ideal fuel for fuel cells. For this reason, steps toward building a hydrogen economy will at once stimulate large-scale uses of all types of fuel cells in the most diverse applications.

The definition provided above implies that hydrogen will be used not merely—and in fact not so much—as the fuel for fuel cells but also—and predominantly—in the most diverse heat engines: internal combustion engines in vehicles, and gas turbines in small and large electric power plants. Competition and contests are expected, for example, between vehicles driven by hydrogen-powered ICEs and vehicles propelled by electric power generated by hydrogen-fed fuel cells. Without having given a detailed description of hydrogen-powered heat engines, this is not the place to discuss the advantages and disadvantages of these two systems. We only make the following remarks.

In a hydrogen economy, just as in a widespread application of fuel cells, two problems exist that so far have not found a definite solution: distribution of hydrogen to the consumer, and hydrogen storage at the consumer. A completely new infrastructure must be created to provide for universal hydrogen distribution, not unlike the existing infrastructures for natural gas and petroleum products. For hydrogen transport over long distances, pipelines are thought to be feasible; they would carry energy much more efficiently than high-tension power lines. For hydrogen transport over shorter distances and its storage at the consumer site, three possibilities have been shown to exist in Chapter 11: (1) compressed gas in steel or lightweight tanks; (2) liquefied in cryogenic containers; and (3) absorbed as hydrides in various metal alloys. The first two possibilities are realistic for stationary systems and for large mobile systems such as ocean vessels. Tanks, being very bulky, appear to be a severe problem when used in lighter vehicles. Liquefied hydrogen incurs excessive losses in such vehicles while they are refueled or parked. Hydride storage is the most promising solution but requires further advances in solid-state physics and chemistry, finding new alloys or other materials capable of reversibly absorbing and releasing large amounts of hydrogen while releasing and absorbing the smallest possible amounts of heat. New hydride storage materials would be a very considerable contribution toward progress in work for the hydrogen economy as well as in work for fuel cells.

In the opinion of Zegers (2006), "the slow progress of fuel-cell commercialization is a major barrier for the introduction of a hydrogen economy." Many aspects of the hydrogen economy and its connection with fuel cells are discussed in a paper by Conte et al. (2001) and in a book by Romm (2005).

17.3 ACTIVITIES IN NORTH AMERICA

In 2005, almost 50% of the more than 7000 fuel cell stacks put in operation were made in the United States and Canada. In 2006, the corresponding fraction of fuel cell stacks from North America dropped to 20%, owing to the very considerable increase in the fraction coming from Japan (Adamson and Crawley, 2006).

In his 2003 State of the Union Address, President George W. Bush launched the Hydrogen Fuel Initiative to ensure a clean environment and the long-term energy security of the United States. President Bush's vision was summed up in the statement that "the first car driven by a child born today could be powered by hydrogen and pollution-free."

In the context of this initiative, the U.S. Department of Energy (DOE) has developed a number of programs for the advancement of fuel cells and of hydrogen energy. The DOE is now the largest funder of fuel cell technology in the United States.

DOE's Office of Fossil Energy is funding several major programs for the development of fuel cell-based systems for distributed power generation. These programs include SOFCs and MCFCs. A new project named FutureGen was initiated early in 2004, with the aim of producing electricity and hydrogen from coal in a virtually emission-free plant. This project, costing $1 billion US dollars, was set up for 10 years. It is intended to work as a government/industry partnership, and include a consortium representing the coal and energy industries. The National Energy Technology Laboratory (NETL) and certain other laboratories under the aegis of DOE's office formed a Solid State Energy Conversion Alliance (SECA) and formulated a program which among other aims should produce a drastic decrease in the cost of producing SOFC-based FC power plants (Williams et al., 2004). Recent policy considerations see an alternative to the zero-emission plant in retrofitting existing coal-based power plants.

DOE's Energy Efficiency and Renewable Energy office and Science Office have developed a detailed R&D plan aiming at the gradual replacement of

ICE-based vehicles operated with gasoline with fuel cell-based electric vehicles operated with hydrogen. The plan reckons with a commercialization of appropriate fuel cells by 2015. In the meanwhile, an appreciable reduction in gasoline consumption should be attained by wider introduction of hybrid vehicles where electric traction sustained by rechargeable batteries is operative in parallel with an ICE. In this connection, an R&D program aiming at the development of improved batteries (particularly lithium-ion batteries) was set up together with the US Advanced Battery Consortium (USABC). The final aim of this program is the introduction of fuel cell–based vehicles (FCV) working with PEMFCs. Apart from work toward further improvements in this fuel cell type and toward a drastic decrease in manufacturing costs, a large part of the efforts should be aimed at solving the key problem of hydrogen storage onboard the electric vehicles (Chalk and Miller, 2006).

In the beginning of 2006, President Bush launched a new Advanced Energy Initiative (AEI) that addresses the challenges of both transportation and power generation. This initiative provides for a 22% increase in DOE's funding of research involving clean energy technologies. In the area of means of transportation, this initiative suggests developing fuel cell–based electric cars, with the argument that they could provide appreciable fuel economy (even 50% higher than that to be achieved with ICE-based cars operating with hydrogen as a fuel). According to DOE's Hydrogen, Fuel Cells & Infrastructure Technologies (HCIT) Program, to achieve this target the cost of the fuel cells should be lowered to $30/kW (as compared to $275/kW in 2002 and $110/kW in 2005), and the operating life should be raised to 5000 hours (compared to 1000 hours in 2003 and 2000 hours in 2005) (Milliken et al., 2007).

A large amount of fuel cell R&D is conducted in Canada; part of it has been mentioned in earlier chapters. Power stations of different size working with membrane fuel cells are produced by the Canadian company Ballard Power, a world leader in this area.

17.4 ACTIVITIES IN EUROPE

In Europe, 27 countries are now united in the European Union. According to the Union charter, questions of scientific research are subject to regulations at the Union level rather than at the level of the individual member states. In 1995, the EU Directorate General for Energy and Transport had produced a document "Research, Development and Demonstration Strategy Up to 2005."

In pursuit of this strategy, a European Hydrogen and Fuel Cell Platform was created. On this platform, specific research targets related to lower costs and higher durability were formulated for PEMFCs, PAFCs, MCFCs, and SOFCs as the major fuel cell varieties. The cost limit for a fuel cell stack was fixed at €200 to 500/kW, the cost limit for the entire power system at €1000 to 1500/kW. The durability target was set at 20,000 to 40,000 hours. It must be pointed out that even from a current perspective, these 1995 targets were highly optimistic. In a 1998 revision of this program, many of these parameters were adjusted.

Apart from work on fuel cells and their peripherals (such as equipment for reforming and gas purification), the platform targets included many hydrogen-energy-related items.

The financing of scientific research is fixed in five-year Framework Programs of the EU. These programs include selected priority projects. Fifty percent of the cost of each project is covered by the EC, the other 50% by partners of the projects. The following are examples of projects from the current sixth Framework Program (Adamson, 2005):

- Elevated-temperature PEMFC systems: €4 million
- Compact DMFC for portable applications: €2.1 million
- Reliable, durable, efficient, and cost-effective SOFC systems: €9 million

In 2007, the European Commission announced that it is to support a project concerned with the development of hydrogen and fuel cell technologies. Over the next six years, €470 million will be provided to fund the Fuel Cells and Hydrogen Joint Technology Initiative (JTI), a sum expected to be matched by industry. It is hoped that the program will "accelerate the development of hydrogen technologies to the point of commercial take-off between 2010 and 2020" (*Fuel Cell Today*, October 17, 2007).

17.5 ACTIVITIES IN OTHER COUNTRIES

Japan

In Japan, mid- and long-term technological projects in which there is an R&D investment risk that cannot be borne on a private-sector level by entrepreneurs and universities are carried out by the Ministry of Economy, Trade and Industry and by the New Energy and Industrial Technology Development Organization (NEDO). In 2006, NEDO developed a Fuel Cell/Hydrogen Technology Roadmap, providing a detailed description of intermediate targets (toward 2007, 2010, 2015) as well as longer-range targets (toward 2020 to 2030) for technical and economic parameters of PEMFCs, DMFCs, SOFCs, and SOFCs, as well as for progress in hydrogen technology. NEDO's program lists all intermediate steps of the work and the financial resources needed. All special features associated with the three basic areas of applications are taken into account: stationary plants, transport media, and portable equipment. A detailed listing is also provided as to which work should be performed in the private sector and which should be government financed (*Fuel Cell Today*, Article 1131, is an English translation of the NEDO Roadmap).

In earlier chapters we pointed out that in Japan in the 1980s and 1990s, a relatively large number of stationary power plants of various sizes, including different fuel cell variants, had been built and operated over extended periods of time. These plants have made a considerable contribution to solving power supply problems in the country. Subsequently, for the beginning of the twenty-first century, the attention of fuel cell R&D has strongly shifted toward power plants for electric vehicles and for portable applications. The development of fuel cells of intermediate and small size is now undertaken by car companies (Honda, Mitsubishi, etc.) and by numerous electronic companies (Hitachi, Toshiba, Casio, etc.). In 2006, of the fuel cell–based power plants built worldwide, 20% were Japanese.

China

At present, about 70% of all power needs in China are covered by coal resources. This may explain the interest displayed by government structures in work toward alternative energy sources. Among this work, research in the areas of fuel cells and hydrogen energy is not a minor item. In 2002 it was learned that the Chinese Academy of Sciences would spend $12 million on a three-year program for the development and commercialization of various types of fuel cells (*Fuel Cell Technology Update*, February 2002). Since that time, the funding of fuel cell work by government bodies has increased continuously. These sources of funding are the Knowledge Innovation Program of the Chinese Academy of Sciences, the High-Tech Research and Development Program of the National Natural Sciences Foundation of China, and other funds and national programs.

Fuel cell work is conducted at many universities, research laboratories, and enterprises in China (including Hong Kong). A large number of the papers published in the scientific and technical literature, and a large number of patents, come from Chinese authors. Efforts to reduce the imports of oil and oil products drastically are also a driving force for work on fuel cells. An additional stimulus is rising air pollution from growing automotive transport in the cities. This problem was of specific relevance to the 2008 Olympic Games in Beijing and to the 2010 World Expo in Shanghai. It was in the context of these problems that the large international 2008 China Hydrogen Energy and Fuel Cell Exhibition was assembled in Shanghai at the end of March 2008.

Early in 2008, China announced it will invest more than $10 billion on renewable energy as part of its commitment to generate 15% of its energy from renewables by 2020, up from 8% today (Butler, 2008a).

South Korea

South Korea issued a document entitled "Long-term Vision for Science and Technology Development toward 2025," listing 40 priority targets. As a result of this document, a 10-year energy technology R&D plan was developed that includes a National Fuel Cell Technology Plan. The road map provided by this plan describes in detail the targets and all intermediate aims (up to the year 2010) associated with fuel cells to be built for transport and other applications

(Adamson, 2006). Korean scientists from various universities and national laboratories often publish their results in international journals and contribute actively to international meetings.

Australia

In 1992, a company called Ceramic Fuel Cells Limited (CFCL) was founded by a number of companies from the electrotechnical sector and other industries as well as certain government structures. It is the basic aim of this company to develop flat solid-oxide fuel cells for distributed power generation. Soon after its work beginning, this firm attained great success in SOFC development and became one of the world leaders in this area. About 80 persons work in this company. The funding volume from the private sector and from government subsidies attained a level of 60 million Australian dollars (Godfrey et al., 2000).

Russia and the Former Soviet Union

In 1962, the Presidium of the Academy of Sciences of the Soviet Union created a Science Council on Fuel Cells, including representatives of a number of academic institutions, universities, and electrotechnical and chemical industry firms. The head of this council was Alexander Frumkin, well-known electrochemist and academician. The council did important work coordinating scientific studies performed at the various organizations and facilitating the exchange of information. National fuel cell meetings were organized periodically. Unfortunately, the council had no funds of its own and could not materially advance the work on fuel cells. It existed until Frumkin's death in 1976.

In the 1970s and 1980s, a 10- or 15-kW fuel cell power plant, Photon, was developed in the Urals Integrated Electrochemical Plant in collaboration with the S. P. Korolev "Energy" Space Corporation for the planned Buran space shuttle. Preliminary tests demonstrated the excellent quality of this AFC-based power plant. Further work was discontinued when the Buran project was shelved.

In today's Russian Federation, research work on fuel cells is conducted in a number of institutes of the Russian Academy of Sciences and in a number of universities. Fuel cell power plants are still developed in the Urals Plant and in the Russian science center known as the Kurchatov Institute and in other institutes of the nuclear industry (Korovin, 2005).

Beyond the countries listed above, fuel cell work is conducted (in part with government support) in, for example, India, Taiwan, Singapore, Malaysia, and certain South American countries.

In conclusion, we wish to quote what Adamson said when comparing the national plans existing in different countries in a 2006 review: "The USA at the federal level has clear R&D targets, but lacks the very long term vision shown for example in Japan."

17.6 THE VOLUME OF PUBLISHED FUEL CELL WORK

As noted earlier, work on fuel cells today is conducted in almost all developed countries of the world. In some of them the work is highly advanced, and power plants based on different variants of fuel cells begin to be produced commercially. In others, this work is still in an initial stage and is limited to academic studies of certain special problems. Within any one country, many scientific and commercial entities are usually involved.

For successful pursuit of this work, a well-adapted system of information exchange on the national and international levels is very important. It is quite natural that this exchange cannot cover 100% of all fuel cell data gathered. Many private companies (particularly the larger ones) have proprietary information that they do not wish to share with the competition and thus is kept confidential or classified. This is usually true for certain technical solutions or know-how, particular formulations, special scientific and technical details, and often enough for the economic aspects. General aspects of the structure and functioning of the various types of fuel cells, and particularly the aspects reflected in this book, belong to the public domain and form a body of broad and widely distributed scientific and technical experience. For its dissemination, various media exist.

Meetings and Expositions

Each year during the annual meeting and during special meetings of the International Society of Electrochemistry (ISE), held in turns in different countries, sessions discussing specific problems, such as electrocatalysis, or sessions discussing general problems of fuel cells are organized. Similar sessions can be found in the semiannual meetings of the Electrochemical Society (ECS) in the United States, also heavily attended by many foreign specialists. Some 3000 to 5000 persons come together annually in the meetings of the ISE and ECS, who in 2007 signed a cooperative agreement and have already held many joint meetings. The biannual international Grove Fuel Cell Symposium, first held in London in 1987, is the largest forum on fuel cells offered in Europe. Although progress in fuel cells is always the central topic, in 2007 during the Tenth Grove Symposium, achievements since the first symposium in 1987 were the focus. Regular international meetings on solid-oxide fuel cells have been held in Japan, the United States, Switzerland, and France. National meetings or seminars are held in many countries. The annual Hanover Fair in Germany has featured special stands of companies presenting their fuel cell achievements. Tokyo will hold its fifth (annual) International Hydrogen and Fuel Cell Expo in February 2009.

Publications

Work on fuel cells and hydrogen energy is published in many scientific and technical journals. The *International Journal of Hydrogen Energy* has already been mentioned. Most fuel cell work is published in the *Journal of Power Sources*. In the 23 issues of this journal in 2007, more than 2500 fuel cell papers were published, including 16 reviews, which represent an average of 200 papers and at least one review every month. A relatively large number of fuel cell papers can also be found in the periodicals *Electrochimica Acta*, *Journal of the Electrochemical Society*, and *Electrochemical and Solid-State Letters*. A relatively small specialized publication is Elsevier's *Fuel Cells Bulletin*, which concentrates on technical and economic news. A large volume of information on the fuel cell industry and on fuel cell applications is available online in the biweekly newsletter *Fuel Cell Today* (www.fuelcelltoday.com). This online source issues detailed special review articles periodically.

Patents

Despite the fact that patents have the prime purpose of protecting intellectual property and commercial rights of individual companies or people, and also those of other organizations, they are a very important vehicle for the spread of information. Patents are bought, sold, and licensed to be used by third parties in the same country or in other countries. In the United States alone, every week about 12 new fuel cell patents are issued. About 500 companies are holders of at least one patent in the field (Khan, 2004). The distribution between countries is highly nonuniform. For instance, of the 1582 general fuel cell patents published in the fourth quarter of 2007, 51% came from Japan, 15% from the United States, 9% from South Korea, 6% from China, and about 1% from European countries (Butler, 2008b).

Fuel cell manufacturing patents are filed mainly under the International Patent Classification Code (IPC) H01M008.

17.7 LEGISLATION AND STANDARDIZATION IN THE FIELD OF FUEL CELLS

Long ago, fuel cells ceased to be a scientific and technical curiosity or mere examples for a possible direct conversion of the chemical energy of different fuels to electrical energy. Different types of fuel cells are now manufactured in tens and hundreds of pieces, and some will soon number several thousand. Production figures counted in the millions are not now beyond reach, in view of portable and transport needs. This calls for formulating documents on legislation and standardization addressing the conditions of fuel cell use (e.g., questions of safety) as well as of fuel cell construction.

National standards have begun to be formulated in a number of countries in the field of fuel cells. Fuel cell trade across international borders calls for the development of documentation that is legally valid in all countries.

In view of rapid progress in fuel cell work and in response to the many inquiries received from industry, the International Electrotechnical Commission

(IEC) in 1996 created Technical Committee 105 (Fuel Cell Technologies). In August 2004, the first International Standard on fuel cells, IEC 62282-2 Fuel Cell Technologies, was published. It covers the minimum safety requirements for modules (not systems) that manufacturers should comply with when producing fuel cells (AFCs, PEMFCs, MCFCs, and SOFCs) destined for use by customers.

Many questions and aspects requiring regulations and standards remain to be answered. These include the mating of fuel cell stacks with their reactant containers (more particularly, standard junctions for cartridges holding methanol solution in miniature DMFCs) and the reporting of parameters (voltage, frequency, etc.) of the power generated by a fuel cell–based power station, to name just two of many more. A need also exists to standardize certain aspects of fuel cell terminology (the authority here again being the IEC). The term *high temperature*, for instance, has a different meaning for different fuel cell types.

An important act in fuel cell standardization are the new (revised) guidelines of the International Civil Aviation Organization, in force as of January 1, 2009, addressing fuel cells and fuel cartridges in both the hand and checked baggage of airline passengers.

REFERENCES

Adamson K.-A., Fuel cell and hydrogen R&D targets and funding: comparative analysis, presentation at the Fuel Cell Seminar, 2006, http://www.fuelcelltoday.com.

Adamson K.-A., European Union fuel cell and hydrogen R&D targets and funding, *Fuel Cell Today*, Mar. 2005.

Adamson K.-A., G. Crawley, *Fuel Cell Today 2006 worldwide survey, Fuel Cell Today*, Jan. 2006.

Butler J., 2007 Legislation quarterly review, Q4, *Fuel Cell Today*. Jan. 2008a.

Butler J., 2007 patent review, Q4, *Fuel Cell Today*, Jan. 2008b.

Chalk S. G., J. F. Miller, *J. Power Sources*, **159**, 73 (2006).

Conte M., A. Jacobazzi, M. Ronchetti, R. Vellone, *J. Power Sources*, **200**, 171 (2001).

Godfrey B., K. Föger, R. Gillespie, et al., *J. Power Sources*, **86**, 68 (2000).

Khan E., *J. Power Sources*, **135**, 212 (2004).

Korovin N. V., *Fuel Cells and Electrochemical Power Units* [in Russian], Power Engineering Institute, Moscow, 2005.

Milliken J., F. Josek, M. Wang, E. Yuzugullu, *J. Power Sources*, **172**, 121 (2007).

Romm J. J., *The Hype About Hydrogen*, Island Press, Washington, DC, 2005.

Williams M. C., J. P. Strakey, S. C. Singhal, *J. Power Sources*, **131**, 79 (2004).

Zegers P., *J. Power Sources*, **154**, 497 (2006).

CHAPTER 18

OUTLOOK

In this concluding chapter, we wish to provide an appreciation of past development work and achievements, point to some misconceptions, take a glance at natural cold combustion, refer to relevant aspects of a future hydrogen economy, and summarize what we believe to be the needs and prospects ahead of us.

18.1 PERIODS OF ALTERNATING HOPE AND DISAPPOINTMENT

At all stages of fuel cell development in the past, we have seen great enthusiasm about particular achievements, associated with optimistic predictions regarding fuel cells as a cure for the major ills of energy conversion and supply. When subsequent, arduous work failed to fulfill the high hopes, a period of calm ensued and efforts quickly waned.

Thus, in 1898, Wilhelm Ostwald proclaimed that the energy supply of the future would be tied to electrochemistry and would be smoke-free. Subsequent work on fuel cells was hardly successful enough—largely, from what we know now, because of problems with materials—and the inevitable period of calm that followed did not end until 1959, when Francis Bacon demonstrated his

Fuel Cells: Problems and Solutions, By Vladimir S. Bagotsky

stack of alkaline fuel cells. His patents were basic for the Pratt & Whitney fuel cells in *Apollo* space flight. This situation led to the first real fuel cell boom. (Remember at this point to appreciate Bacon's effort that he begun work on his device in 1932.)

During the 1960s and 1970s, work and progress were vigorous. Phosphoric acid–based fuel cells were assembled and built up into large power plants, first on the scale 100 to 200 kW, then even on the megawatt scale. However, predictions of continued strong growth and ever-wider use failed, and space flight and an energy crisis could not prevent another 20 years of slack. Few additional power plants were built.

This period ended only in the mid-1990s, when great success was achieved with membrane hydrogen and methanol fuel cells. Lamy, Léger, and Srinivasan, prominent workers in these fields, titled their 2000 review: "Direct methanol fuel cells: from a 20th century electrochemists' dream to a 21st century emerging technology."

Fuel cell development has always been related to progress in structural materials. It has also been related to progress in electrocatalysis, a field that had come into its own as a new field of theoretical and practical electrochemistry during the 1970s. Like fuel cells, these fields have seen alternation of great expectations and quenched illusions.

A starting point for the emergence of electrocatalysis was the discovery that hydrocarbons could be oxidized at low temperatures (this condition had not been part of the Ostwald scenario). Then it was discovered that synergistic effects were operative in the use of ruthenium–platinum catalysts for methanol oxidation and that compounds such as platinum-free metalloporphyrins were useful catalysts for certain electrochemical reactions in fuel cells. Now hopes were expressed that in the future, expensive platinum catalysts could be replaced. Again, in attempts toward commercial realization of these discoveries, considerable difficulties were encountered, leading to a period of disenchantment and pessimism in the 1970s and 1980s. It had been demonstrated beyond doubt that fundamentally, hydrocarbons could be oxidized at low temperatures, but practical rates that could be achieved were unrealistically low. It had also been demonstrated that fuel cells could be made to work without a platinum catalyst, but practical lifetimes that could be achieved were unrealistically low. In real fuel cells, platinum was still needed (even if in much lower quantities). To a certain degree, pessimism in the field of electrocatalysis may also derive from the fact that science is still far from having attained one of the final aims of theoretical work in this field: that of providing a full quantum-mechanical foundation for electrocatalysis.

At present, most workers hold more realistic views of the promise and difficulties in the development of fuel cells as well as in the development of electrocatalysis. We concluded in Section 16.4 that an important near-term application is in portable power supply. This is a realistic expectation, based on need and on available solutions.

18.2 SOME MISCONCEPTIONS*

At the various stages of fuel cell development, some misconceptions became apparent, consisting of an incorrect view or understanding of certain aspects of fuel cell function. Sometimes progress was impeded or undue hopes nourished by such misconceptions. At other times, efforts to overcome or correct them contributed to progress in fuel cell development.

Catalysis is certainly one of the greatest forces in chemical technology and an agency bordering on a marvel in nature. Catalysts are at the heart of fuel cells. This section therefore opens with the reminder that the common definition of catalysts is not quite true when we call them *substances facilitating the reactions between reactants (say, A and B), without themselves undergoing any change.* Unfortunately for a fuel cell, a catalyst, be it one acting via surface sites on a macroscopic surface or one acting via temporary molecular bonds at "molecular" surfaces, is capable of catalyzing just "so many" net reactions, *not an infinitely large number.* Its action is governed by a turnover number, not easy to define or measure but meaning essentially that after so many unit reactions, the particular surface site is consumed. As it were, after so many unit reactions the definition of "not taking part" or "not undergoing change" fails, and the catalyst atom or molecule seems to "fly off," bound up in an undesired side reaction that has a very low but finite probability. Examples are the catalytic converters for passenger car exhaust or the hemoglobin in blood. For fuel cells, this means that *lifetime has a basic limitation*, and that engineers and chemists will strain to obtain catalytic substances and configurations lasting the largest possible number of turnovers (or unit reactions).

Another misconception is linked to conclusions from the *Carnot cycle*, a limitation not imposed on fuel cells. In fact, Carnot's cycle is a thermodynamic concept, stating that work input and work output occurring between different temperatures lead to a basic efficiency limitation that applies to heat engines such as ICEs. When it comes to fuel cells, where such a temperature change does not occur, a *Carnot limitation* is absent, but basic limitations do not simply vanish—they take on a different guise. In fact, current flow at practical rates is due to chemical reactions occurring at the electrodes with a considerable amount of polarization, which constitutes a new, very basic efficiency limitation. This argument is now rooted not in thermodynamics but in kinetics, and for this reason is not directly comparable with the Carnot argument. It is a practical limitation of different origin. In living energy systems such as the living body, nature—working isothermally like a fuel cell—follows cunning multistep catalytic pathways at patient speeds to optimize efficiency. They seem to be too complex to imitate in real-life technical energy conversion. Still, catalysis will be the key to keeping polarization-related limitations low.

A third misconception may arise when the term *direct energy conversion* is used. So other types of energy conversion could be understood as being

* This section was written by Klaus Müller.

indirect! This is not really what one wants to say. By *direct* one wants to say "from chemical to electric energy without any detours involving thermal and mechanical energy." Implicitly, all additional intervening energy forms give rise to energy conversion losses. This is the all-important reason for using the advantages offered by fuel cells. However, diesel electric traction is an example of a system providing higher efficiency despite adding a conversion stage. Thus, operating conditions may affect the efficiency as much as do all additional stages.

A fourth misconception is that of *clean energy*. Emissions, be they CO_2 or NO_x, soot or waste heat, should always be summed up in relevant ways. Evidently, like any electrically powered operation, fuel cell–powered electric operation is clean at the user's site (assuming the fuel of a fuel cell to be harmless), but emissions may arise elsewhere in fuel manufacture or in the production of the electric grid power. So a local ecological sum is required for consumer sites, and a global sum is required for global impact, including potential local consequences of global deterioration.

A fifth and final misconception that we want to mention here is *cheap solar energy*. Strictly, cheap solar energy may be the proper term for a basketful of grapes or corncobs, or for hot water prepared on a roof in the sun. However, the conversion of solar energy to electricity and of electricity to hydrogen as a fuel for fuel cells, to name one route relevant to fuel cells, requires high investments for solar collectors and electrolyzers, both having considerable efficiency limitations. The solar collectors require land and maintenance (cleaning). Deserted areas are good for solar energy capture but may lack the water for electrolyzers, and long-distance electric power lines have higher losses than are generally acknowledged.

18.3 IDEAL FUEL CELLS

At this point we have reviewed the achievements in fuel cell development and pointed out that driven by the balance between requirements and performance, fuel cells are bound to find interesting uses in large numbers. We have also taken a more sober look at the ups and downs, the hopes and disillusionments. We now wish to reassure and position ourselves by a look at nature.

It is not unusual to compare manufactured technical devices with their natural analogs. A well-known example of such a comparison includes certain parameters of sharks and of the electric torpedos used by navies. The two objects are of similar size and move underwater with comparable speed. That implies that the power required for them to move is comparable as well. A shark gets energy by consuming "cheap" organic seafood, whereas complex storage batteries containing heavy metals, sometimes even large amounts of expensive silver, must be used to propel the torpedo. Successful torpedo designers will have to show how close the energy consumed by their torpedos comes to that employed by sharks.

In identical fashion, fuel cell designers often noticed an analogy between processes taking place in fuel cells and processes taking place in the human body or certain other biological systems. In their 1962 book, one of the first modern monographs on fuel cells, Justi and Winsel had a chapter on "Similarities and Differences Between Fuel Cells and Living Beings." In 1967, Bockris and Srinivasan in a Short Note to the British journal *Nature*, wrote to that effect. The essence of this note can also be read in Srinivasan's recent fundamental monograph (2006; Sec. 9.9.7.3).

A fuel cell has the task of using (air) oxygen for direct conversion of the chemical energy of fuels (including organic fuels) to electrical energy while bypassing the intermediate stage of producing thermal energy, which, speaking thermodynamically, is "little useful." The heat produced in cold combustion in a fuel cell as the result of various irreversible processes is not an intermediate form of energy in the conversion process but a real, final product and can be used in addition to the electrical energy produced.

In the human body, as an example, mechanical energy is obtained by an analogous direct conversion from the chemical energy of food products of an organic nature, again by such cold combustion, and in fact, many food items are labeled with an indication of their caloric content. According to Srinivasan, the conversion efficiency has been estimated at about 35%. As an important point, note that mechanical energy, just like electrical energy, belongs to forms of energy that are thermodynamically useful, and as such admit interconversion with almost 100% efficiency.

On average, humans consume food products having an energy content of 3000 kcal or 12.5 MJ/day. Spread out uniformly over 24 hours, this provides an electrical equivalent of about 126 W. The mechanical energy derived by the human body is consumed in part for internal needs: blood circulation; breathing; and functioning of the brain, intestines, and other organs (certain data indicate that a brain runs on the equivalent of just 1 W). Tens of watts at a time may be available to perform external work. Although in some modern societies, people tend to delegate mechanical work to machines and then work out in a gym, we should remember the useful physical work performed by slaves and criminals on galleys or that performed by the Volga boatmen.

Thus, from a purely energetic point of view, and abstracting from all other human activity, one may look at a human as the equivalent of a 50-kg fuel cell working with an efficiency of 35% without interruption for an average period of 70 years, producing (after satisfying internal needs) tens of watts of useful power and having a high degree of autoregulation. For fuel cell engineers, such parameters are a matter of envy and distant dreams.

It is remarkable that the analogy is not merely superficial. In fact, physiological processes taking place in the human body, including metabolism, are often of an electrochemical nature and reminiscent of the processes taking place in fuel cells.

The first steps in the metabolic process are a purely chemical breakdown of nutrients into simpler low-molecular-mass compounds that occur in the digestive tract. The basic oxidation steps occur at the membranes of the cell mitochondria. Hydrogen atoms are split off from the low-molecular-mass compounds under the effect of the enzyme dehydrogenase and are transferred to molecules of the organic compound NAD (nicotinamide-adenine dinucleotide), yielding $NADH_2$. Further oxidation (or dehydrogenation) of this compound occurs purely electrochemically. The electrons and protons are transmitted through a multienzyme system called an *electron transfer chain*. The last step is electron and proton transfer from the enzyme cytochrome oxidase to an oxygen molecule existing in a hemoglobin complex. This oxygen reduction step is entirely analogous to the oxygen reduction reaction at the cathodes of fuel cells (see Bagotsky, 2005, Chap. 30).

The high-energy conversion efficiency found in the human body is the result of many detailed factors, chief among which are: (1) the high activity level and high selectivity of catalytic systems consisting of many different enzymes, and (2) the presence of exceptionally efficient organs for control and management, such as the brain and nervous system, securing a high degree of coordination of all internal physiological processes and reacting rapidly and effectively to any perturbation. It may be a long way to go for the fuel cell designer, but nature does point the direction and destination.

18.4 PROJECTED FUTURE OF FUEL CELLS

Prior to 1960, only a few enthusiasts believed in Ostwald's idea of direct conversion of the chemical energy of natural fuels to electrical energy, and so tried to build fuel cells. For many others, at that time the idea either was altogether utopian or a matter for a very distant future. Decisive change came about after Bacon's demonstration of a working stack of fuel cells and the first practical uses of fuel cells in *Gemini* spacecraft. Now many scientific groups, private companies, and in the end, government bodies became interested. This led to the broad development work described in this book. Numerous projects testing the practical utility of fuel cells in various departments of the economy were performed. Many variants of fuel cells attained a high degree of technical perfection. Industry began to produce certain types of fuel cells (or rather, fuel cell stacks or batteries) in numbers reaching hundreds per year.

Many problems remained without a solution, which hindered further progress. Primarily, such problems were either economic (the need to lower the production costs) or lifetime-related (the need to extend the working life). The problems were not fundamental. A considerable number were not even associated with fuel cells proper but with ancillary units such as systems for reactant storage and feed, product removal, thermal management, humidity management, and the like. The ancillary units of a fuel cell power plant account for about 40% of the total weight, a large part of the cost, and a considerable number of operating defects (King and O'Day, 2000).

For further improvement in fuel cell performance, obvious problems as have been mentioned in this book need to be solved, such as raising the activity of catalysts for methanol oxidation, lowering the sensitivity of oxygen–electrode catalysts toward methanol, and building improved membranes. In addition, a number of fundamental problems need to be solved in the field of electrocatalysis.

1. *Developing new catalysts for the oxygen electrodes.* It has been pointed out repeatedly that considerable energy loss occurs in fuel cell operation due to the irreversibility of the electrochemical reduction of oxygen, a phenomenon also responsible for the considerable departure of the oxygen electrode's open-circuit potential from the thermodynamic value. (Analogous losses are seen in anodic oxygen evolution when electrolytically splitting water.) The metabolic reduction of oxygen in the human body that was mentioned in Section 18.3 occurs under conditions close to reversibility, and without energy losses, despite the fact that it, too, has strong electrochemical components. This is food for thought by research workers in the field of electrocatalysis and reassurance that we are not facing a fundamental impasse here!
2. *Developing catalysts for the complete 12-electron oxidation of ethanol.* Such catalysts are needed for the efficient utilization of biofuels in fuel cells producing electric power, which in the final analysis is derived from solar energy involving the consumption (rather than liberation) of CO_2 as a greenhouse gas.
3. *Developing highly selective catalysts for electrochemical reactions.* The platinum–metal catalysts generally used in fuel cells are highly active but completely nonselective. They catalyze equally well reactions that are desired and reactions that are not needed or that are even detrimental. In contrast, enzymes as natural catalysts are highly active and at the same time highly selective. A given enzyme will catalyze only one particular reaction. If we had selective catalysts for the two electrodes in a fuel cell, we could realize single-compartment fuel cells, that is, a type of fuel cell discussed widely at present in which the fuel and the oxidizer are supplied together to the fuel cell. Such cells would have a much simpler design than that of current cells requiring a separate reactant supply, and they would be easier and cheaper to manufacture than today's fuel cells. Such cells would also be free of the numerous difficulties arising from the need for sealing, that is, isolating the fuel compartment from the oxidizer compartment, which is a real headache for fuel cell makers.

It may be possible, again following leads provided by nature, to achieve adequate solutions to these problems by developing catalysts with surface properties that have been tailored deliberately to provide favorable catalytic action on all intermediate steps that require it. Such catalysts should be polyfunctional and exhibit a certain degree of chemical and structural micro-inhomogeneity. Nanoelectrochemistry (see Bagotsky's 2005 monograph, Chap. 36) and nanoelectrocatalysis (Stimming, 2008) may be possible approaches to synthesizing such surfaces.

REFERENCES

Bagotsky V. S., *Fundamentals of Electrochemistry*, 2nd ed., Wiley, Hoboken, NJ, 2005.

Bockris J. O'M., S. Srinivasan, *Nature*, **215**, 197 (1967).

Justi E. W., A. Winsel, *Fuel Cells: Kalte Verbrennung*, Steiner, Wiesbaden, Germany, 1962.

King J. M., M. J. O'Day, *J. Power Sources*, 86, 16 (2000).

Lamy C., J.-M. Léger, S. Srinivasan, Chap. 3 in: J. O'M. Bockris, B. E. Conway (eds.), *Modern Aspects of Electrochemistry*, Vol. 34, Plenum Press, New York, 2000, pp. 53–117.

Srinivasan S., *Fuel Cells: From Fundamentals to Applications*, Springer-Verlag, New York, 2006.

Stimming U., Invited talk at the Fuel Cell Meeting (75th Meeting of the Electrochemical Society of Japan, Kofu, Japan), 2008.

GENERAL BIBLIOGRAPHY

HISTORICAL BOOKS

O. K. Davtyan, *The Problem of Direct Conversion of the Chemical Energy of Fuels into Electrical Energy* [in Russian], Publishing House of the USSR Academy of Sciences, Moscow, 1947.

G. J. Young (ed.), *Fuel Cells*, Reinhold, New York, Vol. 1, 1960; Vol. 2, 1963.

E. W. Justi, A. Winsel, *Fuel Cells: Kalte Verbrennung*, Steiner, Wiesbaden, Germany, 1962.

W. Mitchell (ed.), *Fuel cells*, Academic Press, New York, 1963.

B. S. Baker (ed.), *Hydrocarbon Fuel Cell Technology*, Academic Press, New York, 1965.

W. Vielstich, *Brennstoffelemente*, Verlag Chemie, Weinheim, Germany, 1965.

H. A. Liebhafsky, E. J. Cairns, *Fuel Cells and Fuel Cell Batteries*, Wiley, New York, 1968.

J. H. Hoare, *The Electrochemistry of Oxygen*, Wiley, New York, 1968.

M. B. Breiter, *Electrochemical Processes in Fuel Cells*, Springer-Verlag, Berlin, 1969.

J. O'M. Bockris, S. Srinivasan, *Fuel Cells: Their Electrochemistry*, McGraw-Hill, New York, 1969.

K. Kordesch (ed.), *Brennstoffbatterien*, Springer-Verlag, Berlin, 1984.

K. Kordesh, G. Simader, *Fuel Cells and Their Application*, VCH, Weinheim, Germany, 1996.

CONTEMPORARY BOOKS

W. Vielstich, A. Lamm, H. Gasteiger (eds.), *Handbook of Fuel Cells: Fundamentals, Technology, Applications*, 4 volumes, Wiley-VCH, New York, 2003.
S.D. Singhal, K. Kendall (eds.), *High-Temperature Solid Oxide Fuel Cells: Fundamentals, Design and Applications*, Elsevier, New York, 2004.
N. P. Bansal, D. Zhu, W. M. Kriven (eds.), *Advances in Solid Oxide Fuel Cells* (a collection of papers presented at the 29th Internatinoal Conference on Advanced Ceramics and Composites, Cocoa Beach, FL), Wiley, Hoboken, NJ, 2005.
F. Barbir, *PEM Fuel Cells: Theory and Practice*, Academic Press, New York, 2005.
R. O'Hayre, S.-W. Cha, W. Colella, F. B., Prinz, *Fuel Cell Fundamentals*, Wiley, Hoboken, NJ, 2005.
B. Sørensen, *Hydrogen and Fuel Cells: Emerging Technologies and Applications*, Academic Press, San Diego, CA, 2005.
S. Srinivasan, *Fuel Cells: From Fundamentals to Applications*, Springer Science + Business Media, New York, 2006.
K. D. Kreuer, T. Nguyen (eds.), *Advances in Fuel Cells*, Vol. 1, Elsevier, New York, 2007.
E. Fontes, *Handbook of Fuel Cell Modelling*, Elsevier, Amsterdam, to be published in 2010.
M. Gasik (ed.), *Materials for Fuel Cells*, Woodhead Publishing Ltd., Cambridge, UK, 2008.
G. Walker (ed.), *Solid-State Hydrogen Storage: Materials and Chemistry*, Woodhead Publishing Ltd., Cambridge, UK, 2008.

PERIODICALS

Electrochemical and Solid-State Letters, The Electrochemical Society, Pennington, NJ.
Electrochemistry Communications, Elsevier, Amsterdam, The Netherlands.
Electrochimica Acta, Elsevier, London.
Fuel Cell Today, online newsletter, http://www.fuelcelltoday.com.
Fuel Cells Bulletin, Elsevier Advanced Technology, Oxford, UK.
International Journal of Hydrogen Energy, International Association of Hydrogen Economy, Miami Beach, FL.
Journal of Applied Electrochemistry, Springer, Dordrecht, The Netherlands.

Journal of Power Sources, Elsevier Seuoia, Lausanne, Switzerland.
Journal of the Electrochemical Society, The Electrochemical Society, Pennington, NJ.

REVIEWS ON GENERAL TOPICS

Butler J., Transport 2008, *Fuel Cell Today*, July 2008.
Butler J., Legislation 2008, Q2, *Fuel Cell Today*, July 2008.
Dicks A. L., The role of carbon in fuel cells, *J. Power Sources*, **156**, 128, 142 (2006).
Huleatt-James N., Hydrogen infrastructure, *Fuel Cell Today*, Feb. 2008.

AUTHOR INDEX

The index lists only authors whose contributions are explicitly discussed in the text. Authors merely appearing in literature references are not listed.

Fuel Cells: Problems and Solutions, By Vladimir S. Bagotsky

SUBJECT INDEX

The index lists only those page numbers where a given term is used for the first time, or where another aspect of the same term is discussed.

Fuel Cells: Problems and Solutions, By Vladimir S. Bagotsky

THE ELECTROCHEMICAL SOCIETY SERIES

Corrosion Handbook
Edited by Herbert H. Uhlig

Modern Electroplating, Third Edition
Edited by Frederick A. Lowenheim

Modern Electroplating, Fourth Edition
Edited by Mordechay Schlesinger and Milan Paunovic

The Electron Microprobe
Edited by T. D. McKinley, K. F. J. Heinrich, and D. B. Wittry

Chemical Physics of Ionic Solutions
Edited by B. E. Conway and R. G. Barradas

High-Temperature Materials and Technology
Edited by Ivor E. Campbell and Edwin M. Sherwood

Alkaline Storage Batteries
S. Uno Falk and Alvin J. Salkind

The Primary Battery (in Two Volumes)
Volume I *Edited by* George W. Heise and N. Corey Cahoon
Volume II *Edited by* N. Corey Cahoon and George W. Heise

Zinc-Silver Oxide Batteries
Edited by Arthur Fleischer and J. J. Lander

Lead-Acid Batteries
Hans Bode
Translated by R. J. Brodd and Karl V. Kordesch

Thin Films-Interdiffusion and Reactions
Edited by J. M. Poate, M. N. Tu, and J. W. Mayer

Lithium Battery Technology
Edited by H. V. Venkatasetty

Quality and Reliability Methods for Primary Batteries
P. Bro and S. C. Levy

Techniques for Characterization of Electrodes and Electrochemical Processes
Edited by Ravi Varma and J. R. Selman

Electrochemical Oxygen Technology
Kim Kinoshita

Synthetic Diamond: Emerging CVD Science and Technology
Edited by Karl E. Spear and John P. Dismukes

Corrosion of Stainless Steels
A. John Sedriks

Semiconductor Wafer Bonding: Science and Technology
Q.-Y. Tong and U. Göscle

Uhlig's Corrosion Handbook, Second Edition
Edited by R. Winston Revie

Atmospheric Corrosion
Christofer Leygraf and Thomas Graedel

Electrochemical Systems, Third Edition
John Newman and Karen E. Thomas-Alyea

Fundamentals of Electrochemistry, Second Edition
V. S. Bagotsky

Fundamentals of Electrochemical Deposition, Second Edition
Milan Paunovic and Mordechay Schlesinger

Electrochemical Impedance Spectroscopy
Mark E. Orazem and Bernard Tribollet

Fuel Cells: Problems and Solutions
Vladimir S. Bagotsky